Pain

Pain

New Essays on Its Nature and the Methodology of Its Study

edited by Murat Aydede

A Bradford Book
The MIT Press
Cambridge, Massachusetts
London, England

MIT Press books may be purchased at special quantity discounts for business or sales promotional use. For information, please email special_sales@mitpress.mit.edu or write to Special Sales Department, The MIT Press, 55 Hayward Street, Cambridge, MA 02142.

This book was set in Stone serif and Stone sans by SNP Best-set Typesetter Ltd., Hong Kong, and was printed and bound in the United States of America.

Library of Congress Cataloging-in-Publication Data

Pain: new essays on its nature and the methodology of its study / edited by Murat Aydede.
    p. cm.
"A Bradford book."
Includes bibliographical references and index.
ISBN 0-262-01221-9 (hc : alks. paper)—ISBN 0-262-51188-6 (pbk. : alk. paper)
1. Pain. I. Aydede, Murat.
RB127.P332495 2006
616'.0472—dc22
                                                                        2005042146

10 9 8 7 6 5 4 3 2 1

In memory of

Arda Denkel (1949–2000)
Ilham Dilman (1930–2003)
Berent Enç (l938–2003)

Contents

Preface

The essays collected in this volume concern the nature of pain and the methodology of its study. They reflect the metaphysical, epistemological, methodological, and scientific—rather than the aesthetic, ethical, or religious—interests and perspectives of their authors. The volume is also a manifestation of the increasing scholarly cooperation between philosophers and various scientists who work on consciousness and emotion.

There are various puzzles about pain, philosophical as well as scientific. Any view on the nature of pain or any methodology proposed for its study has to address these puzzles. Many essays in this book address these puzzles and propose new ways of dealing with them—often quite different, indeed sometimes opposing, ways of dealing with them. It is hoped that the spectrum of views presented here will show that the field of pain research is very active and provides intellectually fertile ground for new ideas for interdisciplinary work, especially in relation to larger issues such as the place of consciousness in the natural order and the methodology of psychological research.

Pain is perhaps the most prominent member of a class of sensations usually known as bodily sensations. In his little book, *Bodily Sensations* (1962), David Armstrong divided these into two categories, transitive and intransitive bodily sensations, and put pain in the latter category. The idea was roughly to capture the fact that reporting some bodily sensations seems to involve the use of a transitive verb whose object is a sensible quality of an external object or event, like temperature or pressure. In reporting transitive bodily sensations, we seem to be doing the same sort of thing when we report seeing the color of an object or hearing a sound. We say, "I felt the warmth of his hand," or "I feel the pressure of the stethoscope on my back." These seem to be straightforward tactile perceptual reports on a par with the visual, auditory, gustatory, or olfactory reports by which we report the perception of something external to our minds.

Intransitive bodily sensations, by contrast, pose a puzzle. When we report them, even though we sometimes seem to be reporting the perception of something, many think that either we are not making a perceptual report, or if we are, what we are perceiving is not something external to our minds. Consider for instance reporting an itch or tingle in part of one's body, or take the standard practice of reporting pain in a bodily location. Even though we sometimes use a sentence where the verb seems transitive, as in "I feel a sharp pain in my right thumb" or "I feel my back itching," reporting an itch, a tingle, or a pain can be done by using intransitive verbs, as in "My back is itching" or "My right foot is tingling," or "My wisdom tooth is aching." We also typically locate pains, itches, and tingles in body parts by saying things like "There is a pain in my thigh" or "There is tingling in my arm," and so forth.

It is part of the commonsense conception of pains, itches, and tingles that they can't exist without someone's feeling them—quite unlike the qualities of physical objects or events we can sense or perceive. In this respect, saying that we feel a pain, an itch, or a tingle seems very unlike saying we feel the temperature or pressure of something. The heat or pressure applied to my skin can exist without my feeling them, and when I do feel them they can be perceived (felt) by others in the same way. Shortly after the dentist injects novocaine into the gum around my tooth to fill a cavity, I can still feel the pressure of the driller but (luckily) not the pain—the pain no longer exists. Or if the novocaine is very strong, I may not feel even the pressure of the driller, but in this case there is no temptation to say that the pressure no longer exists. By contrast, it would seem absurd to say that the pain still exists just as the pressure does but I no longer feel it. Pains, itches, and tingles seem essentially subjective and logically private (i.e., accessible only) to their owners in a way the warmth or coldness or the pressure of an object is not. The latter are often said to be objective and public in their nature. Of course the very *feeling* of the temperature or pressure of something may be subjective and private, but the objects of this feeling (perception) are not.

The essential subjectivity and privacy of pains and other intransitive bodily sensations are two core puzzles about them, and they often go together. Related to these, or because of these, our knowledge of our own pains (and other bodily sensations) seems to be so secure that challenging someone about the accuracy of their pain report (when they are sincere and not confused) seems to be conceptually or logically out of the question. This again contrasts sharply with the case of knowledge about the external objects of our perceptions. We are sometimes wrong about the temperature or pressure of an object that we feel. I touch something a few seconds, and it feels hot. But you tell me that it was in fact very cold. I am not baffled; I take your word for it. The shot hurts: I feel a sharp sudden incisive pain on my upper left arm. You tell me that

that it didn't hurt, it was in fact pleasant. I am baffled! I suspect you must be joking; otherwise you must be deeply confused. That we seem to be incorrigible, or indeed infallible, in our beliefs about our pains and other intransitive bodily sensations is another puzzle about them. How could we have such epistemic authority if pain reports are simply a species of ordinary perceptual report? No perception seems immune to error. Illusion or hallucination is always a possibility: there is always an appearance–reality distinction in genuine perception. But nothing of this sort seems to apply to feeling pains.

It might be said that there are in fact no deep puzzles about pains and other intransitive bodily sensations of the sort alleged above, since even transitive sensations share the same features. I may be in error in feeling (perceiving) the correct *temperature* of an object, but just as in the case of pain, I cannot be mistaken about *feeling* (correctly or incorrectly) its temperature. After all they are all sensations—transitive, intransitive, or otherwise. Ordinary perception always involves conscious sensory experience. When I close my eyes, I don't think the things I was looking at go out of existence. It is rather that my visual experience of them ceases to exist—I no longer *see* them. But all sensory experiences, it may be plausibly said, are subjective, private, and the topic of incorrigible belief. What is so special about pains and other intransitive sensations?

Fair enough. But we don't typically locate sensory experiences on bodily locations, except perhaps in the head. Experiences are mental phenomena; as such, they are in the mind if they are anywhere. If the mind is in the head, sensory experiences can at most be in the head. But we don't think it absurd to locate pains, itches, and tingles in toes, thighs, arms, and so forth. Indeed we don't stare incredulously at the doctor who asks whether we feel a tingling sensation in our feet as she examines us for, say, a suspected neurological disorder. Come to think of it, it is *perfectly OK* with common sense that we may have sensations in body parts other than the head—say, under our right foot! Why? This, then, is another puzzle about pains and other intransitive bodily sensations: how to properly understand the common practice of locating what appear to be essentially subjective and private sensations in various parts of the body.

The proponents of so-called perceptual theories of pain argue that despite the many peculiarities involved, feeling pain is an instance of ordinary perception—typically the perception of some bodily condition, either tissue damage or impending tissue damage or some sort of assault on the biological integrity of bodily tissue. Many naturalistically oriented philosophers in the last half century have sought ways in which pains and other intransitive bodily sensations can be integrated into a unified theory of perception, perception of a mind-independent objective reality with which we interact in ways our scientific theories of perception and action have been uncovering. The motivation behind this drive was to show that pains and other intransitive sensations do not threaten a scientifically acceptable naturalistic understanding of human mind and conscious experience, on the assumption that ordinary perception could be accounted for in naturalistically acceptable ways. Most of the puzzles we have discussed so far seem to point to the fact that there is something special about these intransitive bodily sensations. They seem to be signs or indications of what makes minds metaphysically special in the natural order of things, and many think they are the most salient tip of a metaphysical iceberg that cannot be made to fit into the natural order. Indeed many dualists have used them to make trouble for physicalistic naturalism. Ordinary perception, insofar as it involves conscious experience, already has enough to worry philosophers who want to see minds as part of physical order. But if pains and other intransitive sensations are not even perceptions of an ordinary (bodily) reality, they pose a much more radical and direct threat to such philosophers. At least, this has been one of the main, perhaps the most, important motivations for developing a perceptual theory of pain. Chapter 3 by Christopher Hill develops and defends a perceptual theory of pain, addressing many of the puzzles discussed above. His aim is to show that most of the puzzles can be traced back to our ordinary commonsense conception of pain, which is muddled. Once we see that this is the source of the puzzles, we can see that there is no metaphysical or scientific barrier to understanding pain in perceptual terms.

Traditionally, the proponents of perceptual theories have tended to be direct realists who reject the claim that perception of an external mind-independent reality rests on direct awareness of mental intermediaries (e.g., sense data). Indirect realists, by contrast, insist that our perception of mind-independent reality is indirect, although we are not ordinarily aware of this. The reasons for this traditional alignment between direct realists and proponents of perceptual theories of pain are complex and to some extent historically obscure. But part of the reason may be that direct realists have thought that repudiating mental intermediaries is necessary for defending physicalistic naturalism, that is, for the project of naturalizing the mind in general and perception in particular. The introductory chapter by Murat Aydede contains an extended presentation and critical discussion of the complex dialectic that existed among the various parties to this debate. It also reviews the basic trends in the scientific study of pain in order to reach a preliminary assessment of how the perceptualist–representationalist tradition in philosophy fares when compared with recent developments in pain science.

Chapter 11 by Moreland Perkins presents an exception to this alignment. Perkins develops an elegant indirect realist version of a perceptual theory of pain without giving up physicalism—although his primary concern in this chapter is not to show how his indirect realism is compatible with physicalism.

So-called representationalist theories of perception are attempts to reduce the qualitative phenomenology of experiences (their very feels) to naturalistically acceptable representational relations. These are the most recent incarnation of direct realist views of perception, and they include an explicit attempt to address the worry about how to account for the feels of conscious perceptual experiences in a naturalistically acceptable fashion. These views are sometimes called strong representationalism or representationism (see chapter 6 by Ned Block), so as to distinguish them from the earlier indirect but "representative" views of perception, in which the direct object of perceptual awareness, although mental, is *representative* of a nonmental (physical) reality—hence, direct awareness of this intermediate object or quality constitutes the indirect *perception* of this physical reality. Strong representationalism denies direct awareness of these intermediaries and claims that, on the contrary, we are aware of the feel of the experiences by being directly aware of what our experiences represent, namely aspects of extramental reality with which we (directly) perceptually interact. The strong representational theory of pains is the application of this general view to pains. Chapter 4 by Michael Tye contains the most sophisticated development of this sort of view. He tackles many of the puzzles about pains discussed above that stand in his way. The following four chapters, by Murat Aydede, Ned Block, Barry Maund, and Paul Noordhof are mostly critical discussions of the extent to which Tye is successful in his attempt. They typically contain extended discussion that goes beyond mere commentary on Tye's views. Tye's replies in chapter 9 clarify his previous discussion, and more importantly, further develop certain aspects of his approach to pain, such as strong representationalism about the affective-motivational aspects of pain experiences, the subjectivity of pains, and the problem of spatially locating pains.

Not all physicalists think that the best way to naturalize pains and other conscious experiences is by making them follow the pattern of ordinary perception, or indeed by defending strong representationalism about experiences. Even though they may not reject strong representationalism, and thus may subscribe to a perceptual view of pains, some physicalists think that the best way to see pains and other conscious experiences as part of the natural order—and thus to remove some of the puzzles about them—is to show directly how to identify them with brain processes that have been discovered by scientists. Chapter 19 by Tom Polger and Ken Sufka falls in this category. The authors also discuss various methodological as well as theoretical issues that arise in any attempt to identify what seem to be radically different kinds of phenomena as one and the same. In their argument they use the example of pain and what we scientifically know about pain. For this reason, the chapter also contains an accessible review of some aspects of the science of pain.

Despite the many puzzles involved in our ordinary understanding of pain, the dominant thread in the commonsense as well as the scientific conception of pain seems to be that it is a sensory as well as an emotional experience. If so, whatever the solution to the puzzle of spatially locating pains in body parts might turn out to be, we are reporting sensory experiences when we report pains. In other words, we seem to be making introspective reports when we report them. But this raises the question of how we can do this with such ease. Ordinarily, introspection is taken by many to require sophisticated mental capacities. But feeling pain and knowing one is feeling pain don't seem to be very sophisticated mental activities. Indeed, they seem to be quite basic and primitive. The intuition is that many animals and young infants feel pain, and that in some intuitive sense they know when they do. How is this possible? What is the epistemology of feeling pain, and more generally, what is the epistemology of introspecting conscious experiences? This question is discussed by Fred Dretske in chapter 2.

The intuition that animals feel pain, and that they have some rudimentary knowledge of their pains, can be resisted—indeed, has been famously resisted by Descartes, and more recently by Peter Carruthers. Whether animals feel pain is also a very live and pressing question for pain science, since a lot of pain research is done on animals. But apart from this, there is growing sensitivity to animal welfare issues. Since feeling pain seems to be an essentially subjective and private affair, how we know whether animals feel pain, or feel pain in much the same way that we feel pain, is an extremely difficult question to answer. Chapter 20 by Colin Allen and his colleagues, and parts of chapter 21 by Jaak Panksepp, take up precisely this question and suggest ways in which it can be fruitfully addressed. Allen et al. take a middle road and suggest a new experimental paradigm that they claim has the potential to deliver more significant results in the way of answering the question of animal pain. They discuss the methodology of third-person scientific pain research on animals and humans and suggest ways in which it can deliver first-person results. These two chapters also contain an accessible presentation of some of the relevant scientific research.

Perhaps the most salient fact about pains is that they hurt—they are "painful." Pains have an immediate negative affect. The scientific community has embraced the idea that pain experiences have both a sensory-discriminative and an affective-motivational dimension. The neural mechanisms underlying both dimensions have been largely identified and studied extensively in the last thirty years or so, and we know quite a lot about their interactions. There are curious pain syndromes where the affective aspect of pains is eliminated to various degrees while the patients are still capable of processing the nociceptive sensory information. Some of these patients identify their experience as pain but say that they don't find them painful and are not bothered by them, claims that are consistent with their nonverbal behavior and vital signs. The phenomenology of these patients' pain experiences seems to lack the usual negative affect.

Many chapters of this volume discuss the affective dimension of pains. Chapter 10 by Austen Clark argues that the painfulness of pain is not a quale. The term 'quale' is a technical term used by philosophers to denote the raw feels or phenomenological qualities of experiences. Clark's use of the term is restricted to the *intrinsic* sensory qualities of our experiences. He argues that pain's negative affect is not a quality intrinsic to pain experiences. On his view, it consists rather of mostly hard-wired functional connections of pain experiences to drive states. In contrast, Tye argues in chapters 4 and 9 that the affective quality of pain experiences is purely representational, just like their sensory-discriminative phenomenology—although Tye would agree with Clark that the phenomenal affective content of pain experiences is not an intrinsic feature of them. Aydede criticizes Tye's strong representationalism about pain's affect in chapter 5, to which Tye responds in chapter 9.

Not surprisingly, the perceptual–representational theories of pain developed by philosophers have traditionally emphasized the sensory-discriminative aspect of pains. Most have tended to downplay the affective aspect of pain, treating it as nothing more than the experiencers' immediate cognitive reactions to their experiences. Don Gustafson by contrast argues in chapter 12 that pains are best construed primarily as emotional states rather than perceptual states. He points to the intellectual history of conceptions of pain to support his recommendation. He argues that this view is not only consistent with the scientific facts about pain, it is also suggested by them. Indeed the science of pain has increasingly emphasized the affective-emotional dimension of pain and has uncovered a great deal about its underlying neural and psychological mechanisms. In chapter 21, Jaak Panksepp presents scientific evidence for the claim that emotional pain associated with grief, social stress, and social isolation implicates some of the same mechanisms involved in the affective dimension of physical pain. In fact, he argues, there is good reason to think that emotional pain has evolved out of the affective aspect of physical (in particular, visceral) pain.

The subjectivity and privacy of pain pose something of a dilemma for anyone who wants to scientifically study and conquer it. On the one hand we find scientists insisting on a conception of pain according to which pain is a subjective sensory and emotional experience (see chapter 1), and on the other hand we find the rigorous demands of centuries-old scientific research methodology that insists on the objectivity of evidence and publicly verifiable results. So how is the scientist qua scientist to study objectively something that appears to be essentially subjective and private like pain? Although the problem also arises in the study of conscious experience in general, the problem for the pain researcher is especially pressing, for reasons already discussed in relation to ordinary perceptual experiences versus pain. Chapter 13 by Donald D. Price and Murat Aydede addresses this problem. The authors examine the role of introspection in the study of psychological phenomena and present a theoretical framework within which first-person methods of gathering data can be justified both scientifically and philosophically without compromising the objectivity of scientific methods. They go a step further and propose that pain scientists should involve themselves as subjects in their own experiments. Price and Aydede present a well-defined four-step experimental procedure to guide scientists in involving themselves in their own experiments. They argue that this paradigm, when combined with standard third-person scientific methodologies, is superior to standard scientific practice. They call their paradigm the "experiential-phenomenological approach." In chapter 14, Shaun Gallagher and Morten Overgaard present other recent proposals that are very similar to the experiential-phenomenological approach. This seems to show that first-person methods, which have been traditionally shunned by scientists (especially those under the influence of behaviorism), are making a comeback. In chapter 16, Robert Coghill discusses how he and his research group have been using first-person methods in pain imaging studies with interesting and fruitful results. Eddy Nahmias presents further reasons (philosophical as well as scientific) to be optimistic about the prospects of successfully integrating first-person methods with the standard third-person methodology. Robert D'Amico presents a challenge to Price and Aydede by pointing out that there is a radical epistemic asymmetry between first-person and third-person data. Under certain conditions, introspective reports seem immune to certain kinds of revision, whereas third-person data are always revisable in principle. This seems to show, according to D'Amico, that the conceptual framework guiding first-person mental talk is not logically congruent with the conceptual framework guiding standard empirical evidence. Aydede and Price respond to this criticism in chapter 18 by arguing that the asymmetry exists only in degree, not in kind. They think that both kinds of evidence are revisable in principle.

Acknowledgments

The present volume has grown out of a project that was prompted by the kind encouragement of Natika Newton and Ralph Ellis. I thank them both for their help and encouragement. I would like to thank the contributors to this volume for agreeing to write for the project, for their continuous support, and for their patience with me. I am deeply grateful to my home department at the University of Florida for having provided me with enough research time to focus on this and other projects. Many people assisted generously with the long and arduous process that culminated in the eventual publication of this volume. I would like to thank especially Güven Güzeldere and Philip Robbins for their help with various aspects of this project. I am also grateful to Tom Stone for his quick and sound judgment when I needed it.

1 Introduction
A Critical and Quasi-Historical Essay on Theories of Pain

Murat Aydede

1 The Commonsense Conception of Pain

Consider the ordinary situations where we perceptually interact with our immediate environment. We identify and recognize objects and their sensible features by perceiving them. Our immediate focus and interest are normally the objects of our perception, understood in the broadest possible way (states, events, processes, conditions or their sensible features, etc.). We see the apple on the table; we see its color, size, and shape, we reach and hold it, we feel its surface, we smell it, and as we bite it we taste it. In all this, our perceptual interaction with the apple is geared toward or ensues in conceptually identifying or recognizing the apple and its sensible features. Perception typically results, in other words, in deploying concepts that apply in the first instance to the objects of our perception and their sensible qualities. We perceptually categorize the apple as an apple, its color as red, its shape as roughly spherical by mobilizing (consciously or not) the relevant concepts. In short, the locus of immediate identification or recognition, that is, immediate concept application, in perception is the perceptual object and its sensible qualities. Call this kind of situation the Normal Situation (NS).[1]

Now consider situations in which we feel pain. I reach down to pick up the pencil that I accidentally dropped on the floor behind the desk. In trying to reach the pencil behind the desk, I twist my arm and orient my hand uncomfortably, and then suddenly I feel a sharp jabbing pain in the back of my hand. In frustration, I pull back and start rubbing my hand thinking that I probably got a muscle cramp or pulled a muscle fiber. I feel pain in the back of my right hand, and I think and say so. How do situations like this compare with Normal Situations?

Of course, I *recognize that* it is my *hand*, in particular the *back of my hand*, that hurts. I *realize that* the pain *started at the moment* I twisted my arm in that way. I also *realize that* something bad must have happened, *physically*, at the location I feel pain, though

I am not sure what: my guess is that it is a small muscle cramp, and I *hope that* I didn't pull a muscle string because I know this takes more time to heal and is more painful overall. But what is it that I felt in the back of my hand when I extended my arm in that way? The obvious commonsense answer is: pain, a jabbing pain, and I recognized it as such, that is, as a jabbing pain.

Prima facie, the two situations seem quite parallel. The pain I feel has a certain location and temporal properties as well as other sensible qualities like having a certain intensity and jabbing quality, just as the apple I see on the table has certain sensible qualities such as its color, size, and shape that I see and feel. Compare:

(1) I see a red apple on the table (and identify it as such).

(2) I feel a jabbing pain in the back of my hand (and identify it as such).

For (1) to be true, there must be a red apple on the table that I stand in the seeing (perceiving) relation to (and must apply my concept RED APPLE to it).[2] What needs to be the case for (2) to be true?

It is in trying to give a satisfactory answer to this sort of question that we start to realize that the noted similarities may be superficial and misleading, and that the range of possible answers are seriously constrained by one's philosophical assumptions and commitments elsewhere in one's metaphysics, epistemology, and semantics. Pondering questions of this sort is, then, a way to see how pain phenomena have provided philosophers with a fertile ground to exercise their analytical skills.

Let us start with a few observations to disentangle the dissimilarities. When I feel pain in my hand, it seems that I have a sensory experience that prompts me to attribute something to the back of my hand by deploying my concept PAIN, just as I apply the concept APPLE to the apple I see. Indeed, it seems as if I apply PAIN to something in my hand. Could this be a physical feature or condition of my hand (such as a pulled string, or a muscle spasm, or stimulation of my nociceptors located there, or an impending tissue damage)? Suppose it is a *physical condition* of some sort that occurred when I extended my arm to pick up the pencil, call this condition *PC*, and call the experience, which occasioned my attribution of pain, *E*. Then, if I attributed this condition to the back of my hand by the application of PAIN—that is, by thinking or uttering something like (2)—the following two claims would have to be true:

• I would have a pain if I had *PC* but no *E* (as would be the case if I had absolutely effective analgesics before or during the event),

• I would not have a pain if I had *E* but no *PC* (as in the case of centrally caused pain or phantom-limb pains).

In other words, our identification, that is, our application of PAIN, would track the presence or absence of a physical condition in the relevant body parts where we normally feel pain. But this seems not to be the case. Even though some sort of tissue trauma in a bodily location is *usually* the cause of the pain we feel in that location, our ordinary concept of pain is such that our attribution of pain to that location remains true even when there is nothing physically wrong with that bodily location. Contrariwise, we do not attribute pain to a damaged body part when we don't have a certain kind of experience, that is, when we don't *feel* pain there, even when we can detect the damage by other means. In short, the truth-conditions of (2) do not seem to put any constraints on how things *physically* are with my hand.[3] Indeed, the cases of phantom-limb pain have been well documented. After losing their limbs, some people do feel pain in what might be called their phantom limbs. Sometimes this condition becomes very severe and chronic. These people are in *genuine* pain. They are not hallucinating their pain or having an ersatz pain: they are feeling real pain. Moreover, their pain is not in their stump or somewhere else in their body: they experience pain as if it were in just that location on their limb if they had not lost their actual limb. The phenomenon of so-called referred pain is similar. Even though the cause of the pain felt by people with a heart condition might be in their heart, we don't say that they feel pain in their heart when they sincerely report pain in their left hand. The pain *is* in their left hand even though there is nothing physically wrong with their hand and the cause of the pain is somewhere else. In these cases, again, it seems that our ordinary concept of pain does not track something physical in just those bodily locations we feel pain. Nor does it seem to track the *peripheral cause* of pain experiences.

But once we realize this, the attribution of pain to bodily locations or locating pain in body parts becomes very puzzling. If sentences like (2) are genuine attributions of a property or condition to body parts, then it is totally mysterious what it is that we might be attributing or locating when we sincerely utter sentences like (2).

These observations go hand-in-hand with other interesting and puzzling features of pain that have been traditionally observed and discussed by philosophers: namely, their privacy and subjectivity, and the incorrigibility of their owners' testimony about their own pains. These three features contrast sharply with the features of the objects of perception in Normal Situations.

Pains are said to be *private* to their owners in the strong sense that no one else can epistemically access one's pain in the way one has access to one's own pain, namely by feeling it and coming to know one is feeling it on that basis. This sharply contrasts with the *public* nature of objects of perception in Normal Situations: the very same apple I see on the table can be seen by others in possibly the exact way I see it, so is not private in this sense.

Pains seem also to be *subjective* in the sense that their existence seems to depend on feeling them. Indeed, there is an air of paradox when someone talks about unfelt pains. One is naturally tempted to say that if a pain is not being felt by its owner then it does not exist. Again compare the subjectivity of pains to the *objectivity* of the objects of perception in Normal Situations. The apple I see does not depend on my perceiving it in order to exist: *pace* Berkeley and phenomenalists, its existence is independent of my, or for that matter anyone else's, seeing it.

Not only do people seem to have a special epistemic access to their pains, they seem to have a very special epistemic *authority* with respect to their pain: they seem to be incorrigible, or even infallible, about their pains and pain reports. Necessarily, if it seems to me *that* I am in pain and I believe so on that basis, I am in pain. Similarly, necessarily, if I feel pain, then it seems to me *that* I am in pain and I know this on that basis. There doesn't seem to be any room for a possible gap between the appearance of pain and being in pain. In other words, there doesn't seem to be any appearance–reality distinction applicable to pain. As Kripke elegantly puts it:

> To be in the same epistemic situation that would obtain if one had a pain *is* to have a pain; to be in the same epistemic situation that would obtain in the absence of pain *is* not to have a pain. . . . Pain . . . is not picked out by one of its accidental properties; rather it is picked out by its immediate phenomenological quality. . . . If any phenomenon is picked out in exactly the same way that we pick out pain, then that phenomenon *is* pain. (Kripke 1980, pp. 152–153)

If there is no appearance–reality distinction applicable to pain, then it seems that one cannot be mistaken about one's beliefs about one's pain formed on the basis of feeling pain in the way one can be mistaken about the existence and properties of the apple one sees. In the latter case, appearances can be misleading precisely because the perceptual appearance of an apple might not correspond to what the apple is like in reality. In apparent contrast to pain, Normal Situations always involve the possibility of misperception, and thus miscategorization (i.e., misapplication of concepts).[4]

1.1   Scientists' Conception of Pain

So far we have relied in our analysis on nothing else but our ordinary concept of pain. One might think that, given the clinical importance and the urgency of pain experiences, the science of pain may have come up with a novel and more insightful conception of pain. After all, having a good conceptual grasp of one's subject matter is essential for studying it scientifically. Pain scientists, however, share much of this ordinary conception of pain. This is perhaps not very surprising since the pain scientists are part of the folk. As a matter of pinning down the subject matter of their scientific study, however, the "definition" of 'pain' has always been a vexing issue for scientists, so much so that in the early 1980s the International Association for the Study of Pain (IASP) formed a subcommittee on taxonomy to impose some order on the apparently diverse usages of pain terms in the field. "Pain" itself was not left out and became the first entry in the report. The committee consisted of fourteen internationally prominent pain researchers. Their definition has been widely accepted in the field of pain research. Although the acceptance is not universal, the remaining controversy seems to relate to its formulation and details, not to its substance.[5] This canonical characterization of pain was published in 1986 in IASP's official journal, *Pain*, and went like this:

Pain: An unpleasant sensory and emotional experience associated with actual or potential tissue damage, or described in terms of such damage.

Note: Pain is always subjective. Each individual learns the application of the word through experiences related to injury in early life . . . Experiences which resemble pain, e.g., pricking, but are not unpleasant, should not be called pain. Unpleasant abnormal experiences (dysaesthesia) may also be pain but are not necessarily so because, subjectively, they may not have the usual sensory qualities of pain. Many people report pain in the absence of tissue damage or any likely pathological cause; usually this happens for psychological reasons. There is no way to distinguish their experience from that due to tissue damage if we take the subjective report. If they regard their experience as pain and if they report it in the same ways as pain caused by tissue damage, it should be accepted as pain. This definition avoids tying pain to the stimulus. Activity induced in the nociceptor and nociceptive pathways by a noxious stimulus is not pain, which is always a psychological state, even though we may well appreciate that pain most often has a proximate physical cause. (IASP 1986, p. 250)

What is remarkable about this characterization is that it embodies, and even somewhat insists on, all the features we have discussed above. It also adds that pain experiences are always unpleasant.

2 Traditional Sense-Datum Theories

So far we have talked about pain in two ways. We have said that pain is a kind of sensory and affective *experience*, but we have also talked about pains as if they were the proper *objects* of our experience or perception. This double talk is not accidental. Indeed, it is endemic in folk as well as philosophical discourse about pain. It is in fact a robust symptom of what makes pain philosophically puzzling.

When confronted with the observed puzzling characteristics of pain embodied in this double-talk, the traditional philosophical reaction was, largely, to embrace this talk and make pain the cornerstone of so-called act–object analysis of perception, which usually took the form of a sense-datum theory typically associated with indirect realism.

Normal Situations can be analyzed as involving the perception (act) of a public object. The perceptual act on the part of the subject, in turn, is analyzed as involving an experience that typically induces conceptual categorization, that is, application of concepts to the object of perception and its qualities. This kind of act–object analysis of Normal Situations, however, is neutral about whether the perceptual experience itself can be further analyzed into another kind of act–object pair.

Consider a perfect visual hallucination of an apple on the table. Your visual experience may be said to be qualitatively type-identical to an experience you would have had if you had been actually seeing an apple on the table. Despite the absence of an actual apple, you would naturally continue to talk about an apple and its sensible qualities if you did not suspect that you were having a hallucination. In other words, you would continue to deploy your ordinary concepts as if you were applying them to an actual apple and its qualities. You would locate an "apple" on the table by thinking and saying "there is an apple on the table." You would describe an "object" as red, roughly spherical, and relatively small in size, and so on. In other words, you would attribute certain properties to what you take to be a real object. But, by hypothesis, there is no real object of the right sort located on the table or anywhere else you can visually examine. Even when you knew you were having a hallucination, it would be totally natural, and in some sense even correct, for you to talk about an "apple" and its qualities that you were seeing.

A sense-datum theorist believes that this kind of case needs to be analyzed as involving actually seeing an object that really has the sensible properties that you attribute to it in having the hallucination. This object is not, of course, an apple. But it is an object that is shaped like an apple and is really red. It is a sense-datum, a phenomenal individual that really has the qualities that it visually appears to have. Phenomenal objects, however, are no ordinary objects: they are private, subjective, and the source of incorrigible knowledge, according to sense-datum theorists. Phenomenal objects are, in some sense, internal to one's consciousness: they are not before one's sense organs.

Because this kind of nonveridical cases is phenomenologically indistinguishable from veridical perceptual experiences and because phenomenology is believed to be intrinsic to one's experience or consciousness, sense-datum theorists extend their act–object analysis of nonveridical experiences to cover veridical perception. Hence perception of the external physical reality is always indirect according to sense-datum theorists. One perceives physical objects and their qualities indirectly by directly perceiving sense data internal to one's consciousness that resemble in various degrees the physical objects and their qualities that cause them. According to sense-datum theorists, however, we are rarely, if ever, aware of this indirection in ordinary (veridical) perception in Normal Situations. It is only critical philosophical reflection on abnormal situations like illusions and hallucinations (typically along with other considerations) that reveals that the indirection must occur.

Whatever merits sense-datum theories might have with respect to genuine perception and misperception, its attraction seems undeniable when it comes to its treatment of pains and other "intransitive" bodily sensations like itches, tickles, tingles, and so on.[6] According to many sense-datum theorists, pains are paradigm examples of phenomenal individuals, mental objects with phenomenal qualities whose existence depends on their being sensed or felt, and thus are logically private to their owners who feel them. This position presumably explains the double-talk about pain being both an object and an experience, since the objects postulated are claimed to be very special: they cannot exist without the corresponding acts, that is, without someone's acts of experiencing them.[7] In other words, the concept PAIN plausibly applies to both the object part of the act–object pair in the analysis of experience and to the having of the experience that would naturally be identified with the act–object pair considered together on a sense-datum theory.

The puzzle of locating pains in body parts can be treated in more than one way within this framework. The most straightforward way is simply to take the phenomena at face value and say that pains as mental objects or sense data are literally located where they seem to be located in body parts or even in empty space (but within one's somatosensory field) where one's limb would have been before the amputation. That pains are mental particulars and depend for their existence on being sensed apparently does not *logically* preclude their being capable of having, literally, a spatial location.[8] It is not clear, however, to what extent this idea could be defended without risking incredulous looks. Pains, according to sense-datum theories, are mind-dependent objects, and many have taken this to imply that pains are internal to one's consciousness or experience, and are epistemically transparent to their owners partly because of this.[9] According to this latter view, pains cannot be literally located in physical space, but have a location in a phenomenal space or field that is somehow isomorphic to, or has systematic counterparts in, physical space. In the case of bodily sensations, this phenomenal space is sometimes called one's somatic field just as one can have a visual field that maps onto physical space.[10]

2.1 Difficulties with Sense-Datum Theories

Sense-datum theories seem to be tailormade for the peculiarities of pain experiences that we have discussed above. Indeed, their function seems to codify these peculiarities into a theory rather than explain them. After we are told how the sense-datum theories treat pains and other bodily sensations, our understanding of these don't seem to be deepened or advanced at all. Whatever puzzles we had at the start either seem to remain or be transformed into puzzles about what the theories themselves say or imply. For instance, the question about what it is that we seem to attribute to or locate in our body parts when we claim to have pains in just those parts is answered on one version of the theory by saying that we literally locate mental objects with phenomenal qualities in those parts. Furthermore, only the owner of the body can sense these objects even though they are in public space. On the other version, we are told that even though we seem to locate something in public space, appearances are misleading, we are locating private mental objects in a private phenomenal space. So pains are not, after all, located in body parts, ordinarily understood.

Pointing out the implausibility of these theories, of course, is not refutation. The defenders of these views have provided strong, often ingenious, philosophical reasons and arguments to support them. However, they don't seem to quench our intellectual thirst to understand what is going on when we attribute pains to body parts, nor do they seem to explain how there could be private subjective objects that bestow incorrigibility to their owners. Moreover, these theories seem to commit their defenders to antiphysicalism. A naturalist who is trying to understand pain phenomena within a physicalist framework could hardly admit the existence of phenomenal objects.[11] Indeed, attempts to understand pains and other bodily sensations as species of perception were motivated precisely by such concerns. The so-called perceptual theories of pain are advanced and defended on the hope that pains and other bodily sensations, contrary to first appearances, are species of information gathering that work on the same principles that govern other sensory modalities.

3 Perceptual Theories of Pain

3.1 Introduction and a Road Map

Although the question of whether pains and other intransitive bodily sensations are a form of bodily perception cuts across the question of whether they are best understood on the sense-datum model, historically at least, defending a perceptual theory of pain in the tradition of direct realism was seen as a way of avoiding the unnaturalness and the metaphysical excesses of sense-datum theories. This is most evident in Pitcher 1970, one of the most influential and elegant early direct realist theorists who has argued extensively against sense-datum theories.[12] Pitcher formulates his main motivation in promoting a perceptual view of pain as follows:

The obstacles [to a direct realist version of the perceptual view of pain] are some features of pain that seem to rule out [such a view], since they seem to demand either (a) that pains be mental (or at any rate nonphysical) particulars, or (b) that the awareness of pains be the awareness of subjective "sense-contents" that are not identical with anything in the physical world. My aim in the paper is to show that these obstacles are merely illusory, and there are no features of pains that force on us the mental-particulars view of pain. So although my attack on [this view] is only indirect, I nevertheless regard it as lethal. (Pitcher 1970, p. 369)

Pitcher characterizes (a) as the "act–object" analysis of pains and (b) as the "adverbial" version of it—which we will discuss below. Interestingly, he calls both of them the "mental-particulars view of pain" and treats them as equally ontologically problematic, probably because he thinks that even in the adverbial version we would still be committed to (bundles of) *qualities* that appear as irreducibly mental as any sense data (*qua* mental *objects*) are thought to be.

   This sort of worry (among others) has been echoed in the works of many other perceptual theorists such as Armstrong (1962, 1968), McKenzie (1968), Wilkes (1977), and Fleming (1976). Graham and Stephens (1985), and Stephens and Graham (1987) were also explicit about the ontological concerns (among others) in developing a perceptual view of pain, what they have called a "mixed composite state account" of pain according to which "to be in pain is to be in a complex state whose components include a quale and a certain cognitive and affective attitude towards that quale. Further, we hold that the qualitative component of pain is a sensible quality of the body, not a feature of the mind. . . . One who suffers pain is in a certain sort of psychological state with respect to a sensible quality of his body" (Stephens and Graham 1987, p. 395). Newton (1989) also shares the metaphysical concerns and develops a perceptual account of pain according to which the sensuous pain quality is analyzed on the model of secondary qualities of physical objects. On her view, pain is a sensible secondary quality of body parts whose integrity is threatened by potentially damaging stimuli.

   However, it would be fair to say that the most influential perceptual theorists were Pitcher and Armstrong, who developed their views about the same time and more or less independently of each other. Perhaps part of the reason for their tremendous influence was that defending a perceptual view of pain on a direct realist platform was part of their general physicalist program for naturalizing the mind in general and perception in particular. They had to deal with the intransitive bodily sensations that had always been problematic for materialism. They offered a *systematic* approach from which their account of pain was supposed to fall out naturally.[13] Because Pitcher's and Armstrong's accounts are the most developed, complete, and well-known accounts, we will focus on their views in what follows. Indeed, there is a certain sense in which their accounts, and especially Pitcher (1970), are so well carved and responsive to the problems surrounding such theories that they constitute the ideal targets in discussing the strengths and difficulties of any perceptual views of pain, including the more recent direct realist representationalist accounts developed by Dretske (1999, 2003) and Tye (1997), as we will discuss below.

However, in order to have a better understanding of why philosophers responded to the puzzles surrounding pain and other intransitive bodily sensations in the way they did, it is important to look at the dialectics of the debate as it appeared in the late 1960s and early '70s when the perceptual theories of pain had started to be developed. The opening paragraph of Pitcher's article lays out this dialectic for us quite well as follows:

I shall defend the general thesis that to feel, or to have, a pain, is to engage in a form of sense perception, that when a person has a pain, he is perceiving something. This perceptual view of pain will strike many as bizarre. But sense-datum theorists, at least, ought not to find anything at all odd in it: indeed, I am puzzled why philosophers of that school do not subscribe to the perceptual view of pain *as a matter of course*. Since I am not a sense-datum theorist, however, but a direct realist, I espouse what must at first appear to be an irremediably perverse position—namely, a direct realist version of the perceptual view of pain. (Pitcher 1970, p. 368)

A little later, Pitcher tentatively formulates this view thus: ". . . to be aware of a pain is to perceive—in particular, to *feel*, by means of the stimulation of one's receptors and nerves—a part of one's body that is in a damaged, bruised, irritated, or pathological state, or that is in a state that is dangerously close to being one or more of these kinds of states" (ibid., p. 371).

Why would a perceptual view of pain strike many as bizarre—including the sense-datum theorists? Isn't it evident that potentially injurious stimuli typically cause us to experience pain in just those bodily locations threatened by the stimuli? Add to this observation the discovery that throughout our body there are nerve endings specialized to respond to only potentially injurious stimuli, and the effects of these stimuli are carried through such pathways to specialized areas of the brain, and the conscious experience of pain is *normally* the result of such stimulation. Why, then, can't we regard pain experiences as perceptions of the states of our body tissues just as Pitcher proposes? The situation is even more curious with regard to sense-datum theorists. For sense-datum theories were in fact developed primarily as philosophical accounts of *perception* that would do justice to phenomenology and provide a unitary explanation of both the veridical and nonveridical perception like illusions and hallucinations. Indeed this is how we introduced these theories above, by reference to visual hallucination, and noted that the sense-datum theorists extended their theory to cover all perception. Nevertheless, as the quote from Pitcher implies, the sense-datum theorists had not embraced a perceptual view of pain. To understand why the sense-datum theorists had been—along with almost every one else in fact—reluctant to regard pains and other intransitive bodily sensations as species of perception, we need to return to our initial observations about feeling pain and other bodily sensations and highlight the differences between them and perception in Normal Situations from a slightly different perspective.

Because the dialectic of the debate is rather complex, as can be seen from Pitcher's opening paragraph, it is useful to have a road map for the discussion to come in the remainder of this section. We will first (sec. 3.2) spell out the general reasons for why giving a perceptual account of pain and other intransitive bodily sensations—whether this account be an indirect (sense-datum) or direct realist account—seemed implausible and proved so difficult to develop. Second, (sec. 3.3), we will lay out how a sense-datum theory can accommodate a perceptual view of pain; in other words, we will show how sense-datum theorists could in fact subscribe to a perceptual view of pain as a matter of course, as Pitcher puts it. This will give us one version, the indirect realist version, of the perceptual view of pain. Third, (sec. 3.4), we will discuss the main proposal such a theorist could make in order to solve the main difficulty confronting perceptual theories. After briefly assessing the plausibility of such a proposal, we will, fourth, (sec. 3.5), take up the direct realist version of perceptual theories of pain and other intransitive bodily sensations, and show why such theories might indeed strike one as perverse, much more so than the indirect realist versions of the theory. As we will see, the direct realists will have the same (inadequate) resources to meet the main difficulty but will have additional problems. We will also have a discussion of so-called adverbialism and intentionalism in addressing the direct realist attempts to meet some of these difficulties. Then, in the following section (4), we will take up the more modern versions of direct realist attempts to give a perceptual account of pain: these are known as representationalist theories of pain. We will see that these views do not seem to fare any better than their predecessors.

3.2   The Main Difficulty for Perceptual Views of Pain

Here is a line of argument that brings out the intuitions behind a general resistance against a perceptual view of intransitive bodily sensations. When I feel a sudden itch on my back, or a tickle under my armpit, or experience an intense orgasm, or a jabbing pain in the back of my right hand, what is it that I am perceiving? What is it in the relevant parts of my body that my experience represents? The answer seems to be: "I have absolutely no idea" (and, moreover, for the most part, "I don't care!"). In many cases I can identify the cause, but not always. Sometimes a particular spot on my back suddenly and intensely starts to itch for no apparent reason at all. I have absolutely no idea what, if anything, I am perceiving, let alone misperceiving. If this is the proper thing to say, then itching is a very odd form of perception if it turns out to be such. When I seem to see something, I can tell a lot about what it is that I seem to see—and not just in objectual terms: I can also talk about patches of colors, shapes, and so on, in my visual field that directly relate to the object of my perception. In other words, I can readily *conceptually articulate* what I take myself to see and what its (sensible) properties are, and in principle others can do the same regarding the same object. This is not surprising: in perception we gather information about our immediate environment which we use in crucial ways in navigating that environment. If I have no idea about the object of my perception and its properties in itching except the awareness of a certain kind of feeling (sense datum?), and an immediate desire to relieve the accompanying discomfort by scratching, then what is the point of viewing itching as a form of perception? The same goes for other intransitive bodily experiences including pain of course.

Recall that Normal Situations are those in which we are prompted by the perceptual experience involved to deploy concepts that apply, in the first instance, to the objects of our perception and to their sensible qualities. In seeing an apple on the table, we are having a visual experience as of an apple and identify or recognize *it*, not the experience but the apple, as such. Ordinarily, we are rarely—if ever—interested in our experiences in perception. However, as the line of thought above reveals, situations in which we feel pain or an itch or a tickle seem never Normal, in that the concepts we are immediately prompted to apply don't seem to apply to public objects of perception at all: the concepts we are immediately induced to apply are none other than the concepts PAIN, TICKLE, or TINGLE, and so on. What these concepts apply to, however, are not public and objective, at least in the sense that they don't seem to track a physical condition of those body parts to which they are applied or attributed, nor do they seem to track the peripheral causes of these experiences—as indeed emphasized by the IASP definition of 'pain.' As we have seen in the beginning section (1), that is wherein the mystery of intransitive bodily sensations lies that attracts philosophers' attention. Pain is the most dramatic and pronounced example of all, which has also the most serious practical (and clinical, as the case may be) urgency of all.

So there is an asymmetry in what we might call the primary locus of concept application or recognition. Whereas conceptual identification tracks public objects and their sensible properties in perception in NS, conceptual identification or recognition that occurs in having intransitive bodily sensations seems to track experiences or their features, not what these experiences may be tracking—if they do track anything public (see figure 1.1). A consequence of this asymmetry is that to the extent we are incorrigible about recognizing our experiences and their subjective and private objects—if there be any such—to that extent we are incorrigible in recognizing our pains and other intransitive bodily sensations, and this is very unlike having perception in Normal Situations, which always involve the genuine possibility of misperception and thus misidentification (misapplication of concepts) of public objects.

Perception in Normal Situations (e.g., vision):

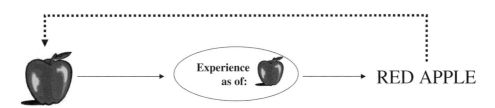

Intransitive Bodily Sensations (e.g., pain):

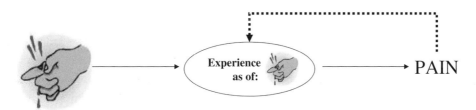

Figure 1.1
The asymmetry in concept application even when it is assumed that the structure of information flow is symmetrical in the contrast cases. (But the assumption that the information flow is identical and that the pain experience represents tissue damage may be resisted, indeed, might even be rejected on the basis of scientific considerations—see below.) Sense-datum theories analyze the experiences as themselves consisting of an act–object pair, in which case the little pictures inside the experience circles may be taken to represent the sense data involved in the pair, the immediate experiencing of which would be the act. The pair as a whole can then be taken as the indirect perception or indirect experience of the public object.

Tempted by these observations, suppose we identify genuine perception with perception in Normal Situations. Such an identification seems quite natural given the job perception is supposed to accomplish as a gatherer of information about one's immediate environment (including bodily environment). Then feeling pain and having other such bodily sensations will not count as species of genuine perception, and this is *the main difficulty* for perceptual views of pain. So the perceptual theorist has to explain or explain away why there is such an asymmetry in concept application in such a way that the result would still justify taking pains and other intransitive bodily sensations as species of genuine perception.

3.3   Indirect Realist (Sense-Datum) Theories as Perceptual Theories of Pain

Given this difficulty, the sense-datum theorists (indeed, any theorist) are justified, at least prima facie, in not subscribing to a perceptual view of pain. The fact that these theorists analyze genuine perception in NS by postulating sense data that resemble their causes to various degrees does not require that they should analyze every phenomenon involving sense data as a species of perception. This, in fact, seems to be an advantage of sense-datum theories given that feeling pain and other bodily sensations seem so unlike genuine perception. The dissimilarities between pain and perception in NS, however, are perfectly general and seem to justify one to believe in the nonperceptual character of feeling pain whether or not one is a sense-datum theorist.

In the next section, we will examine why a direct realist version of a perceptual view of pain might strike many as more perverse than an indirect realist version of it. To see this, however, we first need to make a case for how a sense-datum theorist as an indirect realist could in fact subscribe to a perceptual view of pain and other such bodily sensations.

The crucial move is to acknowledge that the concepts PAIN, ITCH, TICKLE, and so on, apply to the relevant perceptual experiences or to their internal private objects, sense data, but insist that these, namely, sense data or acts of sensing these sense data, *constitute* the perception of disordered bodily conditions of the relevant parts we attribute these sensations to. So pains are experiences in which we indirectly sense bodily damage by directly sensing the relevant sort of sense data, just as we indirectly see the apple on the table and its sensible properties by directly "seeing" an apple-shaped sense datum and its qualities. The difference is not in the structure of the perceptual process or the flow of information involved—the two cases are quite parallel. Rather, the difference is in the locus of recognition or concept application just as we have observed above. But a sense-datum theorist defending a perceptual view of pain would insist that this difference, in and of itself, has no bearing on whether the relevant sense datum involved or the experience considered in general as the act–object pair (i.e., as the act of directly sensing the sense datum) is genuinely a *perceptual* act.

This amounts to denying that genuine perception is perception in NS where the recognitional locus is the public object of perception. If feeling pain is an act of direct sensing of a certain sort of sense datum, and if this sensing is in fact the indirect but genuine perception of a disordered state of a body part, then, according to the perceptual sense-datum theorist, misperception is certainly possible in having a pain, just as hallucinations are possible in visual perception. Namely, if this act of sensing or the relevant sense datum (= pain) represents a certain body part as being damaged or traumatized when this is not the case as in referred pains, phantom limb pains, and most typically centrally caused chronic pain syndromes, then the relevant act of sensing or its datum *is* a misrepresentation, a misperception. In such cases, it is not that we are "misperceiving" our pain experience (no, we correctly recognize our experience as pain), rather the experience itself misrepresents, and so we are indirectly misperceiving the condition that our relevant body parts are in.

For this move to work, the perceptual sense-datum theorist must offer an adequate account for why the locus of concept application is different in the two cases despite the same information flow structure. Recall that the main intuition behind restricting genuine perception to NS was that perception is essentially an activity of gathering information about one's immediate external environment, that is, environment external to one's mind and experiences, where this environment involves one's body or events internal to one's body. It is this information that will serve one's practical interests of often vital importance: the information about the structure of one's experience and its internal objects is useful only insofar as it is a guide to one's external environment. By anchoring perceptual recognition (i.e., conceptual identification directly induced by perception) in the external environment, genuine perception and the conceptual mechanisms that it feeds into efficiently serve us to negotiate this environment of which we are a part. Genuine perceptual experience, in other words, is transparent largely because the locus of recognition is the public object of experience and its sensible qualities. But anchoring the locus of recognition in phenomena internal to one's mind or experience will not immediately serve this vital function or will not serve it well, since the indirection involved will not be automatic or very accurate: we will be directed to relevant body parts only insofar as we succeed in using the information in our experience as a guide to its causes. In other words, pain and other intransitive bodily sensations will not be *transparent* and reliable guides to the physical conditions that they represent and typically carry information about.[14]

Given all this, internal anchoring, at a minimum, must by default be interpreted as a sign that the processes involved are not perceptual. It is the burden of the perceptual theorist (the direct as well as the indirect realist) to dislodge this default assumption and offer an explanation of why there is an anomaly in the locus of concept application in bodily sensations even though there is no difference or anomaly in the information flow.

3.4 A Common Response to the Main Difficulty

The most natural explanation, on the part of a sense-datum theorist, is that unlike other perceptual modalities, the acts of sensing a sense datum involved in feeling pain have a very pronounced negative affective quality: pains are unpleasant, awful, hurtful. This negative hedonic quality, on a sense-datum theory, may attach to either the sense data involved in pain experiences, or to the act of sensing them. Either way, it is this negative affect that explains why the locus of concept application is the acts of sensing or its internal objects, rather than the external objects of perception, that is, traumatized body parts. In other words, it is this negative quality that turns the recognitional focus onto itself or the datum it attaches to. If this is correct, then, of course, pains are equally unpleasant even when they misrepresent. This is why we pick out our acts of sensing (or their direct internal objects) rather than their external objects irrespective of their informational etiology: whether or not they are veridical, they equally hurt.

Although we have presented this sort of explanation as given by a sense-datum perceptual theorist, it is the kind of response that can be and has been co-opted, *mutatis mutandis*, by *all* perceptual theorists, as we will see below.

This seems to be a plausible explanation on a first pass. Indeed, a perceptual theorist can even run, plausibly, an evolutionary story about why these sensing acts should feel unpleasant: they represent or signal a property of body parts that tends to hinder survival. However, can the same sort of explanation be given for other intransitive bodily sensations like itches, tickles, tingles, and orgasms? Perhaps it can be done for experiencing orgasms, which are usually intensely pleasurable and have obvious evolutionary benefits. But what about others? Itches have an unpleasant quality to them that normally makes one want to scratch the spot where one itches to stop it. But tickles and tingles can be pleasant at times and unpleasant at others, as well as affectively neutral at still other times. Still, the concept of a tickle or tingle is very much like the concept of a pain in that they apply to the acts of sensing or experiencing, rather than to what external conditions these acts may be representing—if they represent anything. Indeed, if they were genuine perception of bodily conditions, they would be very odd forms of perception: nobody seems to have the slightest idea what they are perceptions of, which is manifested in the fact that we don't have concepts that apply to what these experiences may be representing. The concepts we have are none other than the concepts TICKLE, TINGLE, which, like PAIN, don't seem to track physical conditions of the body parts that they are applied to—you can run the little thought experiments we conducted with respect to pain/PAIN in the first section. Besides, gustatory and olfactory experiences can be pleasant, unpleasant, or affectively neutral, yet the acts of sensing involved in the exercise of these sensory modalities seem to be largely transparent in that we apply the relevant concepts like SWEET, BITTER, and so on, to the external objects of these experiences in the first instance, and only derivatively or incidentally to the acts of sensing or to the experiences themselves.

These observations cast doubt on the plausibility of the explanation offered by the perceptual theorist for the asymmetry in concept application. However, the perceptual theorist can still claim that whatever the explanation might be in the case of other intransitive bodily sensations, the explanation offered for pain is essentially correct, and such a theorist may thus conclude that she has discharged the burden of proof in claiming that experiencing pain is engaging in genuine perception. This reply has some initial plausibility since pain experiences have almost always a pronounced negative affect. Nevertheless, if one suspects—as one should—that the intransitivity of certain kinds of bodily sensations as a whole must have a unified explanation, one would be wise to conclude that the offered explanation cannot be the whole story and thus is not adequate all by itself.

There is, however, quite substantial scientific evidence that there are abnormal pain phenomena where the sensory and affective aspects of pain experiences are disassociated from each other—see section 5 below. The most typical case is known as the pain asymbolia syndrome, where people who suffer from it have pain experiences without the negative affect. Interestingly, these people still identify their experience as pain, but show no bodily, emotional, and behavioral signs typically associated with the unpleasant aspect of pains. They are feeling a pain that doesn't hurt! If pains are not necessarily unpleasant, as this syndrome seems to show, it's an interesting and open question whether feeling pains without its negative affect would still retain its intransitive nature where it's conceived as an experience that doesn't normally admit a perceptual object beyond itself. (For considerations that it would, see Aydede and Güzeldere 2005.)

3.5 Direct Realist Perceptual Views of Pain
We are now in a position to appreciate why defending such a view on a direct realist version of the perceptual theory would appear to be more perverse than defending it

on an indirect realist version. As we have seen, a perceptual sense-datum theorist claims that one indirectly perceives one's disorderly bodily states by directly sensing a certain kind of sense datum, and either identifies pain with this sense datum or identifies feeling pain with the act of sensing this sense datum (which pair we might just call the pain experience); this then becomes the locus of concept application without regard to the informational etiology of the experience. In other words, the indirect realist *does have a locus for concept application*; it is the experience itself or its internal object whose direct perception constitutes the indirect perception of a bodily condition.

If one wants to run the perceptual theory on a direct realist version of it, however, one immediately runs into the difficulty of finding an appropriate locus for concept application, of finding an object, state, or event to which we apply PAIN, ITCH, TICKLE, and so on. This is because the mark of any direct realism in the theory of perception is the repudiation of perceptual intermediaries that mediate genuine perception in Normal Situations: when I see a red apple on the table, there is no object or quality distinct from the apple and its redness such that I see the apple in virtue of seeing it, or more generally, in virtue of perceiving or being aware of it. On this view, when I see an apple, I directly perceive, or am directly acquainted with, the apple and its qualities, like its redness. This view works well in veridical perception in NS: the locus of concept application is always the public object of perception, like the apple and its properties. It also explains why the concept application is the way it is even when one hallucinates or has illusions: as we have discussed above, even when one hallucinates a red apple, one applies the concept RED APPLE to what one either takes or is tempted to take to be a public object and its publicly available features, rarely (if ever) to what one believes to be a private mental object.

So what does a direct realist perceptual theorist propose that we do when we attribute pains to body parts and report being in pain? Contrary to the initial impressions one might get from reading the passage quoted from Pitcher (1970) above, pain is not identified with the disorderly condition of one's body. According to early direct realists like Armstrong and Pitcher, to feel pain in the back of one's hand is to be in an experiential state that constitutes the perception of a disorderly physical state of the back of one's hand. Thus reporting pain is reporting one's experience to that effect. Even if there is nothing physically wrong with one's hand so that the experience misrepresents, it is still true that one is having an experience to that effect. Pitcher offers an ingenious comparison between feeling pain in a part of one's body and catching a glimpse of something in order to explain the subjectivity of pains:

I maintain that the concept of a pain is remarkably similar to that of a glimpse. Just as there can be no uncaught glimpses, there can be no unfelt pains. Glimpses must be caught in order to exist, not because they have troublesome nonphysical status, but simply because we deem it a necessary condition for the existence of a glimpse that an act (or state) of catching it is going on. In exactly the same way, pains must be felt in order to exist, not because they have troublesome nonphysical status, but simply because we make it a necessary condition for the existence of a pain that an act (or state) of feeling it is going on. So the to-be-is-to-be-felt quality of a pain corresponds precisely to the to-be-is-to-be-caught quality of a glimpse. (Pitcher 1970, p. 377)

And a little further, he says the following about the privacy of pains comparing them to glimpses:

[T]he criterion for the identity of a pain is not the identity of the bodily part whose disordered state is felt, but rather the identity of the act (or state) of feeling it. Pains, then, are not interestingly private: I mean, they are private because they are particulars enjoying a special nonphysical status. They are boringly private, because their privacy really amounts only to the following triviality: each person can perform only his own acts of feeling something (or can be only in his own states of feeling something). So the undoubted privacy of pains corresponds precisely to the (perhaps doubtful) privacy of glimpses. (Ibid., p. 378)

Direct realists reject the act–object analysis of experience advanced by sense-datum theorists. In visually hallucinating a red apple on the table, one does not directly see a private mental particular; rather one is having a visual experience that is like an experience that is normally brought about when one actually sees a real red apple on the table. Direct realists, in other words, typically insist that such cases should not be analyzed as a perceiver standing in a certain perceptual relation to a private mental object. Rather the analysis involves only one particular, the perceiver herself, and her being in certain sorts of (perceptual, experiential) states or conditions that are typically brought about under certain circumstances in which one genuinely perceives something. In standard cases, when one is having a veridical perception, the experiential state of the perceiver is brought about by the actual object of her perception, and the perceiver's state is qualitatively differentiated by the causal influences of the sensible qualities of the public object.[15] In nonstandard cases like in hallucinations and illusions, the same kinds of states are brought about by different causal routes, and the qualitative differentiation of one's experiential state in such nonveridical cases is the result of deviant causal influences.

3.5.1 Adverbialism and intentionalism

This sort of analysis of experience is sometimes known as adverbialism in the literature, because in perceiving a red object one is said to be in a state of perceiving something "red-ly." The theoretical import of this way of talking is that perceiving something that is red is a *manner* of perceiving that object that would be distinct from the manner of perceiving it if the object were blue, in which case one would be perceiving it "blue-ly." Similarly when one hallucinates a red object, there is only one object, the perceiver who is sensing in a certain manner, namely, red-ly. In other words, 'red' is said to qualify not a private object but rather a state or activity of a person, that state being a manner of perceiving or sensing physical objects that are red. Compare the following analogies:

(3)   Judy is dancing a waltz.

(4)   The smile on John's face was mischievous.

It would be a mistake to analyze these sentences respectively as follows:

(3a)   There is an object that is a waltz such that Judy is standing in the dancing relation to it.

(4a)   There is a unique object which is a smile and mischievous such that John has stood in the having-on-the-face relation to it.

Rather it is obvious that they should be rendered something like:

(3b)   Judy is dancing waltz-ly.

(4b)   John was smiling mischievously.

The ontology of (3a) and (4a) seems costlier than that of (3b) and (4b) in that the former require two mysterious particulars *in addition to* and *distinct from* Judy and John, whereas the latter require, at least prima facie, only two particulars of a familiar sort, Judy and John, and their being engaged in an activity of a certain kind or in a certain manner. The point of adverbialism in philosophy of mind and perception is to use this maneuver with the hope that whatever the true formal semantics of adverbs will turn out to be it will not commit us to strange sorts of particulars: on the whole we feel confident about what is going on ontologically when we express a state of affairs by uttering a sentence in which one adverbially qualifies a verb.[16]

So similarly when one is having a bluish afterimage, one is not standing in the seeing relation to a certain mysterious particular that is blue in the way sense-datum theories demand. Rather, on adverbialism, one is in a state of experiencing in a certain manner, that is, blue-ly; this is usually unpacked as being in a sensing state of a certain sort, of the sort one is typically in when one is actually seeing something blue. In other words, 'blue' in reporting a blue afterimage does not qualify a physical or mental particular that is actually blue, rather it qualifies an activity of the person qua standard perceiver.

Adverbialism has various versions and is fraught with many difficulties. Moreover, not all of them are motivated by the same set of concerns. Nevertheless, for our purposes, we can take adverbialism as an attempt to get rid of mysterious mental objects in favor of metaphysically less costly states or activities of persons or manners of perceiving that qualify persons qua subjects of experience. So a pain experience, for a direct realist, is a specific manner in which tissue damage is (somatosensorially) perceived in a bodily region. When we report pain we report the occurrence of experiences understood this way—adverbially.[17]

Pitcher seems to apply this sort of adverbialist analysis to pain: "The important respects in which the nonstandard cases resemble the standard ones are obvious . . . in both, it seems to the person just as if he were feeling the disordered state of a certain part or region of his body; it feels to him, as we might also put it, as though there is something wrong in that part of region of his body" (Pitcher 1970, p. 384). Since the concept of pain applies to the act of feeling the disturbance rather than the disturbance itself, the locus of concept application is the experience. The difference between the direct realist and the indirect realist defender of a perceptual view of pain is that while the latter analyzes experience in act–object terms, the former doesn't. For the direct realist, even though the experience may be necessary for us to perceive the world, we do not in any way need to directly or otherwise perceive our experiences or their alleged internal objects in order to perceive the public world around us. We perceive the world with or through the experiences we have but we do not perceive our experiences in order to perceive their public objects.[18]

But we still need to know what exactly is going on when I truly say that I feel a pain in the back of my hand. How should utterances of this sort be analyzed on a direct realist version of the perceptual view of pain? How does the main adverbialist tenet help us understand the semantics of such utterances? Armstrong proposes the following analysis:

'I have a pain in my hand' may be rendered somewhat as follows: 'It feels to me that a certain sort of disturbance is occurring in my hand, a perception that evokes in me the peremptory desire that the perception should cease'. . . .

The force of the word 'feels' in this formula is no more and no less than the force of the word 'feels' in 'My hand feels hot,' where this latter sentence is so used that it neither asserts nor excludes my hand being hot in physical reality. . . . Now while there is no distinction between felt pain and physical pain, there is a distinction between *feeling* that there is a certain sort of disturbance in the hand, and there actually *being* such a difference. Normally, of course, the place where there feels to be such a disturbance *is* a place where there actually is a disturbance, but in unusual cases, such as that of the "phantom limb," or cases of "referred pain," there feels to be disturbance in a place where there is no such disturbance.

The problem of the "location" of physical pains is therefore solved in exactly the same way as that of the location of "transitive" sensations, such as sensations of pressure. . . . [T]he "location" of the pain is . . . an intentional location. There feels to be a disturbance of a certain sort in the hand, whether or not there is actually such a disturbance.

This account of the location of pain enables us to resolve a troublesome dilemma. Consider the following two statements: 'The pain is in my hand' and 'The pain is in my mind.' Ordinary usage makes us want to assent to the first, while a moment's philosophical reflection makes us want to assent to the second. Yet they seem to be in conflict with each other. But once we see that the location of the pain in the hand is an intentional location, that is, that it is simply the place where a disturbance feels to be, but need not actually be, it is clear that the two statements are perfectly compatible. (Armstrong 1968, pp. 314–316)

Armstrong and Pitcher seem to propose what we might call an "error theory" of pain attribution. Even though the surface structure of sentences like

(2)   I feel a jabbing pain in the back of my hand

suggests that there is an attribution of pain to a bodily location, or more strictly, that I stand in the feeling relation to a pain that is located in a part of my body, the proposed analysis says that this is not what is going on. What I do when I utter (2) is an attribution of a different sort: I attribute to *myself* a *feeling state* that has an intentional content to the effect that a certain region of my body (the back of my hand) is in a physical condition of a certain sort (in a physically damaged or disordered state or in a state close to being damaged). The error is due to the fact that the pain is not in my hand; the pain, being a state of feeling or experience, is "in my mind." It is the physical disturbance that is in my hand which my feeling state represents (in a confused and indistinct way, as Descartes would put it). A moment's philosophical reflection, however, according to Armstrong, makes us realize that in uttering (2) I actually attribute an intentional feeling *state* to *myself* which in turn attributes a physical disturbance to my hand. The colloquial ways of speaking just jumble the pain with the disturbance and thus mislead us. Pains, on this view, are experiences, not objects of our experiences. Moreover, since these experiences have intentional content, they have accuracy conditions: they can be correct or incorrect; they can veridically represent or misrepresent. But even when they misrepresent, these experiences are pain experiences. So I can be in genuine pain, even though there is nothing physically wrong with my hand. The incorrigibility mentioned before reduces to one's incorrigibility about one's occurrent experiences. To the extent to which we are incorrigible in discriminating and conceptually identifying our own experiences, to that extent we cannot be wrong about our own pains. Hence the locus of concept application are the pain experiences, even though it might seem to us as if we were applying the concept of pain to bodily locations.

3.5.2 Difficulties with the direct realist perceptual view   The direct realist version of the perceptual view of pain appears to have the same resources that the indirect realist version has for answering the crucial question of why we are conceptually "focused" on the experience rather than its object, the bodily disturbance. In other words, the question of why the perception involved in feeling pain does not conform to ordinary perception in Normal Situations gets answered by saying that the perceptual state of feeling pain is of such a nature that it immediately evokes in the person having it the "peremptory desire that the perception should cease." Pitcher writes, for instance:[19]

[I]n the nonstandard cases, the person usually has an immediate inclination to change his "state of awareness," an immediate desire to want it stop, just as a person who has a pain in the normal cases usually has. In other words, the person's state of awareness in the nonstandard cases is usually every bit as unpleasant, intolerable, or whatever, as the state of awareness of someone usually is who, in standard cases, has a pain. This feature of his state of consciousness is obviously an extremely important one—indeed, probably, in most cases, its most important aspect— and as such exerts a powerful force toward assimilation: it makes us unhesitatingly subsume the nonstandard cases under the concept of pain (or of feeling pain). Conscious states having this feature are clearly of great concern to us, so that it is only to be expected that we should gather them all together under a single concept. (Pitcher 1970, p. 385)

Pitcher talks about a specific "feature of the state of consciousness" of someone in pain. This feature seems to be identified with the unpleasant (intolerable) aspect of pain experiences. Furthermore, this feature is present in the experience, whether or not this experience is veridical ("nonstandard"), and is causally efficacious in making the person undergoing it want to stop the experience. It is the feature that anchors the concept of pain to the experiences rather than their public objects.

If we were not, in some sense, aware of this "extremely important" feature, we would not, presumably, subsume the nonveridical experiences along with veridical ones under the concept of pain. Indeed it is the lack of this feature that explains, for instance, why we don't have an *ordinary* unitary concept of "seeing" that subsumes both visual hallucinations or illusions and veridical visual perceptions.

As we have noted above, this sort of explanation of the crucial asymmetry between pain and perceptions in NS is not adequate: it is at best incomplete, and at worse, wrong. We have other intransitive bodily sensations (tickles, tingles) where we have the same sort of asymmetry, but these don't exhibit the same kind of strong negative or positive affect. We also have cases of genuine perception in NS (smelling, tasting) where we have often affect but not the same kind of internal concept anchoring. So a perceptual theorist still has some important explaning to do.

However, there is a further issue about the direct realist's appeal to this sort of explanation. Recall that the basic tenet of direct realism is that we do not perceive our experiences or anything like their alleged internal objects (sense data) in order to perceive their public objects. We perceive the world with or through the experiences we have, but we need not and in fact don't perceive them to do that. However, there is a question about to what extent it is legitimate for Pitcher to appeal to a feature of conscious pain experiences in the explanation he offers. He obviously talks about occurrent phenomenology one enjoys in having pain when he talks about this unpleasant feature. How do we become aware of this feature? In fact, there is a more fundamental question: how do we become aware of our pains so that we correctly discriminate and conceptually categorize them by thinking or uttering sentences like (2)? If feeling pain is having a perception of one's body part, the correct mode here should be external perception, but what we seem to have is rather introspection in coming to know one feels pain. For if introspection is a special first-person mode of knowing one's mental states and experiences in particular, and feeling pain is undergoing a conscious experience, then coming to know one is in pain is to engage in introspection. Therefore, if pain is a form of bodily perception, it must be a very different kind of perception: it seems to involve introspection in an *essential* way. For if we didn't know we were in pain, that is, we didn't introspect, we could not perceive the physical disturbance in a bodily region. Recall that according to the "error theory" in reporting pain we primarily report that we are in a certain feeling state, that is, we self-attribute a mental state, and only indirectly report a bodily disturbance as the intentional content of that state. So in the general case of intransitive bodily sensations and in pain in particular, it seems that we need to "perceive" (become aware of) our experiences and their phenomenal features after all in order to perceive bodily disturbances they represent.[20] So it seems that it is simply not true that we do not need to be aware of our experiences in order to be aware of the external bodily conditions they represent.[21]

Early direct realist perceptual theorists like Pitcher and Armstrong played down the importance of phenomenology.[22] Indeed, in one of the quotes from Pitcher earlier, we have seen that he regards as lethal the objection that "the awareness of pains be the awareness of subjective 'sense-contents' that are not identical with anything in the physical world." This was in line with their naturalist program in the theory of perception and mind. Both Pitcher and Armstrong developed a theory of perception as a form of belief acquisition where phenomenology did not play any significant role and beliefs are treated as functional states. Indeed, they were vehemently criticized on this ground: beliefs qua conceptually structured cognitive states don't seem to have any essential phenomenology attached to them. When perception was explained essentially as a form of belief acquisition, there was no place for the phenomenology of perceptual experiences in the theory. This is an especially acute problem when it comes to the negative affective feature of pain experiences that these theorists had to advert to in order to explain the asymmetry in concept application in pain and in other perceptions in NS. Armstrong, for instance, talks about pain as a "perception that evokes in [one] the peremptory desire that the perception should cease" to accommodate the pain's robust affective phenomenology. Pain as a feeling state does, of course, evoke this desire in one. But Armstrong offers this as an *analysis* of pain attribution sentences. Understood this way, however, it seems lacking: one would naturally be looking to see mention of a *qualitative feature* of one's experience as the *reason* (or at least, cause) for the desire—if we are to understand this talk of desire in its ordinary sense. No such feature is ever mentioned.

The robust episodic phenomenology of affect has posed serious difficulties with these theories also because they didn't seem assimilable to perception at all. They could perhaps be aspects of perceptual states, but it was not clear how a direct realist could give a perceptual account of these aspects themselves. The general tendency was to analyze them as certain functional aspects of perceptual states as Armstrong's analysis seems to propose indeed.

4   Modern Representationalist Theories of Pain

More modern direct realist views about pain usually start with the acknowledgement that the qualitative phenomenology of all perceptual experiences are essential to their being perceptual.[23] That is because these theories are openly advanced generally as *representational* theories of qualitative phenomenology (qualia). Unlike earlier direct realists, most modern direct realists are realist about qualia. Although representationalism in this form is a general theory about the phenomenal qualities of *any* experiences, some of their defenders, aware of the problematic character of bodily sensations, have attempted to explicitly apply the theory to pain. Among the defenders of representationalism who have done this are Harman (1990), Dretske (1995, 1999, 2003), Tye (1996, 1997), Byrne (2001), and Seager (2002).[24] But the most prominent and elaborate defenders of the theory are Dretske and Tye.

Representationalism about qualia in this context needs to be understood in a robustly reductionist sense: the claim is that the representational content of an experience metaphysically constitutes and exhausts the nature of its phenomenological content. As we have seen earlier, indirect realism, especially in the form of sense-datum theories, were also advanced as representationalist theories (perhaps excluding the intransitive bodily sensations). Indeed, the old name for indirect realism was "representative realism." On these theories, in having a perceptual experience we are directly acquainted with qualia, understood either as intrinsic qualities of experiences or as qualities of phenomenal individuals like sense data. But, at least in perception in NS, these qualities *represent* objective sensible properties of public objects in virtue of either resembling them or by being regularly caused by their instantiations—or both. In other words, on an indirect realist approach, they are distinct existences: qualia or sense-data come to represent public objects and their sensible properties in virtue of some contingent relations between them (e.g., resemblance and causation). However, in her acknowledgment of phenomenology, the reductionist direct realist cannot have such actual phenomenal objects somehow internal to one's mind or experience, nor is it likely that she can endorse the existence of qualities *intrinsic* to experiences that we can become aware of in introspection. If qualia are to be retained in one's direct realist picture of perception, these qualia need to be reduced to representational content of perceptual states. Like earlier direct realists, representationalists tend to be naturalist or physicalist.

The essential move here is the strict identification of representational properties with phenomenal ones. Even though any reductionist account would in principle do (reductionism per se does not entail naturalism), given direct realism and the naturalistic motivation behind it, reductionist accounts tend to be naturalist or physicalist. Therefore, the reductionist move brings with it an attempt to give a naturalist account of these representational properties. Accordingly, experiences are conceived as brain states with a certain functional profile. What makes these brain states experiences, in other words, is that they play a certain role in the mental economy of their possessors in virtue of their representational content. These theories therefore come with a naturalist account of how these states acquire the representational content they have. The standard story is either an ideal causal covariation theory (informational semantics) or a teleological psychosemantics or both, which are externalist theories.[25]

When applied to pain, qualia representationalism is known as the representationalist theory of pain. Tye states the theory succinctly:

Pains . . . are *sensory* representations of tissue damage. To feel a pain, one need not have the resources to conceptualize what the pain represents. . . . One need not be able to say or think that such-and-such tissue damage is occurring. Still, the content of the pertinent sensory representation is what gives the pain its representational character. . . . A twinge of pain is a pain that represents a mild, brief disturbance. A throbbing pain is one that represents a rapidly pulsing disturbance. Aches represent disorders that occur *inside* the body, rather than on the surface. . . . (Tye 1997, p. 333)

In brief, the representational view of pain is the direct realist perceptual view of pain with a naturalist representationalist (hence, reductionist but realist) understanding of the *entire* phenomenology of pain experiences.

4.1   Standard Difficulties with Representationalism

The theory is quite controversial and has many opponents. It would probably be fair to say that the prevailing wisdom seems to be that even though pains may have representational content, or better, even though some aspects of pain phenomenology may be constituted by pains' representational content, a pain's phenomenology is not exhausted by its representational content. Many modern opponents put their objections in terms of representational–intentional versus purely sensational properties of experiences. McGinn, for instance, makes the point quite strongly:[26] "[B]odily sensations do not have an intentional object in the way perceptual experiences do. We distinguish between a visual experience and what it is an experience of; but we do not make this distinction in respect of pains. Or again, visual experiences represent the world as being a certain way, but pains have no such representational content" (McGinn 1997, pp. 8–9). This is perhaps a bit too strong, but it certainly reflects a widespread sentiment about representationalism about pain and other intransitive bodily sensations.

It appears, then, that the representational theory of pain has, at least structurally, the same weaknesses and strengths of the early perceptual direct realist theories. Just like its predecessors, it confronts the problem of explaining why pain experiences do not seem representational or perceptual at all, or as we put it earlier, why the concept of pain (hence our ordinary understanding) works the way it does, why there is an asymmetry in concept application when the direct realist representationalism in fact crucially predicts otherwise. However, in modern theorists' discussions we don't even see an explicit acknowledgement of this problem, let alone a wholehearted attempt to deal with it as the early theorists did. Recall that it is this problem (what we have called above "the main difficulty with pain") that makes any perceptual view of pain intuitively implausible.

The affective phenomenology of pain experiences is given—if at all—at most a cursory treatment, and a cognitivist/functionalist, not a representationalist, one at that, while advocating a full-blown representationalism for *all* qualia of any kind. Tye, for instance, writes:

How is [the painfulness of pains] to be accounted for within the above proposal? To begin with, it should be noted that we often speak of bodily damage as painful. When it is said that a cut in a finger or a burn or a bruise is painful or hurts, what is meant is (roughly) that it is *causing* a feeling, namely the very feeling the person is undergoing, and that this feeling elicits an immediate dislike for itself together with anxiety about, or concern for, the state of the bodily region where the disturbance feels located. Of course, pains do not themselves normally cause feelings that cause dislike: they *are* such feelings, at least in typical cases. So, pains are not painful in the above sense. Still, they are painful in a slightly weaker sense: they typically elicit the *cognitive* reactions described above. Moreover, when we introspect our pains we are aware of their sensory contents as painful. This is why if I have a pain in my leg I am intuitively aware of something in my leg as painful (and not in my head, which is where, in my view, the experience itself is). My pain represents damage in my leg, and I then cognitively classify that damage as painful (via the application of the concept *painful* in introspection). (Tye 1997, pp. 332–333, emphases in the original; see also Tye 1996, pp. 111–116 and 134–136)

As indicated before, this proposal mainly suffers from intuitive implausibility: it certainly doesn't appear that the hurting aspect of pain experiences is just a matter of our *cognitive reactions* to them as *ordinarily* understood. Cognitive reactions qua cognitive don't seem to have any qualitative phenomenology to them.[27] But more importantly, this move appears to mislocate the problem. The question is: in what does the painfulness, the hurting quality, of pains consist? Tye's answer seems to be: in our cognitive–conative reaction to the experience, something like having a desire for it to stop, for instance. But one would like to think that it is because the experience is painful that one desires it to stop, not the other way round.

The response therefore seems to deny that there is anything qualitative, intrinsic, or common to pains that we can identify as *their* hurting, painful phenomenology. The only qualitative content of pain experiences is, according to Tye, their representational content, that is, tissue damage along with its spatiotemporal properties. Furthermore, it is not at all clear that we need to have a concept, PAINFUL, that we apply to the experience in introspection in order to be aware of our pains as painful.[28]

4.2 Representationalism and the Difficulty of Introspecting Pains

But representationalism about pain and other intransitive bodily sensations has *additional* problems—at least externalist versions of the theory. Because phenomenal content is exhausted by their representational content, coming to know about one's experiences and their phenomenal content is coming to know about their representational content. But their representational content is what they represent. Knowing that, however, requires one to possess the concepts that one would need to use to articulate what one's experiences represent—in addition to the concepts required to articulate through what modality one is having these experiences. According to many representationalists, experiences have nonconceptual representational content, which is to say that they are not conceptually structured states. So one can have experiences even without possessing the concepts that apply to the objects they represent. But, as we have seen, perception in Normal Situations results in conceptual identification of the external objects and their properties that these perceptual experiences represent: the locus of concept application in NS is outside the mind. However, to the extent to which qualia are introspectable features of our experiences, that is, to the extent to which the nonconceptual representational content of these states are accessible to us, to that extent we need concepts to articulate this phenomenal content, and the concepts are none other than the concepts that apply to the objective content of the experiences. So without having the concepts that apply to the objects of our experiences we cannot introspect their phenomenology. This is not to say that without the concepts we apply in perception in NS, we cannot perceive. All it says is that without such concepts we cannot introspect our perceptual experiences.[29]

This view of introspection has two disturbing consequences for representational theories of pain and other intransitive bodily sensations. One is that if the affective aspect of the phenomenology of pain experiences is not representational, as it seems not to be, then we cannot introspectively be aware of this aspect. But this is absurd: we are all too painfully aware of this aspect of pains. Indeed this "important feature of our experiences" was the ground for the early theorists' attempt to explain the pains' intransitivity and the asymmetry in concept application we discussed above. But even if the affective aspect of pain is representational, it is not at all clear how we can become aware of this aspect given the view of introspection just outlined. For we need to have the concepts that apply to what this negative affect represents in order to introspect the affect itself. But it seems that we don't have such concepts and, as a matter of fact, we don't need to have them to come to know about the unpleasantness of our pains.[30]

The second difficulty is equally serious. Suppose my liver starts to hurt and I come to fully realize this. Supposing that my coming to fully realize that I have a pain in my liver (or liver area) is at least partly the result of engaging introspection, then the process involved in my realization, according to representationalism and the theory of introspection it naturally comes with, is supposed to go like this. I have a bodily perception of my liver and its damaged condition (suppose I have a serious damage in my liver), and I am able to conceptually articulate it as perception in NS requires; so for instance, I can think and say to myself that something is physically wrong in my liver area. This is so far a perception of a bodily condition first, and then a conceptual recognition of that condition that results from this perception. Then I come to realize that I have this perception (let's say, while having it), which is a feeling state, that is, the pain experience. According to the theory, the necessary condition for my being able to realize this is that I can conceptually articulate the nonconceptual intentional content (= the representational content) of this experience, namely, that there is something physically wrong in my liver area. It is only because I can do that, or something like it, can I come to realize that I am in pain, that is, can I self-attribute a pain experience. But it is clear that the only concept I need to have to be able to realize that I have a pain in my liver area is the concept of pain itself (and perhaps the relevant spatiotemporal—locating—concepts involved). However, the concept of pain is *not* the concept of a physical disturbance, or if we want to be even more minimalist, the concept of something being physically wrong—this is what the intransitivity of pain consists in. It is not even the concept of an experience that (mis)represents physical disturbance. It may be that every experience we classify as pain may turn out to represent some sort of physical disturbance; this may even be necessary in some suitable sense. Still, it is evident that the concept of pain minimally necessary for someone's realizing one has a pain is neither the concept of a kind of physical disturbance, nor the concept of an experience that represents a kind of physical disturbance.

5   General Assessment of Perceptual/Representational Views of Pain in the Light of Scientific Pain Research

One would like to think that naturalistically motivated philosophical approaches to pain would be in harmony with what we scientifically know about pain. Interestingly, however, the theoretical trends in scientific pain research and in philosophy seem to have gone by and large in opposite directions in the last forty years or so. While the science of pain has increasingly conceived of pain as less like perception of an objective reality and more like emotions by first drawing the sensory–affective distinction and then emphasizing more and more its affective aspect,[31] the trend in philosophy, on the other hand, has been in the other direction: as naturalism has started to become an orthodoxy in the second part of the twentieth century, philosophers have increasingly sought for ways in which they could assimilate pain to ordinary perception like vision, audition, and so on. To assess the prospects of perceptual/representational approaches to pain, it will be instructive to have a brief look at the forces that have shaped the trend in the scientific research community. This will also give us a chance to outline the underlying subpersonal mechanisms of pain processing.[32]

Up until the 1960s, the so-called specificity theory of pain had been more or less the dominant view in the medical and scientific community. This view was inspired by Descartes's discussion of a pain pathway in analogy to the idea that "pulling at one end of a rope one makes to strike at the same instant a bell which hangs at the other end" (Descartes 1664), and gained scientific credence by Müller's (1842) "doctrine of specific nerve energies." This doctrine was subsequently modified by von Frey's (1894) account of the cutaneous sensations according to which pain was subserved by the peripheral free nerve endings specialized to respond to all and only nociceptive stimuli, which were then faithfully transmitted to the pain center in the brain. Head and Rivers (1920) suggested that a certain area in the thalamus was the pain center. Although there were other competing theories in the scientific community, the specificity theory dominated the field.[33] This was partly due to the demands of the urgent practicalities of clinical settings and the sociology of medical sciences. A simplistic theory promised a better and more effective intervention. However, by the 1960s it had been getting increasingly clear that observed facts about pain could not be accounted for on the basis of such a simplistic theory.

5.1 Two Kinds of Observed Data for the Science of Pain to Account For

There were basically two sets of observed data that prompted a search for a better theory. One set included data about the high variability of the relationship between nociceptive stimuli and the pain experience. The second set was about disassociation effects.

5.1.1 The variable link between the stimulus and pain experience
The same kind of injury or the same stimuli can elicit pain of widely different intensity and quality in different people or in the same person at different times depending on the circumstances they perceive themselves to be in, their past experiences, and the previous states of their bodies—and sometimes elicit no experience at all. Conversely, sometimes pain occurs with no peripheral stimuli at all, or occurs as a result of totally unexpected stimuli. For instance, given that 40 percent of all Americans suffer from chronic pain at some period in their lives, prolonged pain in the absence of stimuli is common. Sometimes, innocuous stimuli elicit pain (allodynia). Sometimes, nociceptive stimuli elicit a disproportionately increased pain (hyperalgesia). Sometimes one feels pain in areas in one's body quite remote from the location of stimulus, and sometimes one continues to feel pain even after the stimulus is terminated (referred pain, spatial and temporal summation, persistent pain serving a recuperation–healing function). This great variability between stimulus and the pain experience is one of the scientific puzzles about pain, one that the specificity theory could not explain. Recall that the modern IASP definition of pain explicitly warns against tying pain to peripheral stimulus, just as the commonsense conception of pain distinguishes the causes of pain from the pain itself.[34]

5.1.2 The disassociation of pain affect from its sensory dimension
It had been well known that certain surgical procedures, some pain syndromes, and drugs reduced or removed the unpleasantness of pain while preserving its sensory-discriminative aspects. These data typically came from patients who had undergone prefrontal

lobotomy (Freeman et al. 1942; Freeman and Wattz 1946; Freeman and Watts 1950; Hardy et al. 1952; Barber 1959; Bouckoms 1994) or cingulotomy (Foltz and White 1962a,b; White and Sweet 1969) as a last resort for their intractable chronic pain (as frequently involved in phantom limb pain, neuralgia, causalgia, severe psychogenic, and cancer pains), from patients under the effects of hypnotic suggestion (Barber 1964; Rainville et al. 1997; Rainville et al. 1999), nitrous oxide (laughing gas), and some opium derivatives like morphine (Barber 1959). These patients by and large agreed that when they were in pain, they could recognize and identify it as such, but did not feel or seem bothered by it or distressed in ways characteristic to having pain experiences. Although it is usually not recognized in the scientific and philosophical literature, there are, however, important differences among the phenomena afflicting these patients, which are manifested in patients' reports and behavior. For instance, pain asymbolia also typically produces a kind of disassociation—a rather strong kind—sometimes similar to cingulotomy patients' but interestingly different from lobotomy patients' (Rubins and Friedman 1948; Hurt and Ballantyne 1974; Berthier et al. 1988; Berthier et al. 1990; Devinsky et al. 1995; Weinstein et al. 1995).

In fact, there is evidence that pain asymbolia may be the only form of genuine disassociation (Grahek 2001). These patients, for instance, didn't react to even momentary pains like pinpricks, small cuts, or burns. Experimental pain stimuli failed to produce any recognizable affective reactions. Nevertheless, the patients insisted that the stimuli caused pain—they identified their experiences as pain (Rubins and Friedman 1948; Berthier et al. 1988; Berthier et al. 1990; Dong et al. 1994; Weinstein et al. 1995). The lobotomy and morphine patients, on the other hand, showed the usual affective reactions and symptoms when they were stimulated momentarily by normally painful stimuli. But they didn't seem to care or be bothered by their standing persistent or chronic pains. Probably, they still felt the negative affect but didn't mind it, whereas the pain asymbolia patients didn't even feel the momentary negative affect. These two cases also need to be distinguished from so-called congenital insensitivity to pain, a condition where the patients don't even report any pain experience upon various kinds of nociception—these patients don't live long (McMurray 1955, 1975; Baxter and Olszewski 1960; Sternbach 1963; Brand and Yancey 1993).[35]

It was partly the accumulation of this sort of (mostly) subjectively obtained abnormal data indicating dissociable phenomenological components of pain experiences (along with the observed variability between stimulus and pain experience discussed above) that contributed to the demise of the specificity theory and led to a search for a better theory capable of accounting for the data.

5.1.3 Endogenous modulatory mechanisms for the variable link Although a number of scientists had already proposed important modifications of or alternatives to the specificity theory that prepared the way for Melzack and Wall's gate-control theory (see figure 1.2), it was the introduction of this theory in 1965 (Melzack and Wall 1965) that initiated what might be truly termed a revolution in scientific pain research. The theory in its initial form was geared mainly toward explaining the first set of data about the variability of the link between nociceptive stimuli and pain.

According to gate-control theory, noxious stimuli from the peripheral nociceptors are carried to the spinal cord through basically two types of fibers (see figure 1.3): large myelinated fibers with faster conduction velocity (A-beta fibers) and small fibers (S) with slow conduction velocity (small myelinated A-delta and unmyelinated C-fibers— although A-beta fibers are involved in the noxious stimuli, they are not specific to noxious stimuli). Before entering the gray matter of the spinal cord the axons of A-beta fibers branch out and project to the thalamus in the brain. The other branch and the small fibers enter the subtantia gelatinosa (SG) (laminae I and II) and laminae IV and V, in the dorsal horn of the gray matter of the spinal cord. The gating mechanism was postulated to be somewhere in these laminae. This gate was conceived as a neural mechanism (see figure 1.4) that acts like a modulating or regulating system that controls the amount of nerve-impulse transmission from the periphery to the spinal cord transmission cells (T-cells), that is, to the second-order nerve cells that transmit the modulated output of the gate to the higher brain structures. Under normal circumstances, the theory proposed, it is necessary for the systems in the brain to interpret the incoming signals as pain that the output of this gate reach or exceed a certain critical level. This output is regulated in the gate by various excitatory and inhibitory factors. From a theoretical viewpoint, the most interesting of these is the descending inhibitory signals from the brain.

The proposal of a modulatory gating mechanism has since then generated a great deal of scientific research, which uncovered a great variety of endogenous modulatory (inhibitory as well as excitatory) mechanisms both local (segmental) within the dorsal horn of the spinal cord and descending from various areas in the brain stem and the brain. There are also mechanisms within the brain stem and the brain. The general thrust of the gate-control theory was the postulation of endogenous modulation of the nociceptive transmission to higher areas in the brain in order to explain the great variability of the link between the stimulus and the pain experience. This much has certainly been confirmed by later scientific research—although this research has uncovered many more modulatory mechanisms of various kinds not envisaged by the gate-control theory as even a quick look at the recent work could readily reveal.[36]

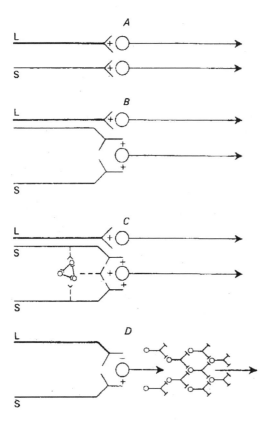

Figure 1.2

Schematic representation of the development of conceptual models of pain mechanisms before the introduction of gate-control theory. ***A***: The specificity theory espoused by von Frey (1894), according to which large fibers (L) transmitted touch, and small fibers (S) transmitted pain impulses in separate, specific, straight-through pathways to touch and pain centers in the brain. ***B***: Goldscheider's (1894) summation theory, showing convergence of small fibers onto a dorsal horn cell (touch was assumed to be carried by large fibers). ***C***: Livingston's (1943) conceptual model of reverberatory circuits underlying chronic pathological pain states where constant nociceptive signals from the periphery generate prolonged activity in the self-exciting chain of neurons in the dorsal horn, which then transmit abnormally patterned volleys of nerve impulses to the brain. ***D***: Noordenbos's (1959) sensory interaction theory where large fibers inhibit (−) and small fibers excite (+) central transmission neurons, which consist of a multisynaptic system that projects to the brain.

Note that these models propose increasingly more sophisticated modulatory mechanisms that sever the reliability of the correlations between stimuli and pain sensation. What we see here is an increasing theoretical sophistication in attempts to explain the first set of observed data. (Adopted from Melzack and Wall 1988, caption modified).

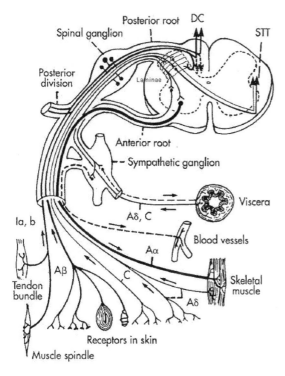

Figure 1.3

A schematic showing peripheral nerve fibers entering the spinal cord. The laminar structure of the dorsal horn is indicated. There are about 6 such laminae within the dorsal horn, 10 in the whole ipsilateral white matter (i.e., in the one wing of the butterfly shaped structure). The first two laminae at the beginning of the dorsal horn are called subtantia gelatinosa (GS). Note the branching out of the fast A-beta fibers into the dorsal column (DC) without entering the horn. Note also the free endings of nociceptive A-delta and C-fibers. The reflex arc is also shown. STT, the spinothalamic tract (a major ascending nociceptive pathway—T-cells). (Modified from Bonica 1990.)

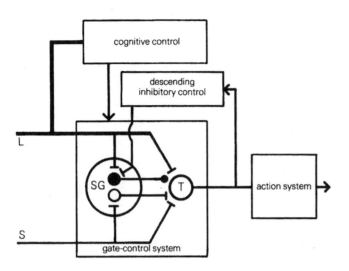

Figure 1.4
This is the schematic of the gate–control theory as modified later by Melzack and Wall (1983). The new model included excitatory (white circle) and inhibitory (black circle) links from SG to T-cells as well as descending inhibitory control from some brain-stem structures. The round knob at the end of the inhibitory link implies that its actions may be presynaptic, postsynaptic, or both. All connections are excitatory, except the inhibitory link from SG to T-cells. (Slightly modified from Melzack and Wall 1983.)

Nociception seems to be unique among the sensory systems in general in incorporating a distinct endogenous modulatory control system.

5.1.4 Functionally distinct underlying mechanisms for the disassociation effects As an attempt to explain the second set of observed data about the affective aspects of pain and its disassociation from its sensory aspects, Melzack and Casey (1968) expanded the gate-control theory by postulating multiple parallel central processing systems of noxious stimuli coming from the peripheral nervous system. According to this proposal, after the noxious stimuli are modulated in the gate in the spinal cord, they are projected through various pathways to two different brain areas to be centrally processed. One of the systems is phylogenetically older: the reticular formation, limbic system, and hypothalamus. In humans this system heavily interacts with the prefrontal cortex. Melzack and Casey called this stream the motivational-affective system. Indeed, the limbic system had long been known to be responsible essentially for emotional and motivational processes. They called the other the sensory-discriminative system, which involved the ventrobasal thalamic nuclei at which the noxious stimuli arrive through the ascending neospinothalamic tract and go directly to the somatosensory cortex, the basic sensory component of the system. These two systems were also monitored and controlled by what Melzack and Wall had called "a central control trigger" that had usually already been aroused by the signals carried through the aforementioned fast conducting A-fibers that branch out before entering the gray matter in the dorsal horn. The location of this system was generally thought to be in the frontal cortex. The behavioral output in the broadest sense was supposed to be a varying function of these three systems. The functional organization of the underlying structures proposed by the theory can be seen in broad outlines in figure 1.5.

Melzack and Casey's speculations opened up a whole new chapter in modern pain research: more elaborate models with more detailed and precise anatomical mappings have emerged. The Melzack–Casey model was speculative and lacked detailed functional and anatomical specifications. However, even so, as revealed by later scientific research it was on the right track by postulating functionally distinct central systems for what appeared to be phenomenologically distinguishable and dissociable components of pain experience.

Let us briefly look at the current evidence. There are several ascending pathways that carry nociceptive signals to brain structures. Almost all of these (second-order) neurons, after synapsing with the primary afferents in the dorsal horn, cross the anterior white commissure and climb in the anterolateral quadrant (ALQ) of the spinal

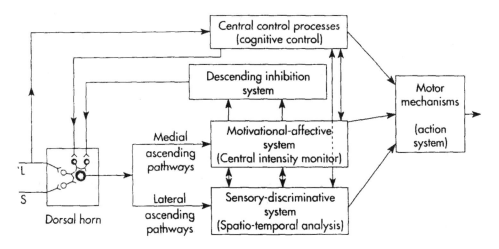

Figure 1.5
This is a modified version (Melzack and Wall 1983) of a diagram that first appeared in Melzack and Casey 1968. It also shows part of the original gate-control mechanism in the dorsal horn (Melzack and Wall 1965).

cord. Three major nociceptive tracts play a particularly important role: the spinothalamic tract (STT), the spinoreticular tract (SRT), and the spinomesencephalic tract (SMT). STT itself may be divided into two major tracts: the more laterally projecting neospinothalamic tract (nSTT), which terminates in the ventroposterolateral thalamic nucleus (VPL) where the nerves synapse with the third-order fibers that project mainly to the somatosensory areas I and II (SS1, SS2). The vast majority of nSTT neurons are wide-dynamic-range neurons (WDR), which originate in laminae I and II and have large myelinated axons. WDR neurons are fast conducting neurons that respond to a variety of peripheral stimuli including noxious stimuli, and encode their intensity very precisely. They have usually smaller receptive fields and seem to be somatotopically organized. The medially projecting part of the STT is usually called the paleospinothalamic tract (pSTT). The neurons of this tract have a variety of more diffuse central projections: some end in the medial and intralaminar thalamic nuclei (MIT), some diffusely project to different areas in the reticular formation (RT), pariaqueductal gray (PAG), and the hypothalamus (H), where they synapse with third-order neurons that project to the limbic forebrain structures (LFS)—such as the singular cortex (SC), anterior cingulate cortex (ACC), amygdala (A) and the orbitofrontal cortex—as well as to other areas of the brain, including weak diffuse projections to SS1 and SS2. The SRT and SMT, along with pSTT, are considered as the *medial* projection system or simply the medial system, whereas nSTT as the *lateral* system (see figure 1.6).

Like pSTT, SRT and SMT also diffusely project to areas in the reticular formation as well as to the parabrachial (PB) nucleus of the dorsolateral pons, to PAG, H, and MIT— although, as a rule, as they ascend higher in the brain stem, their projections become weaker and even more diffuse than those of pSTT. The neurons making up SRT and SMR comprise mostly the nociceptive neurons (NS) that are smaller in diameter and slower, and respond only to peripheral *nociceptive* stimuli. They are less precise and have larger receptive fields. (The trigeminal pathways that carry signals from facial structures have a similar medial and lateral organization—see figure 1.6.) There are other ascending systems. Some of them seem to be made up of several neurons synapsing each other as they climb up to areas in the brain stem. The existence of such diffuse pathways is probably one of the reasons why cutting major ascending pathways in the anterolateral quadrant seems not very effective in the removal of chronic pain.

Since Melzack and Casey (1968), it has usually been assumed in the scientific and especially in the medical community that the lateral and medial systems serve different functions.[37] In particular, the lateral pathways terminating in the somatosensory cortical areas (SS1, SS2) are assumed to serve the capacity of the pain system to distinguish between different sensory properties of noxious stimuli, such as bodily

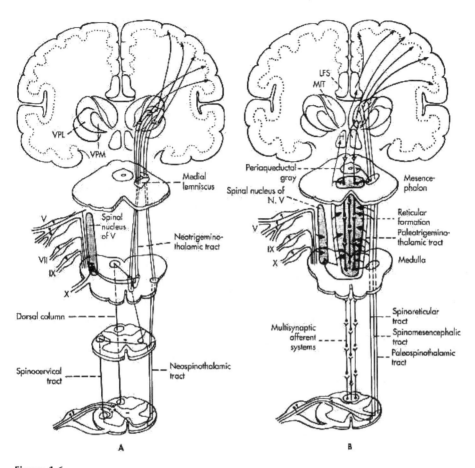

Figure 1.6

A: The lateral system, showing both nSTT and the neotrigeminothalamic tract which innervates facial structures. **B**: The medial system comprising pSTT, SRT, and SMT, as well as the paleotrigeminothalamic tract (from Bonica 1990).

location, intensity, and quality (thermal, incisive, traction, extension, etc.). In contrast, the medial pathways—due to the connections to insular and cingulate cortices and to limbic structures in general, which are known to play a crucial role in emotional behavior—are assumed to serve the capacity of pain to generate the appropriate affect-laden responses to nociception (such as unpleasantness, urgent desire for stimulus cessation, interruption of attention). In brief, it has been assumed that while, roughly, the sensory–discriminative aspects of pain are subserved by the lateral system, the affective–emotional aspects are subserved by the medial system. Moreover, since Melzack and Casey (1968), these aspects have usually been thought to be processed in parallel (see figure 1.5 above). This was thought to be an advance over the specificity theory's relegation of unpleasantness to a mere serial mental reaction to pain sensations proper.[38]

There are no doubt important truths in the common wisdom incorporated in this picture. The medial system have important direct connections to many areas in the limbic–forebrain structures that are known to be directly involved in affective-motivational phenomena. They are also involved in the arousal and the regulation/adjustment of autonomic responses to incoming stimuli. The evidence for the role SS1 and SS2 play in the sensory-discriminative aspects of pain is decisive. Nevertheless, things seem to be more complicated than what this neat picture tells us. In particular, there is evidence that there are also serial processes in the production of pain affect. The nociceptive signals reaching SS1 and SS2 seem to proceed through a cortico-limbic pathway to the posterior parietal cortex (PPC) which are then integrated with and analyzed in light of information coming from other sensory systems such as visual and auditory streams. Damage to this area produces profound defects in learning the affective significance of the stimuli being presented to monkeys and humans.[39] The integrated output is then sent to insular cortex where this serial stream converges with the stream coming from the medial pathways. The processing proceeds to the anterior cingulate cortex (ACC), which seems to regulate attention among other things. Among the most robust results in pain brain-imaging studies is the fact that ACC reliably lights up in correlation with the intensity of the unpleasantness of pain.[40] ACC has strong connections to sensorimotor areas and the prefrontal cortex, which is thought to be involved in setting response priorities, planning and action execution. Interestingly, part of the stream coming from ACC-FC feeds back to IC and to the amygdala, which receive direct input from the medial pathways (see figure 1.7).[41]

To summarize, there is considerable evidence that the affective-motivational aspects of pain are processed both in series with its sensory-discriminative aspects and in parallel. On the basis of our present knowledge of pain mechanisms and the pathways feeding to them we can give plausible explanations of various pain disassociation

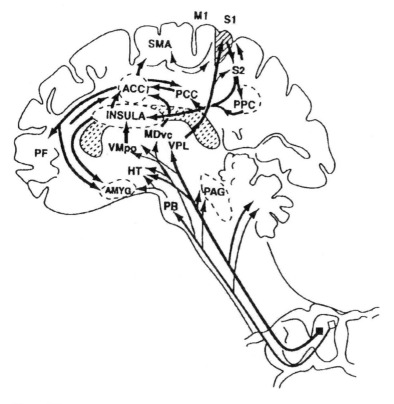

Figure 1.7
Outlines of ascending pathways, subcortical and cortical structures involved in processing pain affect both in series with and parallel to sensory–discriminative processes in pain. VMpo, ventromedial part of the posterior nuclear complex; MDvc, ventrocaudal part of the medial dorsal nucleus; PCC, posterior cingulate cortex; HT, hypothalamus; PPC, posterior parietal cortex; SMA, supplementary motor area; AMYG, amygdala (adapted from Price 2000).

effects. Although they implicate different structures or different stages in the pain processing, the essential and invariant point in all these explanations is that there are functionally and anatomically different—although overlapping—underlying brain mechanisms and these mechanisms can therefore be selectively impaired in ways characteristics to disassociation effects.

5.2 Prospects for the Philosophical Perceptual/Representational Approaches to Pain

Now that we have the general outlines of a story of what empirical and conceptual forces have been shaping the science of pain, we may ask if the recent trends in pain science are any indication that the naturalistic trends that have been shaping the philosophical theories of pain in the last forty years or so have been on the wrong track. As we have seen, one of the main motivations behind the perceptual/representational views of pain in philosophy is the belief (or hope) that perception as a species of mental representation playing a certain informational/functional role within the mental and behavioral economy of organisms can be accounted for entirely in physicalistic terms. Of course, this is a controversial claim. There are many who think that perception involving as it does conscious phenomenal experience cannot be a purely physical phenomenon. They think that there are explanatory as well as metaphysical gaps between conscious and physical phenomena that cannot be bridged.[42] However, this much can probably be agreed upon generally: it is a perfectly plausible theoretical strategy to pursue an understanding of pain and other intransitive bodily sensations in perceptual/representational terms, because this strategy, if it works, minimizes the diversity of mental phenomena, and thus potentially offers the prospects of a more unified theory of mind, and if this theory turns out to be in harmony with the rest of our sciences and their fundamental metaphysical and methodological assumptions, so much the better. Indeed, it was the plausibility of this strategy and the belief that we will eventually succeed in understanding perception in purely naturalistic terms that have prompted many philosophers to advance perceptual/representational theories of pain. Many in fact believe that philosophy has made significant progress in the second half of the last century in developing the conceptual tools for a better and more naturalistic understanding of perception and the mind in general—for example, the notion of a mental representation and its (broadly) computational processing.

But does the scientific trend toward understanding pain as a subjective experience less like a perception and more like an emotion with quite a variable link to injurious stimuli undermine the philosophical project? There is no simple answer. We may say, "yes it does," if we take the perceptual/representational theories as making the strong claim that pain is strictly nothing but a perception just like other standard perceptions. We may say, "no it doesn't" if we take their claim in a weaker sense to the effect that pain *involves* perception, indeed involves perception in the very same sense and manner that other sensory modalities involve genuine perception—it is just that there is so much more going on when we are in pain. Nothing in the scientific understanding of pain and its underlying mechanisms seems to show that pain involves no perception at all—genuine perception specific to pain processing. On the contrary, as we have seen, there are physiologically specialized systems that process nociceptive stimuli from the moment they effect peripheral receptors to the central processing of these signals in the spinal cord and the brain. Pain processing may not involve sense organs (e.g., the eyes, ears, nose, and the tongue) in the traditional sense of involving *physical (bodily) structures* that are designed to channel the appropriate energy forms to peripheral nerve endings for their transduction. But sense-organs in this sense are not necessary for genuine perception. What is necessary is that there be specific energy forms that are transduced and transmitted by specialized nerve fibers and their end-organs (dendrites), and are processed selectively by specialized central mechanisms. This is what happens in *all* classical five sense modalities, including touch that doesn't involve sense organs in the classical sense. Pain may be classified as a sub-modality of touch.

The modern representationalist theories seem to be making the stronger claim, however. Recall that they are in fact application of a more general naturalism about qualia to the specific case of pain. We have already seen some of the major difficulties facing these theories. What has recently happened in the sciences of pain, as we have just briefly outlined, seems to confirm the seriousness of these difficulties. The robust affective-motivational aspect of pain is less amenable to a representationalist treatment. But there is a physicalistic and very natural alternative: it is plausible to claim that this aspect of pain can be accounted for in a *psychofunctionalist* way. According to this proposal, even though the sensory-discriminative aspect of pain can perhaps be handled representationally,[43] the affective aspect reduces to the way in which the sensory-discriminative information is processed, not for analysis to extract information about the proximal or distal properties of the stimuli, but rather for its significance for the effector or motor systems, to set motivational parameters for action on the basis of stimuli's informational content.[44] There is in fact strong supporting evidence for such a thesis in the evolutionary stories of different organisms at different developmental hierarchies. The neuroscientific evidence about the affective brain seems also to support this idea in general.[45] Such a proposal should not be surprising, after all psychofunctionalism in philosophy as a reductionist proposal has always found its natural home in drive states that are intrinsically strong motivators. There is, then, little wonder why philosophers of mind have typically used the example of pain in the discussion of behaviorist or functionalist proposals about the metaphysics of mind in general.

This is a view that treats pain as both a representational and a psychofunctional state. Such a view still needs to provide a good answer to the main difficulty that we have seen afflicts all perceptual/representationalist views. Why is there an asymmetry in concept application, or in the focus of conceptual categorization? Pointing out that pain has a deeply pronounced negative affect is not entirely adequate even when we have an adequate account of what this affect consists in. We also need a new account of introspection; for if pain affect doesn't consist in representing a certain condition but is rather a psychofunctional affair, then its introspection cannot avail itself to the use of a concept that applies to what this affect represents as is required by pure qualia representationalism—this is probably why pure representationalists usually tend to shy away from psychofunctionalism (how do we introspect a state or an aspect of a state whose nature is psychofunctional?).

These are the major questions that an adequate perceptual account of pain ought to give adequate answers to. Thus, despite significant advances in our philosophical and scientific understanding of pain in the last forty years or so, there is still a lot of work to be done to develop a fully satisfactory account of pain.[46]

Notes

1. It is natural to see perception as a process that involves categorization and to that extent as a process that is itself (at least in part) conceptual. This is consistent with the claim that perception starts with raw sensations. Sometimes, "perception" is used to refer to raw sensations. The essential and uncontroversial point, however, is that perception ends up with (perhaps modality specific) conceptual categorization when the information in the perceptual experience is picked up for use by the cognitive belief forming mechanisms—whether or not this categorization is part of the perceptual process itself.

2. In what follows, concepts will be denoted by capitalized words that name them. So, for instance, "RED" denotes the concept of red, which in turn expresses the property of being red. In other words, RED expresses the property of being red, or *redness* in short (similarly with other concepts and properties). Even though I assume for convenience a representationalist framework for concepts according to which concepts are mental representations realized in the brain (which is the psychologist's preferred reading), nothing very important hangs on this: the reader may substitute his or her own preferred interpretation of how concepts are to be understood. For instance, concepts may be merely certain sorts of mental or behavioral capacities that are functionally characterized.

3. But see Hill, this volume. Hill thinks that the possession conditions for the concept of pain and its reference come apart—so that our ordinary concept of pain is incoherent.

4. D'Amico (this volume) takes the epistemic asymmetry (he uses "unrevisability" rather than "incorrigibility") to show that our mental (introspective) discourse and ordinary perceptual discourse are governed by fundamentally different conceptual schemes. He then argues that this shows these cannot be happily juxtaposed in the way required by the methodology proposed by Price and Aydede (this volume) for the study of pain. Aydede and Price (this volume) deny that there is such a radical epistemic asymmetry *in kind*.

5. Indeed even those who have been critical of this definition consider it to be an important advance over a conception of pain that the note, appended to the main definition, warns against. For critical discussion see Melzack and Wall 1988, Price 1988, 1999, Fields 1999, Hardcastle 1999, and Aydede and Güzeldere 2002.

6. The term 'intransitive' is from Armstrong (1962, 1968). Armstrong contrasts these with "transitive" bodily sensations such as feeling the temperature or smoothness of an object, where the experience has a straightforward nonmental perceptual object. See the preface of this volume.

7. Broad (1959) defends one of the most sophisticated versions of a sense-datum theory. Interestingly, however, pains and aches, according to Broad, are less amenable to an act–object analysis. He writes: "It is by no means obvious that a sensation of headache involves an act of sensing and a 'headachy' object; on the contrary, it seems on the whole more plausible to describe the whole experience as a 'headachy' state of mind. In fact the distinction of act and object seems here to have vanished; and, as there is clearly *something* mental in feeling a headache, just as there is in sensing a red patch, it seems plausible to hold that a sensation of headache is an unanalyzable mental fact, within which no distinction of act and object can be found" (1959, pp. 254–255). Perhaps this is so with headaches and other aches that are spatiotemporally very diffuse. But it is not clear whether it is indeed more plausible to make the same claim about acute pains that are temporally and spatially well delineated, such as a sudden jabbing pain in the back of one's right hand or a pain caused by a pinprick in one's left thigh.

8. See Jackson 1977 for a defense of this line. Jackson 1977 is an unabashed and technically sophisticated defense of a sense-datum view in general and of pains in particular. See also Addis 1986 for a similar line. Jackson no longer defends a sense-datum theory: he has recently rejected antiphysicalism and accepted a physicalistic representationalist view similar to the one we will discuss below. See Jackson 1998, 2000.

9. See for instance Perkins 1983, and this volume. Perkins is not a sense-datum theorist, although he is an indirect realist. He dispenses with phenomenal objects in favor of phenomenal qualities that one's experience instantiates or somehow incorporates—they are part of one's consciousness. But he still takes the act–object model to be the proper analysis of sensory experience except that the act is now directed toward not a mental object but rather an instance of a quality. Perkins is the only indirect realist I know of who keenly advocates a perceptual view of pain that he claims requires physicalism.

10. See H. H. Price 1950, which seems to defend this sort of view. Price introduces the notion of a "sense-field" according to which "sense-data, though they have places in their own sense-fields, are *nowhere*" (p. 248). However, he does not treat pains as sense data on their own, but rather as "outstanding parts of a sense-datum (namely, of the total somatic one) which display a particularly striking sort of sensible qualities" (p. 232). For Price, when it comes to somatic sensations and feelings, there is only one sense datum with its own unique somatic sense field such that all bodily feelings and sensations are qualifications or modifications of it. For early classical discussions of sense-datum theories, see Moore 1903, 1939; Russell 1912.

11. See Lycan 1987a,b for useful discussion of this point.

12. See also Pitcher 1971.

13. The ensuing debate was quite lively. For criticisms of these early perceptual views of pain, see Vesey 1964a,b; Margolis 1976; Mayberry 1978, 1979; Everitt 1988; and Grahek 1991. Grice

(1962) had already argued that there is a fundamental distinction between bodily sensations and standard perception. See Armstrong 1964 and Pitcher 1978 for replies to some of these criticisms. Holborow 1969 and Pitcher 1969 are criticisms of another perceptual theorist's work, McKenzie (1968), who claims that "pain can be accepted as a sense in the way in which smell is a sense" (p. 189). For the most recent and radical incarnation of a perceptual view of pain, see Hill, this volume.

14. This is why Armstrong (1962, 1968) calls pains, itches, and tickles (among others) *intransitive* bodily sensations and distinguishes them from *transitive* bodily sensations like feeling the temperature of an object or the roughness or smoothness of objects' surfaces in contact with skin. The latter are transitive precisely because we attribute the sensible qualities to the objects themselves by the deployment of relevant concepts whose extensions are these objects in the first instance (or body parts if we feel, say, their temperature).

15. Compare the sense-datum analyses, which postulate a private internal particular that actually *instantiates* the qualities also attributed to the external public object that is the cause of the internal object's being sensed by the perceiver.

16. The proper formal semantics of adverbs is still a controversial topic in linguistics. But adverbialism is advanced primarily as an ontological thesis in the philosophy of perception, although the issues interact with each other. Also, strictly speaking, it might not be true that adverbialism about perceptual experiences may make do with only one individual, the perceiver. One prominent version of adverbialism is the event modification theory, which quantifies over primitive events and takes adverbs to modify these events—see Davidson 1980. On this view, there are two individuals: the perceiver and the event of her perceiving an object. But many defenders of this view might take this commitment to be innocuous when compared to a commitment of phenomenal objects like sense data. For early defenses of adverbialism, see Ducasse 1952 and Sellars 1975. Aune 1967 can be interpreted as a variant of adverbialism explicitly applied to pain. Chisholm 1957 defends a view (the "theory of appearing") that has close affinities with adverbialism. Kraut 1982, Lycan 1987a, and Tye 1984a are more recent and technically more sophisticated defenses of adverbialism in general. Tye (1984b) and Douglas (1998) defend adverbialism about pain specifically. The latter is an adverbialist reply to Langsam (1995), who attempts to give an explanation of why we think and talk about pains as mental objects. For powerful criticisms of adverbialism, see Jackson 1975, 1977, Robinson 1994, and Foster 2000.

17. Adverbialism of this sort can be effectively combined with intentionalism or representationalism about experience (Kraut 1982 is especially helpful for an understanding of the intimate connection between them). An adverbialist has to somehow characterize these ways or manners of perceiving/sensing for the purposes of distinguishing among them, and the typical way in which she does that is by appealing to the standard or canonical conditions under which those perceptual events are brought about (recall how direct realists want to handle visual hallucinations: one is having a visual experience that is like an experience that is normally brought about when one actually sees a real red apple on the table). Thus it may be possible and reasonable to argue that those canonical conditions are what the specific perceptual events or activities of the persons represent. For instance, it is plausible to claim that a specific perceptual

activity constitutes the perception of red (= the perceptual event representing the instantiation of red) because it is the kind of psychological event regularly (canonically) caused by red surfaces—indeed one might expect that the psychophysics of sensory modalities would detail these canonical or standard conditions in objective terms. (The development of Tye's position over the years from adverbialist direct realism into an information-theoretic representationalism follows this path.)

18. Now a direct realist may in fact go on to claim, if he wishes, that we do engage in some sort of internal perception of our experiences in introspection, but this seems compatible with his direct realism in that introspecting (= perception of experiences) is not required for perceiving the objective world. Two prominent direct realists, Armstrong (1968) and Lycan (1987a, 1996), hold this position—but see below.

19. Compare Armstrong (1962, p. 125) where he writes:

Now it may well be asked why, in this important field of perception [intransitive bodily sensations], ordinary discourse restricts itself to sense-impression statements, instead of making perceptual claims about what is actually going on in our body. Several reasons may be advanced. In the first place, we have seen that the [conative] reactions characteristically evoked by these bodily impressions are determined almost solely by the impression, whether or not it corresponds to physical reality. Since the reactions are an important part of our concept of almost all the intransitive bodily sensations, it is natural to talk about what feels to us to be the case, whether or not it is the case.

20. That is, to the extent to which introspecting is a perception-like activity. But even if it is not, we still need to do something equivalent—something like an information uptake—to become aware of our pain experiences.

21. Indeed, it is precisely the intransitivity of pain and the indirection that this implies that Perkins (1983, this volume) uses to make an initial intuitive case for indirect realism, which he later argues presents the basic argumentative framework for other senses as well.

22. For instance, when Pitcher first introduces the notion of unpleasantness as the reason why our concept of pain works the way it does, he writes in a parenthetical note: "(When I refer to the act, or state, of feeling pain as an *experience*, I do not, of course, mean that it is an exclusively mental happening or anything of the sort: I mean 'experience' in the sense that riding a bicycle or lying on a rug before a fire is an experience—that is, merely as something that we do or undergo)" (Pitcher 1970, pp. 379–380). Perhaps feeling pain is not indeed an exclusively mental happening. But it is certainly an episode of conscious experience in a much more robust sense than riding a bike is—as he himself seems forced to acknowledge later in the same paper.

23. The main bulk of the present introductory chapter was written in 2002, parts of which were intended as an entry for the *Stanford Encyclopedia of Philosophy*. For this reason it does not cover the later theoretical developments. Indeed, the essays by Tye and Hill in this volume can be seen as fresh and important contributions to the literature that continue the debate where my introductory essay stops. Instead of incorporating the new moves contained in Tye's and Hill's

chapters into the present exposition, I wanted to leave the latter as is so these new chapters could be better appreciated within the larger dialectic partly provided in this introduction. The reader is therefore encouraged to read Hill's and Tye's essays in this volume as fresh and prompt rebuttals of some the critical points made in this section. Dretske's new essay is primarily on the epistemology of pain and introspection in general, rather than a direct defense of representationalism about pain.

24. Lycan 1987a (pp. 60–61) contains a brief statement of a representationalist account of pain; however, Lycan does not claim that the affective aspects of pain can also be handled representationally. He seems to have in mind a mixed representationalist-cum-psychofunctionalist theory about pain and other bodily sensations. See below.

25. See Dretske 1981, Fodor 1987, and Aydede and Güzeldere 2005 for a strict informational psychosemantics (the latter contains an extended discussion of pain in the context of informational theory). See Dretske 1988, 1995 and Tye 1996 for a combination of informational and teleological psychosemantics. See Millikan 1984 and Papineau 1987 for teleological versions.

A naturalist functional role psychosemantics (a form of internalism) is not out of the question. See Carruthers 2000 (ch. 9) for a representationalist account of pain that comes close to this—what he calls a consumer's psychosemantics. Rey 1997 (ch. 11) contains an account of sensory experience according to which phenomenal properties are identified with narrow functional roles of sensory states, which roles are then identified with the narrow representational content of these sensations. See also White 1986 for a similar narrow functionalist account.

26. See also Block 1996, Peacocke 1983, and Searle 1983 for similar views. Shoemaker also seems to believe that many experiences like taste and smell (and a fortiori pain) do have a phenomenology that cannot be captured entirely by their representational contents—see his discussion about taste in his 1996 (pp. 102–104), which can be applied, *mutatis mutandis*, to pain without any difficulty.

27. Tye actually accepts this; see his 1996 (pp. 4–5).

28. From the text it is not clear what the concept is supposed to be applied to. Tye says that he applies it to the damage in his leg, but then he does this in introspection. Do we perceive or introspect our legs? Maybe both are involved, but we are not told how in the passage. (Tye's chapter and his replies to commentators in this volume address this question in more detail.)

29. This view of introspection has close affinities to what Dretske (1995) calls a *displaced perception* view of introspection. But Dretske's view is stronger than what is presented in the main text in requiring an inferential linkage between one's perceptual beliefs about the public object represented by one's perception and the introspective belief that this is what one's experience represents. But this stronger and to my mind implausible requirement is not forced upon a representationalist, although all qualia representationalists *are* committed to the claim that possessing the relevant concepts are necessary for introspective *knowledge* of one's experiences. See Aydede 2003 for a criticism of Dretske's view of introspection. Dretske, with his new essay in this volume, seems to have dropped his earlier view of how introspection works—instead he is actively probing the theoretical space for a more plausible view of introspection consistent with his externalist representationalism.

Also, even if introspection might begin with an internal monitoring device as Armstrong (1968) and Lycan (1995) envisage it, introspective *knowledge* of phenomenal qualities still needs the possession of those concepts that apply to what these qualities represent, according to externalist qualia representationalism.

30. See Aydede 2001 for an elaboration of this objection. (Again: see Tye's contributions to this volume.)

31. See Gustafson's essay in this volume for a proposal that seems to take this trend and extend it to its logical extreme: it is more fruitful to conceive pain experiences as essentially emotional rather than perceptual states.

32. See also the chapters by Polger and Sufka, Price and Aydede, and Allen et al., as well as the chapter by Panksepp in this volume for useful and accessible presentations of aspects of underlying pain mechanisms.

33. See Dallenbach 1939, Botterell et al. 1942, Bonica and Loeser 2001, and Melzack and Wall 1988 for more detailed historical accounts of the development of scientific pain theories. (Gustafson's essay in this volume also contains a brief history of conceptions of pain.)

34. See the first two chapters of Melzack and Wall 1988, which present in great detail a powerful and convincing case for the existence of a hugely variable link between pain and injury. Melzack and Wall were very self-conscious about what they needed to account for in developing their gate-control theory—they go over the history and philosophy of their own discipline in great detail and insight. Melzack and Wall 1962, 1965 are also very useful.

35. See Price 1999, 2000, 2002 for an illuminating discussion of such cases and underlying mechanisms. Price draws a distinction between immediate unpleasantness of a pain experience and what he calls pain's secondary affect, which consists of a conscious cognitive appraisal of the consequences of the pain and its cause, which involves various emotional reactions (e.g., worry, panic, anxiety, arousal, depression) that typically last longer than the onset of pain. He thinks that pain asymbolia involves cases where both the immediate unpleasantness and the secondary affect are absent, whereas in other cases, although immediate unpleasantness is present, the secondary affect is absent.

There is strong evidence that the disassociation between the sensory versus affective components of pain also goes in the other direction: in addition to cases where the intensity of the sensory component can be reduced without affecting the unpleasantness of the experience (Gracely et al. 1979), there is at least one well documented and studied case where the patient experiences something very unpleasant upon receiving nociceptive stimuli without being capable of identifying his experience as pain (Ploner et al. 1999). It is worth quoting from this study to illustrate the prima facie counterintuitive results, which provided strong support for the bidimensional nature of pain experience: "[At higher intensities of cutaneous laser stimulation] the patient spontaneously described a 'clearly unpleasant' intensity-dependent feeling emerging from an ill-localized area 'somewhere between fingertips and shoulder' that he wanted to avoid. The fully cooperative and eloquent patient was completely unable to further describe quality, localization, and intensity of the perceived stimulus. Suggestions from a given list containing 'warm,' 'hot,' 'cold,' 'touch,' 'burning,' 'pinprick-like,' 'slight pain,' 'moderate pain,' and 'intense pain' were denied" (Ploner et al. 1999, p. 213). In personal communication, Price has indicated an important feature of the findings: the unpleasantness reported by the patient has arisen only when the laser stimulus intensity has reached 350 mJ, 150 mJ more than the normal pain threshold established for the normal right hand. As emphasized by Price, this seems to indicate that the disassociation of affect from sensation is not just a matter of a parallel system being shot down, rather it leaves room for a serial interpretation of the interaction between affect and sensation. Price 2002 contains a useful discussion of such cases.

36. See Sufka and Price 2002 for a generally positive reevaluation of gate-control theory after nearly forty years after the publication of Melzack and Wall 1965. See Jones 1992, Sandkühler 1996, and Fields and Basbaum 1999 for a thorough and up-to-date examination of endogenous pain modulation mechanisms. For more accessible general surveys of the modulatory and control systems, see Fields 1987, Melzack and Wall 1988, Price 1999, and Hardcastle 1999.

37. In fact, the idea of a dual afferent system goes back to Head and Rivers (1920). They postulated an "epicritic" system responsible for fine and precise tactile sensory discrimination and a "propopathic" system responsible for generating sensations in response to nociceptive stimulation.

38. It is interesting to note that Marshall (1892, 1894a,b), a philosopher and a psychologist, argued against the conception of negative affect being in series with pain sensations proper, which was a common assumption at the time. Marshall thought of the unpleasantness of pain sensation as a particular kind of negative tone surrounding a specialized sensory response to nociceptive stimuli; thus he conceived of it as a process parallel to pain sensations. In fact, he was arguing against the specificity theory by defending what was called the "affect theory of pain." Marshall generally conceived of (negative and positive) affect as a hedonic tone surrounding all kinds of sensations proper. Such a view seems to have been defended by some of the nineteenth-century introspectionists like E. B. Titchener who wrote: "The pain of a toothache is localized at a particular place, 'in the tooth'; but the unpleasantness of it suffuses the whole of present experience, is as wide as consciousness. The word 'pain' . . . often means the whole toothache experience." (The quotation is from Melzack [1961, p. 47], who does not specify the reference.) See also James 1895 and Duncker 1941 for similar views. Trigg 1970 contains one of the most sophisticated philosophical accounts of pain that is sensitive to these issues.

39. See Greenspan and Winfield 1992, Greenspan et al. 1999, Dong et al. 1996, and Dong et al. 1994. Also, for a fascinating survey of the evidence for the importance of the posterior parietal areas for immediate affect, which is thought to be implicated in patients with pain asymbolia, see Grahek 2001. Grahek, after going over various kinds of pain disassociation effects, concludes that the only cases where absolute disassociation occurs are some forms of pain asymbolia where extensive damage is found in either insular cortex or the posterior parietal areas, especially in area 7b.

40. See the influential studies conducted by Rainville and his colleagues (Rainville et al. 1997, Rainville et al. 1999). By using hypnosis, they were able to selectively influence ratings of pain unpleasantness and pain sensation intensity. They were able to show that unpleasantness can be

varied while sensory intensity was kept constant. But they couldn't demonstrate that the pain sensation can be varied without changing its unpleasantness. These results seem to show that pain affect is causally dependent on (and therefore in series with) the sensory processing of pain intensity. On the basis of these results, they propose a serial model of pain affect resulting from pain sensation processing.

41. Since his 1988, Price has been consistently advocating a serial processing of pain affect. In a series of recent articles (Price 2000, 2002) and in his recent book (Price 1999), he presented the underlying mechanisms and much of the evidence for serial processing of pain affect. He thinks that although affect in general is processed both in parallel and in series by the two major pathways, consciously felt affect results when the nociceptive stream coming from SS1 and SS2 (major sites at the end of the lateral pathway) reaches IC and ACC via PPC and PCC—see figure 1.7. I am much indebted to his work and his patient tutelage.

42. See Robinson 1982, Jackson 1982, 1986, and Chalmers 1996, among others. (The chapter by Polger and Sufka in this volume addresses this problem; they attempt to show that the explanatory gap said to exist between conscious experiences and brain processes can be closed, and they use the case of pain to show that.)

43. *Perhaps*, because the huge variability of the link between nociceptive stimuli and the pain experience can be taken as evidence that even this aspect may not be representational, or may not be *entirely* representational. A pure representationalist needs to argue for her case in the face of apparent scientific counterevidence.

44. See Clark's essay in this volume for an elegant and powerful development of this sort of proposal. However, he thinks that the (negative or positive) affect of experiences is not a quale. But it is not entirely clear whether by this he means to reject the view that the affective aspect of experiences is not part of their overall qualitative phenomenology. Aydede 2000 also contains a psychofunctionalist proposal with respect to pleasure states.

45. See particularly the work of Kent Berridge and his colleagues at the University of Michigan (Berridge and Pecina 1995; Berridge 1996; Berridge and Robinson 1998; Berridge 1999).

46. For a detailed attempt to give such an account answering what we have called above the "main difficulty" and incorporating an appropriate account of introspection, see Aydede and Güzeldere 2005. See also Lycan 1987a, 1996.

References

Addis, L. 1986. "Pains and Other Secondary Mental Entities." *Philosophy and Phenomenological Research* 47(1): 59–74.

Armstrong, D. M. 1962. *Bodily Sensations*. London: Routledge and Kegan Paul.

———. 1964. "Vesey on Bodily Sensations." *Australasian Journal of Philosophy* 42: 247–248.

———. 1968. *A Materialist Theory of the Mind*. New York: Humanities Press.

Aune, B. 1967. *Knowledge, Mind, and Nature: An Introduction to Theory of Knowledge and the Philosophy of Mind*. New York: Random House.

Aydede, M. 2000. "An Analysis of Pleasure vis-à-vis Pain." *Philosophy and Phenomenological Research* 61(3): 537–570.

———. 2001. "Naturalism, Introspection, and Direct Realism about Pain." *Consciousness and Emotion* 2(1): 29–73.

———. 2003. "Is Introspection Inferential?" Pp. 55–64, in *Privileged Access: Philosophical Accounts of Self-Knowledge*, ed. B. Gertler. Hampshire, U.K.: Ashgate.

Aydede, M., and G. Güzeldere. 2002. "Some Foundational Issues in the Scientific Study of Pain." *Philosophy of Science* 69 (suppl.): 265–283.

———. 2005. "Cognitive Architecture, Concepts, and Introspection: An Information-Theoretic Solution to the Problem of Phenomenal Consciousness." *Noûs* 39(2): 197–255.

Barber, T. 1964. "The Effects of 'Hypnosis' on Pain: A Critical Review of Experimental and Clinical Findings." *Journal of the American Society of Psychosomatic Dentistry and Medicine* 11(2).

Barber, T. X. 1959. "Toward a Theory of Pain: Relief of Chronic Pain by Prefrontal Leucotomy, Opiates, Placebos, and Hypnosis." *Psychological Bulletin* 56.

Baxter, D. W., and J. Olszewski. 1960. "Congenital Universal Insensitivity to Pain." *Brain* 83.

Berridge, K. C. 1996. "Food Reward: Brain Substrates of Wanting and Liking." *Neuroscience and Biobehavioral Reviews* 20(1): 1–25.

———. 1999. "Pleasure, Pain, Desire, and Dread: Hidden Core Processes of Emotion." Pp. 525–557, in *Well-Being: The Foundations of Hedonic Psychology*, ed. D. Kahneman, E. Diener, and N. Schwarz. New York: Russell Sage Foundation.

Berridge, K. C., and S. Pecina. 1995. "Benzodiazepines, Appetite, and Taste Palatability." *Neuroscience & Biobehavioral Reviews* 19(1): 121–131.

Berridge, K. C., and T. E. Robinson. 1998. "What is the Role of Dopamine in Reward: Hedonic Impact, Reward Learning, or Incentive Salience?" *Brain Research Reviews* 28(3): 309–369.

Berthier, M. L., S. E. Starkstein, and R. Leiguarda. 1988. "Pain Asymbolia: A Sensory–Limbic Disconnection Syndrome." *Annals of Neurology* 24: 41–49.

Berthier, M. L., S. E. Starkstein, M. H. Nogues, and R. G. Robinson. 1990. "Bilateral Sensory Seizures in a Patient with Pain Asymbolia." *Annals of Neurology* 27(1): 190.

Block, N. 1996. "Mental Paint and Mental Latex." Pp. 19–49, in *Perception*, ed. E. Villanueva. Atascadero, Calif.: Ridgeview.

Bonica, J. J., ed. 1990. *The Management of Pain*. Philadelphia: Lea and Febiger.

Bonica, J. J., and J. D. Loeser, eds. 2001. "History of Pain Concepts and Therapies." Pp. 3–16, in *Bonica's Management of Pain*, ed. J. J. Bonica and J. D. Loeser. Philadelphia: Lippincott, Williams, and Wilkins.

Botterell, E. H., et al. 1942. *Sensation and Perception in the History of Experimental Psychology*. New York: Appleton-Century-Crofts.

Bouckoms, A. J. 1994. "Limbic Surgery for Pain." Pp. 1171–1187, in *Textbook of Pain*, ed. P. D. Wall and R. Melzack. Edinburgh: Churchill Livingston.

Brand, P. W., and P. Yancey. 1993. *Pain: The Gift Nobody Wants*. New York: HarperCollins.

Broad, C. D. 1959. *Scientific Thought*. Patterson, N.J.: Littlefield Adams.

Byrne, A. 2001. "Intentionalism Defended." *Philosophical Review* 110(2): 199–240.

Carruthers, P. 2000. *Phenomenal Consciousness: A Naturalistic Theory*. Cambridge: Cambridge University Press.

Chalmers, D. J. 1996. *The Conscious Mind: In Search of a Fundamental Theory*. New York: Oxford University Press.

Chisholm, R. M. 1957. *Perceiving: A Philosophical Study*. Ithaca: Cornell University Press.

Dallenbach, K. M. 1939. "Pain: History and Present Status." *American Journal of Psychology* 52: 331–347.

Davidson, D. 1980. *Essays on Actions and Events*. Oxford, New York: Oxford University Press.

Descartes, R. 1664. *L'homme*. Paris: Jacques Le Gras.

Devinsky, O., M. G. Morrell, and B. A. Vogt. 1995. "Contributions of Anterior Cingulate Cortex to Behaviour." *Brain* 118(1): 279–306.

Dong, W. K., E. H. Chudler, K. Sugiyama, and V. J. Roberts. 1994. "Somatosensory, Multisensory, and Task-Related Neurons in Cortical Area 7b (PF) of Unanesthetized Monkeys." *Journal of Neurophysiology* 72(2): 542–564.

Dong, W. K., T. Hayashi, V. J. Roberts, and B. M. Fusco. 1996. "Behavioral Outcome of Posterior Parietal Cortex Injury in the Monkey." *Pain* 64(3): 579–587.

Douglas, G. 1998. "Why Pains Are Not Mental Objects." *Philosophical Studies* 91(2): 127–148.

Dretske, F. 1981. *Knowledge and the Flow of Information*. Cambridge, Mass.: MIT Press.

———. 1988. *Explaining Behavior: Reasons in a World of Causes*. Cambridge, Mass.: MIT Press.

———. 1995. *Naturalizing the Mind*. Cambridge, Mass.: MIT Press.

———. 1999. "The Mind's Awareness of Itself." *Philosophical Studies* 95(1–2): 103–124.

———. 2003. "How Do You Know You Are Not A Zombie?" Pp. 1–13, in *Privileged Access: Philosophical Accounts of Self-Knowledge*, ed. B. Gertler. Hampshire, U.K.: Ashgate.

Ducasse, C. J. 1952. "Moore's Refutation of Idealism." In *The Philosophy of G. E. Moore*, ed. P. A. Schilpp. New York: Tudor.

Duncker, K. 1941. "On Pleasure, Emotion, and Striving." *Philosophy and Phenomenological Research* 1(4): 391–430.

Everitt, N. 1988. "Pain and Perception." *Proceedings of the Aristotelian Society* 89: 113–124.

Fields, H. L. 1987. *Pain*. New York: McGraw-Hill.

———. 1999. "Pain: An Unpleasant Topic." *Pain* 6 (suppl.): 61–69.

Fields, H. L., and A. I. Basbaum. 1999. "Central Nervous System Mechanisms of Pain Modulation." Pp. 309–329, in *Textbook of Pain*, ed. R. Melzack and P. D. Wall. Edinburgh: Churchill Livingstone.

Fleming, N. 1976. "The Objectivity of Pain." *Mind* 85(340): 522–541.

Fodor, J. 1987. *Psychosemantics: The Problem of Meaning in the Philosophy of Mind*. Cambridge, Mass.: MIT Press.

Foltz, E. L., and L. E. White. 1962a. "The Role of Rostral Cingulumonotomy in 'Pain' Relief." *International Journal of Neurology* 6: 353–373.

Foltz, E. L., and L. E. White. 1962b. "Pain 'Relief' by Frontal Cingulotomy." *Journal of Neurosurgery* 19: 89–100.

Foster, J. 2000. *The Nature of Perception*. Oxford: Oxford University Press.

Freeman, W., and J. W. Watts. 1950. *Psychosurgery, in the Treatment of Mental Disorders and Intractable Pain*. Springfield, Ill.: Charles.

Freeman, W., J. W. Watts, and T. Hunt. 1942. *Psychosurgery; Intelligence, Emotion and Social Behavior Following Prefrontal Lobotomy for Mental Disorders*. Springfield, Ill.: C. C. Thomas.

Freeman, W., and J. W. Wattz. 1946. "Pain of Organic Disease Relieved by Prefrontal Lobotomy." *Proceedings of the Royal Academy of Medicine* 39: 44–47.

Goldscheider, A. 1894. *Ueber den Schmerz im Physiologischer und Klinischer Hinsicht*. Berlin: Hirschwald.

Gracely, R. H., R. Dubner, and P. A. McGrath. 1979. "Narcotic Analgesia: Fentanyl Reduces the Intensity But Not the Unpleasantness of Painful Tooth Pulp Sensations." *Science* 203: 1261–1263.

Graham, G., and G. L. Stephens. 1985. "Are Qualia a Pain in the Neck for Functionalists?" *American Philosophical Quarterly* 22: 73–80.

Grahek, N. 1991. "Objective and Subjective Aspects of Pain." *Philosophical Psychology* 4: 249–266.

———. 2001. *Feeling Pain and Being in Pain*. Oldenburg, Denmark: BIS-Verlag, University of Oldenburg.

Greenspan, J. D., R. R. Lee, and F. A. Lenz. 1999. "Pain Sensitivity as a Function of Lesion Location in the Parasylvian Cortex." *Pain* 81: 273–282.

Greenspan, J. D., and J. A. Winfield. 1992. "Reversible Pain and Tactile Deficits Associated with a Cerebral Tumor Compressing the Posterior Insula and Parietal Operculum." *Pain* 50(1).

Grice, H. P. 1962. "Some Remarks about the Senses." Pp. 133–151, in *Analytical Philosophy*, ed. R. J. Butler. Oxford, U.K.: Blackwell.

Hardcastle, V. G. 1999. *The Myth of Pain*. Cambridge, Mass.: MIT Press.

Hardy, J. D., H. J. Wolff, and H. Goodell. 1952. *Pain Sensations and Reactions*. Baltimore: Williams and Wilkins.

Harman, G. 1990. "The Intrinsic Quality of Experience." Pp. 31–52, in *Philosophical Perspectives: Action Theory and Philosophy of Mind*, ed. E. Villanueva. Atascadero, Calif.: Ridgeview.

Head, H., and W. H. R. Rivers. 1920. *Studies in Neurology*. London: H. Frowde, Hodder and Stoughton.

Holborow, L. C. 1969. "Against Projecting Pains." *Analysis* 29: 105–108.

Hurt, R. W., and H. T. Ballantyne. 1974. "Stereotactic Anterior Cingulate Lesions for Persistent Pain: A Report on 68 Cases." *Clinical Neurosurgery* 21: 334–351.

IASP. 1986. "Pain Terms: A List with Definitions and Notes on Pain." *Pain* 3 (suppl.): 216–221.

Jackson, F. 1975. "On the Adverbial Analysis of Visual Experience." *Metaphilosophy* 6: 127–135.

———. 1977. *Perception: A Representative Theory*. Cambridge: Cambridge University Press.

———. 1982. "Epiphenomenal Qualia." *Philosophical Quarterly* 32(127): 127–136.

———. 1986. "What Mary Didn't Know." *Journal of Philosophy* 83(5): 291–295.

———. 1998. "Mind, Method, and Conditionals: Selected Essays." *International Journal of Philosophical Studies* 8(2): 259–261.

———. 2000. "Some Reflections on Representations." Draft 5, presented at the NYU seminar on April 25, 2000.

James, W. 1895. "The Physical Basis of Emotion." *Psychological Review* 1: 516–529.

Jones, S. L. 1992. "Descending Control of Nociception." In *The Initial Processing of Pain and Its Descending Control: Spinal and Trigeminal Systems*, ed. A. R. Light. New York: Karger.

Kraut, R. 1982. "Sensory States and Sensory Objects." *Noûs* 16(2): 277–293.

Kripke, S. A. 1980. *Naming and Necessity*. Cambridge, Mass.: Harvard University Press.

Langsam, H. 1995. "Why Pains Are Mental Objects." *Journal of Philosophy* 92(6): 303–313.

Livingston, W. K. 1943. *Pain Mechanisms*. New York: Macmillan.

Lycan, W. G. 1987a. *Consciousness*. Cambridge, Mass.: MIT Press.

———. 1987b. "Phenomenal Objects: A Backhanded Defense." Pp. 513–526, in *Philosophical Perspectives*, ed. J. E. Tomberlin. Atascadero, Calif.: Ridgeview.

———. 1995. "Consciousness as Internal Monitoring, I: The Third Philosophical Perspectives Lecture." *Philosophical Perspectives* 9 (AI, Connectionism and Philosophical Psychology): 1–14.

———. 1996. *Consciousness and Experience*. Cambridge, Mass.: MIT Press.

Margolis, J. 1976. "Pain and Perception." *International Studies in Philosophy* 8: 3–12.

Marshall, H. R. 1892. "Pleasure–Pain and Sensation." *Philosophical Review* 1(6): 625–648.

———. 1894a. *Pain, Pleasure, and Aesthetics: An Essay Concerning the Psychology of Pain and Pleasure*. London and New York: Macmillan.

———. 1894b. "Pleasure–Pain." *Mind* 3(12): 533–535.

Mayberry, T. C. 1978. "The Perceptual Theory of Pain." *Philosophical Investigations* 1: 31–40.

———. 1979. "The Perceptual Theory of Pain: Another Look." *Philosophical Investigations* 2: 53–55.

McGinn, C. 1997. *The Character of Mind: An Introduction to the Philosophy of Mind*. Oxford: Oxford University Press.

McKenzie, J. C. 1968. "The Externalization of Pains." *Analysis* 28: 189–193.

McMurray, G. A. 1955. "Congenital Insensitivity to Pain and Its Implications for Motivational Theory." *Canadian Journal of Psychology* 9.

———. 1975. "Theories of Pain and Congenital Universal Insensitivity to Pain." *Canadian Journal of Psychology* 29(4).

Melzack, R. 1961. "The Perception of Pain." *Scientific American* 204(2): 41–49.

Melzack, R., and K. L. Casey. 1968. "Sensory, Motivational, and Central Control Determinants of Pain: A New Conceptual Model." Pp. 223–243, in *The Skin Senses*, ed. D. Kenshalo. Springfield, Ill.: Charles C. Thomas.

Melzack, R., and P. D. Wall. 1962. "On the Nature of Cutaneous Sensory Mechanisms." *Brain* 85: 331–356.

———. 1965. "Pain Mechanisms: A New Theory." *Science* 150(3699): 971–979.

———. 1983. *The Challenge of Pain*. New York: Basic Books.

———. 1988. *The Challenge of Pain*. London and New York: Penguin.

Millikan, R. G. 1984. *Language, Thought, and Other Biological Categories: New Foundations for Realism*. Cambridge, Mass.: MIT Press.

Moore, G. E. 1903. "The Refutation of Idealism." *Mind* 12: 433–453.

———. 1939. "Proof of an External World." *Proceedings of the British Academy* 25: 273–300.

Müller, J. 1842. *Elements of Physiology*. London: Taylor.

Newton, N. 1989. "On Viewing Pain as a Secondary Quality." *Noûs* 23(5): 569–598.

Noordenbos, W. 1959. *Pain*. Amsterdam, New York: Elsevier.

Papineau, D. 1987. *Reality and Representation*. London: Basil Blackwell.

Peacocke, C. 1983. *Sense and Content: Experience, Thought, and Their Relations*. Oxford: Oxford University Press.

Perkins, M. 1983. *Sensing the World*. Indianapolis, Ind.: Hackett.

Pitcher, G. 1969. "Mckenzie on Pains." *Analysis* 29: 103–105.

———. 1970. "Pain Perception." *Philosophical Review* 79(3): 368–393.

———. 1971. *A Theory of Perception*. Princeton, N.J.: Princeton University Press.

———. 1978. "The Perceptual Theory of Pain: A Response to Thomas Mayberry's 'The Perceptual Theory of Pain.'" *Philosophical Investigations* 1: 44–46.

Ploner, M., H. J. Freund, and A. Schnitzler. 1999. "Pain Affect without Pain Sensation in a Patient with a Postcentral Lesion." *Pain* 81(1/2): 211–214.

Price, D. D. 1988. *Psychological and Neural Mechanisms of Pain*. New York: Raven Press.

———. 1999. *Psychological Mechanisms of Pain and Analgesia*. Seattle: IASP Press.

———. 2000. "Psychological and Neural Mechanisms of the Affective Dimension of Pain." *Science* 288(9): 1769–1772.

———. 2002. "Central Neural Mechanisms that Interrelate Sensory and Affective Dimensions of Pain." *Molecular Interventions* 2: 392–403.

Price, H. H. 1950. *Perception*. London: Methuen.

Rainville, P., B. Carrier, R. K. Hofbauer, M. C. Bushnell, and G. H. Duncan. 1999. "Dissociation of Sensory and Affective Dimensions of Pain Using Hypnotic Modulation." *Pain Forum* 82(2): 159–171.

Rainville, P., G. H. Duncan, D. D. Price, B. Carrier, and M. C. Bushnell. 1997. "Pain Affect Encoded in Human Anterior Cingulate but Not Somatosensory Cortex." *Science* 277(5328): 968–971.

Rey, G. 1997. *Contemporary Philosophy of Mind: A Contentiously Classical Approach*. Cambridge, Mass.: Basil Blackwell.

Robinson, H. 1982. *Matter and Sense: A Critique of Contemporary Materialism*. Cambridge: Cambridge University Press.

———. 1994. *Perception*. London, New York: Routledge.

Rubins, J. L., and E. D. Friedman. 1948. "Asymbolia for Pain." *Archives of Neurology and Psychiatry* 60: 554–573.

Russell, B. 1912. *The Problems of Philosophy*. London: Home University Library.

Sandkühler, J. 1996. "The Organization and Function of Endogenous Antinociceptive Systems." *Progress in Neurobiology* 50: 49–81.

Seager, W. 2002. "Emotional Introspection." *Consciousness and Cognition* 11(4): 666–687.

Searle, J. R. 1983. *Intentionality: An Essay in the Philosophy of Mind*. Cambridge: Cambridge University Press.

Sellars, W. 1975. "The Adverbial Theory of the Objects of Sensation." *Metaphilosophy* 6: 144–160.

Shoemaker, S. 1996. "Qualities and Qualia: What's in the Mind?" In his *The First-Person Perspective and Other Essays*. Cambridge: Cambridge University Press.

Stephens, G. L., and G. Graham. 1987. "Minding Your P's and Q's: Pain and Sensible Qualities." *Noûs* 21(3): 395–405.

Sternbach, R. A. 1963. "Congenital Insensitivity to Pain: A Critique." *Psychological Bulletin* 60(3): 252–264.

Sufka, K. J., and D. D. Price. 2002. "Gate Control Theory Reconsidered." *Brain and Mind* 3: 277–290.

Trigg, R. 1970. *Pain and Emotion*. Oxford: Clarendon Press.

Tye, M. 1984a. "The Adverbial Approach to Visual Experience." *Philosophical Review* 93(2): 195–225.

———. 1984b. "Pain and the Adverbial Theory." *American Philosophical Quarterly* 21: 319–328.

———. 1996. *Ten Problems of Consciousness: A Representational Theory of the Phenomenal Mind*. Cambridge, Mass.: MIT Press.

———. 1997. "A Representational Theory of Pains and Their Phenomenal Character." In *The Nature of Consciousness: Philosophical Debates*, ed. N. Block, O. Flanagan, and G. Güzeldere. Cambridge, Mass.: MIT Press.

Vesey, G. N. A. 1964a. "Armstrong on Bodily Sensations." *Philosophy* 39: 177–181.

———. 1964b. "Bodily Sensations." *Australasian Journal of Philosophy* 42: 232–247.

von Frey, M. 1894. "Beitrage zur Physiologie der Shmerzsinnes." *Ber. Verhandl. konig. sachs. Ges. Wiss.* 46(185).

Weinstein, E. A., R. L. Kahn, and W. H. Slate. 1995. "Withdrawal, Inattention, and Pain Asymbolia." *Archives of Neurology and Psychiatry* 74(3): 235–248.

White, J. C., and W. H. Sweet. 1969. *Pain and the Neurosurgeon: A Forty-Year Experience*. Springfield, Ill.: Charles C. Thomas.

White, S. L. 1986. "Curse of the Qualia." *Synthese* 68: 333–367.

Wilkes, K. V. 1977. *Physicalism*. London: Routledge.

2   The Epistemology of Pain

Fred Dretske

Many people think of pain and other bodily sensations (tickles, itches, nausea) as feelings one is necessarily conscious of. Some think there can be pains one doesn't feel, pains one is (for a certain interval) not conscious of ("I was so distracted I forgot about my headache"), but others agree with Thomas Reid (1785, 1, 1, 12) and Saul Kripke (1980, p. 151) that unfelt pains are like invisible rainbows: they can't exist. If you are so distracted you aren't aware of your headache, then, for that period of time, you are not in pain. Your head doesn't hurt. It doesn't *ache*. For these people (I'm one of them) you can't be in pain without feeling it, and feeling it requires awareness of it. The pain (we have learned) may not bother you, it may lack the normal affective (as opposed to sensory) dimension of pain, but it can't be *completely* insensible. That wouldn't be pain.

Whether or not we are necessarily conscious of our own pains is clearly relevant to the epistemology of pain. If there are—or could be—pains of which we are not aware, then there obviously is an epistemological problem about pain. How do we know we are not always in pain without being aware of it? Pain becomes something like cancer. You can have it without being aware of it. Not only that, you can *not* be in pain and mistakenly think you are. As a result of misleading external signs, you mistakenly think you are in pain that you haven't yet felt. The epistemology of pain begins, surprisingly, to look like the epistemology of ordinary physical affairs. So much for Descartes and the alleged first-person authority about our own mental life.

One could go in this direction, but I prefer not to. I prefer to simplify things and minimize problems. I therefore focus on pains (if there are any other kind) of which we are aware, pains, as I will say, that really hurt. For those who conceive of pain as something of which one is necessarily aware, then, I am concerned, simply, with pain itself: when you are in pain, how do you know you are? For those who think there are, or could be, pains one does not feel, I am concerned with a subset of pains—those one is aware of, the ones that really hurt. When it hurts, how do you know it does?

If this is the topic, what is the problem? If we are talking about things of which we are necessarily aware, the topic is things that, when they exist, we know they exist. There may be a question about *how* we know we are in pain, but that we know it is assumed at the outset. Isn't it?

No it isn't. To suppose it is is to confuse awareness of things[1] with awareness of facts. One can be aware of an armadillo, a thing, without being aware of the fact that it is an armadillo, without knowing or believing it is an armadillo—without, in fact, knowing what an armadillo is. There is a sensory, a phenomenal, form of awareness— seeing or in some way perceptually experiencing an armadillo—and a conceptual form of awareness—knowing or believing that it (what you are experiencing) is an armadillo. We use the word "awareness" (or "consciousness") for both. We are aware *of* objects (and their properties) on the one hand, and we are aware *that* certain things are so on the other. If one fails to distinguish these two forms of awareness, awareness of *x* with awareness that it is *x*, one will mistakenly infer that simply being in pain (requiring, as I am assuming, awareness of the pain) requires awareness of the fact that one is in pain and, therefore, knowledge. Not so. I am assuming that if it really hurts, you must feel the pain, yes, and feeling the pain is awareness of it, but this is the kind of awareness (thing-awareness) one can have without fact-awareness of what one is aware of—that it is pain—or that one is aware of it. Chickens and maybe even fish, we may suppose, feel pain, but in sup- posing this we needn't suppose that these animals have the concept PAIN. We needn't suppose they understand what pain is sufficiently well to believe (hence know, hence be aware) that they are in pain. They are aware of their pain, yes. It really hurts. That is why they squawk, squirm, and wiggle. That is why they exhibit behavior symptomatic of pain. But this does not mean they believe or know that they are in pain. It is the chicken's feeling pain, its awareness of its pain, not its belief or aware- ness that it is in pain that explains its behavior. The same is true of human infants.[2] They cry when they are hungry not because they think (much less know) they are hungry. They cry because they *are* hungry or, if you prefer, because they *feel* hungry, but they can be and feel hungry without knowing it is hunger they feel or that they are feeling it.

So I do not begin by assuming that if it hurts, you know it does. In fact, as a general rule, this is false. Maybe we adult human beings always know when it hurts, but chick- ens and fish (probably) don't. Human infants probably don't either. I'm asking, instead, why we adult human beings always seem to know it and if we really do know it, *how* we know it. What is it about pain—unlike, say, cancer—that confers episte- mological authority on its possessors?

When I see a pencil, the pencil doesn't depend on my awareness of it. Its existence, and its existence as a pencil, doesn't depend on my seeing it. When I stop looking at it, the pencil continues to exist in much the way it did when I saw it. Pencils are indifferent to my attentions. There is the pencil, the physical object of awareness, on the one hand, and there is my awareness of it, a mental act of awareness, on the other. Remove awareness of the pencil, this relational or extrinsic property, from the pencil, and one is left with the pencil, an unchanged object of awareness.

This can't be the way it is with pain since pain, at least the sort I am concentrating on here, unlike pencils, is something one is necessarily aware of. Remove the act of awareness from this object of awareness and this object ceases to be pain. It stops hurting.

So in this respect pains are unlike pencils. Unlike a pencil, the stabbing sensation in your lower back cannot continue to exist, at least not as pain, when you cease to be aware of it. If it continues to exist at all, it continues to exist as something else. When the marital relation is removed (by divorce, say) from a husband, the relation that makes him a husband, you are left with a man, a man who is no longer a husband. If you remove the awareness relation from a stabbing sensation in your lower back, the relation without which it doesn't hurt, what are you left with? A stabbing event (?) in your lower back that doesn't hurt? that isn't painful? What sort of thing could this be?[3]

To understand the sort of thing it might be think about something I will call a *crock*.[4] A crock (stipulate) is a rock you are visually aware of. It is a rock you see. A crock is a rock that stands in this perceptual relation to you. Remove that relation, as you do when you close your eyes, and the crock ceases to be a crock. It remains a rock, but not a crock. A crock is like a husband, a person whose existence as a husband, but not as a man, depends on the existence of a certain extrinsic relationship. Crocks are like that. They look just like rocks. They have all the same intrinsic (nonrelational) properties of rocks. They look like rocks because they *are* rocks, a special kind of rock, to be sure (one you are aware of), but a rock nonetheless.

Should we think of pains like crocks? When you cease to be aware of it, does your pain cease to exist as pain, but continue to exist as something else, something that has all the same intrinsic (and other relational) properties of pain but which requires your awareness of it to recover its status as pain in the way a rock has all the same intrinsic (and most relational) properties of a crock and requires only your awareness of it to regain its status as a crock? Is there something—let us call it *protopain*—that stands to pain the way rocks stand to crocks? Is a stabbing pain in your lower back merely a stabbing (?) protopain in your lower back that you happen to be aware of? Under anesthesia, is there still protopain in your lower back, something with all the same intrinsic properties you were aware of when feeling pain, but something that, thanks to the anesthetic, no longer hurts because you are no longer aware of it?[5]

If we understand pain on this model—the crock model—we have a problem. How serious the problem is depends on how much of a problem it is to know one is aware of something. To appreciate the problem, or at least threat of a problem, think about how you might go about identifying crocks. When you see a crock, it is visually indistinguishable from an ordinary rock. Rocks and crocks look alike. They have the same intrinsic, the same observable properties. They differ only in one of their relational properties. They are, in this respect, like identical twins. This means that when you see a crock, there is nothing in what you are aware of, nothing in what you see, that tells you it is a crock you see and not just a plain old rock. So how do you figure out whether the rock you see is a crock and not just a rock?

I expect you to say: I know the crocks I see are crocks because I know they are rocks (I can see this much) and I know I see them. So I know they are crocks. Assuming there is no problem about recognizing rocks, this will work as long as there is no problem in knowing you see them. But how do you find out that you see, that you are visually aware of, the rock? There is, as already noted, nothing in what you see (the crock) that tells you that you see it. The crock would be exactly the same in all observable respects if you didn't see it—if it wasn't a crock. Just as husbands differ from men who are not husbands in having certain hidden (to direct observation) qualities, qualities one cannot observe by examining the husband, crocks differ from rocks in having a certain hidden quality, one that can't be observed by examining the crock. When you observe a crock, the relational property of being observed by you is not itself observed by you. Just as you must look elsewhere (marriage certificates, etc.) to find out whether the man you see is a husband, you must look elsewhere to find out whether the rock you see is a crock.

But where does one look? If one can't look at the crock to tell whether it is a crock, where does one look? Inward? Is introspection the answer? You look at the rock to see whether it is a rock, but you look inward, at yourself, so to speak, to find out whether it is that special kind of rock we are calling a crock. If we understand pain in your lower back on the crock model—as a condition (in the lower back?) you are aware of—we won't be able to say how you know you have lower back pain until we understand how you find out that this condition in your lower back is not just a condition in your lower back, but a condition in your back that really hurts, a back condition that you are aware of.

This is beginning to sound awfully strange. The reason it sounds so strange is that although awareness (at least awareness of objects) is a genuine relation between a person and an object, it is, we keep being told by philosophers, an epistemologically *transparent*[6] or *self-intimating* relation. When *S* is aware of something, *S* knows automatically, without the need for evidence, reasons, or justification, that he is aware of it. *S* can be married to someone and not realize he is (maybe he has amnesia or he was drunk when he got married), but he can't be aware of something and not know he is. If this were so, there would be no problem about knowing the rocks you see are crocks since you can easily (let us pretend) see that they are rocks and you would know immediately, without need for additional evidence, in virtue of the transparency of awareness, that you see (are visually aware) of them. So anything one sees to be a rock is known, without further ado, to be a crock. The fact that makes a rock a crock—the fact that one is aware of it—is a transparent, self-intimating fact for the person who is aware of it. That is why there is no epistemological problem about pain over and above the familiar problem of distinguishing it from nearby (but not quite painful) sensations—for example, aggressive itches. That is why we don't have a problem distinguishing pain from protopain. Whatever it is, exactly, we are aware of when we are in pain, we always know, in virtue of the transparency of awareness, that we are aware of it. When we have a pain in our back, therefore, we always know it is real pain and not just protopain.

This nifty solution to our problem doesn't work, but it comes pretty close. It doesn't work because awareness is not transparent or self-intimating in this way. Animals and very young children are aware of things, but, lacking an understanding of what awareness is, they don't realize, they don't know, they are aware of things. A chicken is visually aware of rocks and other chickens without knowing it is. That is why animals and young children—even if they know what rocks are (they probably don't even know this much)—don't know the rocks they see are crocks. They don't know they see them. They don't know they are aware of them. If awareness is a transparent, self-intimating relation, it is so for only a select class of people. It certainly isn't so for children and animals.

So if we are going to appeal to the transparency of awareness in the epistemology of pain, we must be careful to restrict its transparency to those who understand what awareness is, to those capable of holding beliefs and making judgments about their own (and, of course, others) awareness of things.[7] We need, that is, a principle something like:

(**T**) If *S* understands what awareness is (i.e., is capable of holding beliefs and making judgments to the effect that she is aware of things) and *S* is aware of *x*, then *S* knows she is aware of *x*.

This sounds plausible enough, but we have to be careful here with the variable "*x*." What, exactly, does it mean to say that *S* knows she is aware of *x*? If *x* is a rock, must *S* know she is aware of a rock? Clearly not. *S* can see a rock and not know it is a rock and, therefore, not know she is visually aware of a rock. She thinks, mistakenly, it is a piece of cardboard. She might not even know it is a physical object. She thinks she is hallucinating. So what, exactly, does principle **T** tell us *S* knows about the *x* she is aware of?

Nothing, except that she is aware of it. Awareness of objects makes these objects available to the person who is aware of them as objects of *de re* belief, as things (a *this* or a *that*) he or she can have beliefs about. Since, however, none of these additional beliefs you have about *x* need be true for you to have them, you needn't know anything about *x* other than that you are aware of it. To illustrate, consider the following example. *S* sees six rocks on a shelf. She sees them long enough and clearly enough to see all six. When *S* looks away for a moment, another rock is added. When *S* looks back, she, once again, observes the rocks long enough and clearly enough to see all seven. She doesn't, however, notice the difference. She doesn't realize there is an additional rock on the shelf. She sees—and is, therefore, aware of—an additional rock on the shelf, but she doesn't know she is. *S* is aware of something (an additional rock) she doesn't know she is aware of.

Does this possibility show, contrary to **T**, that one can be aware of an object and not know it? No. It only shows that one can be aware of something additional without knowing one is aware of it under the description "something additional." Maybe, though, one knows one is aware of the additional rock under the description "the leftmost rock" or, simply, as "one of the rocks I see" or, perhaps (if she doesn't know it is a rock), as "one of the things I see." If all seven rocks are really seen the second time, why not say the perceiver knows she is aware of each and every rock she sees. She just doesn't know they are rocks, how many there are, or that there are more of them this time than last time. But she does know, of each and every rock she sees, that she is aware of it.

If we accept this way of understanding "*S* knows she is aware of *x*," there may still be a problem about the intended reference of "*x*" in our formulation of transparency principle **T**. Suppose *S* hallucinates a talking rabbit with the conviction that she really sees and hears a talking rabbit. *S* mistakenly thinks she is aware of a talking white rabbit. She isn't. There are no white rabbits, let alone talking white rabbits, in *S*'s vicinity. What, then, is *S* aware of? More puzzling still (if we assume she is aware of something), what is it that (according to **T**) *S* *knows* she is aware of? Is there something, something she can (perhaps mentally) pick out or refer to as *that*, that she knows she is aware of? If so, what is it? Is it something in her head? A mental image? If so, does this image talk? Or does it merely appear to be talking? Does it have long ears? Or only appear to have long ears? Is *S*, then, aware of something that has (or appears to have) long white ears and talks (or sounds as though it is talking) like Bugs Bunny?

Knowing what lies ahead on this road (viz., sense data) many philosophers think the best way to understand hallucinations (dreams, etc.) is that in such experiences one is not aware of an object at all—certainly nothing that is white, rabbit-shaped, and talks like Bugs Bunny. Nor is one aware of something that only appears to have these properties. It only seems as though one is. Although there appears to be an object having these qualities, there actually is no object, certainly nothing in one's head, that has or even appears to have[8] the qualities one experiences something as having.[9]

This way of analyzing hallucination, however, seems to threaten **T**. *S* thinks she is aware of something—a talking rabbit, in fact—but she isn't. She isn't aware of anything. So while hallucinating, *S*'s belief that she is aware of something is false. This seems to show that, sometimes at least, *S* can't tell the difference between being aware of something and not being aware of something. Why, then, suppose, as **T** directs, that *S* always knows when she is aware of something?

What we need to understand in order to sidestep this kind of objection to **T** is that the "*x*" *S* knows she is aware of needs to be interpreted liberally. It needn't be a physical object. It needn't be a mental object (a sense datum) either. It can be a property or a set of (appropriately "bound" together) properties the subject experiences something as having. In hallucinating a talking white rabbit *S* is conscious of various sensory qualities: colors, shapes, tones, movements, orientations, and textures. These are qualities *S* experiences (perceptually represents) something as having, qualities *S* is conscious of in having this hallucinatory experience. According to the intended interpretation of **T**, it is such qualities, not some putative object that has (or appears to have) these qualities, that *S* is aware of and (in accordance with **T**) knows she is aware of. The difference between a hallucination of a talking white rabbit and a veridical perception of one isn't—or needn't be—the phenomenal (sensory) qualities one is aware of. These can be exactly the same. The experiences can be subjectively indistinguishable. In one case one is aware of something that has the qualities, in the other case not. But in both cases the subject is aware of, and in accordance with the intended interpretation of **T**, knows she is aware of, the qualities that make the experiences that kind of experience.

So much by way of propping up **T**. What we are left with may appear contrived and suspiciously ad hoc. Nonetheless, it or something close to it seems to do the job. It explains how one can, without additional epistemic effort (beyond what it takes to identify rocks), know that crocks are crocks. More significantly for present purposes, it also explains why someone who understands what awareness is, someone who is cognitively developed enough to *think* she is in pain, can't be aware of protopain (the cluster of qualities she is aware of when in pain) without knowing she is aware of them and, therefore, without knowing she is in pain.

Is **T** true? If it is, *why* is it true? What is it about awareness, or perhaps the concept of awareness, or perhaps the having of this concept, that yields these striking epistemological benefits?

The fact that, according to **T**, *S* must not only be aware of *x*, but also understand what awareness is (an understanding animals and infants lack) in order to know—*gratis*, as it were—she is aware of *x* tells us something important. It tells us the knowledge isn't constitutive of awareness. It tells us that awareness of *x* doesn't consist of knowing one is aware of it. The truth of **T**—if indeed it is true—isn't what Fricker (1998) calls an artifact of grammar. There are some mental relations we bear to objects in which it seems plausible to say knowledge is a component of the relationship. Memory of persons, places, and things is like that. For *S* to remember her cousin (an object), *S* needn't remember that he is her cousin (maybe she never knew this), but she must at least remember (hence know) some facts about her cousin—that, he looked so-and-so, for instance, or that he wore a baseball cap.[10] Memory of persons and things, it seems reasonable to say, consists in the retention (and, therefore, possession) of such knowledge about them. Awareness of objects and persons, though, isn't like that. You—or, if not you, then chickens and children—can be aware of objects without knowing they are. So the knowledge attributed in **T** is not the result of some trivial, semantic fact about what it means to be aware of something. It isn't like the necessity of knowing something about the people you remember. If **T** is true, it is true for some other, some deeper, reason.[11]

Perhaps, though, it goes the other way around. Although a (lower level) awareness of something (a rock) doesn't have a (higher level) belief that one is aware of it (the rock) as a constituent, maybe the higher level belief that you are aware of it (a belief animals and young children lack) has awareness as a constituent. Maybe, that is, awareness of *x* is a relation that holds between *x* and whoever thinks it holds.[12] If this were so, then a belief that you are aware of something would always be true. According to some theories of knowledge, then, such a belief would always count as knowledge. Whoever thinks they are aware of something knows they are because thinking it is so makes it so. So they can't be wrong. So they know.

This possibility would be worth exploring if it really explained what we are trying to explain—namely, why, when we are aware of something, we know we are. But it doesn't. The fact (if it were a fact) that awareness of something is, somehow, a constituent of the (higher order) belief that one is aware of something would explain why the higher order belief, if we have it, is always true—why, if we believe we are aware of something, we are. But it would not explain what we are trying to explain, the converse: why, if we are aware of something, we always believe (thus, know) we are. The proffered explanation leaves open the possibility that, when we are aware of something, we seldom, if ever, believe we are and, therefore, the possibility that, when aware of something, we seldom, if ever, know we are.

So if **T** is, somehow, an artifact of grammar, a truth vouchsafed in virtue of the concept of awareness, it is not in virtue of the belief being a constituent of the awareness or vice versa. If you always know when you are in pain, you know it for reasons other than that the belief (that you are in pain) is a constituent of the pain or the pain is a part of the belief. The pain and the belief that you are in pain are distinct existences. The problem is to understand why then, despite their distinctness, they are, for those who understand what pain is, apparently inseparable.

Chris Peacocke (1992—and earlier, Gareth Evans, 1982, p. 206) provides a way of understanding our possession of concepts in which the truth of **T** can be understood as somehow (to use Fricker's language) an artifact of grammar without supposing that it is to be understood in terms of the knowledge being a constituent of the awareness or vice versa. Concepts not only have what Peacocke (1992, p. 29) calls attribution conditions—conditions that must be satisfied for the concept to be correctly attributed to something. They also have possession conditions, conditions that must be satisfied for one to have the concept. To have a perceptual (what Peacocke calls a *sensational*—1992, p. 7) concept for the color red, for instance, he says that a person must, given normal circumstances, be able to tell, just by looking, that something is or isn't red. She must know the concept applies, or doesn't apply, to the things she sees. Possessing the concept RED requires this cognitive, this recognitional, ability. Those who lack this ability do not have the concept RED.[13]

Adapting this idea to the case of awareness, it might be supposed that a comparable cognitive ability is part of the possession conditions for AWARENESS. Although (as the case of animals and young children indicate) knowledge isn't part of the attribution (truth) conditions for awareness (*S* can be aware of something and not know she is), an ability to tell, in your own case, authoritatively, that you are aware of something may be a possession condition for this concept. You don't really have the concept, you don't really understand what it means to be aware of something, if you can't tell, when you are aware of something, that you are aware of it. This is why, in the antecedent of **T**, an understanding (of what awareness is) is required. Awareness is transparent for those who possess the concept of awareness because its transparency, the ability to tell, straight off, that one is aware of something, is a requirement for *thinking* one is aware of something. If you can think you are aware of something, then, when you are aware of something, you know you are in that special authoritative way required for possession of the concept.

This strikes me as a plausible—if not the only possible—explanation of why **T** is true. Regrettably, though, it doesn't take us very far. It is, in fact, simply a restatement of what we were hoping to explain—namely, **T**: that those who understand what awareness is know, in virtue of having this understanding, when they are aware of something. It does not tell us what we were hoping to find out—the *source* of the epistemological ability required for possession of this concept. If, to have the concept AWARENESS, I have to know I'm aware of everything I'm aware of, how do I acquire this concept? What is it that gives me the infallible (or, if not infallible, then near-infallible) powers needed to possess this concept and, thereby, a capacity to *think* I'm aware of something?[14]

One doesn't explain infallible—or, if not infallible, then authoritative—application of a concept by saying that infallibility (or authoritativeness) in its application is a condition for possessing the concept. That might be so, but that doesn't help us understand where the authority comes from. It merely transforms an epistemological question—how do we know something is *X*—into a developmental question—how do we manage to believe that something is *X*. If, in our skeptical moods, we are suspicious about infallibility or first-person authority, then requiring it as a necessary condition for possessing a concept does nothing to alleviate our skepticism. It merely displaces the skepticism to a question about whether we in fact have the concept—whether we ever, in fact, *believe* we are aware of something. It is like trying to solve an epistemological problem about knowing you are married by imposing infallibility in believing you are married as a requirement for having the concept MARRIED. You can do this, I suppose, but all you really manage to achieve by this maneuver is a kind of conditional infallibility: if you think you are married, you know you are. But the old question remains in a modified form: do you think you are? Given the beefed-up requirements on possessing the concept MARRIED, it now becomes very hard—for skeptics impossible—to think you are married.

We began by asking how one knows one is in pain. Since pain, at least the kind of pain we are here concerned with, is a feeling one is necessarily aware of (it doesn't hurt if you are not aware of it), this led us to ask how one knows one is aware of something. We concluded, tentatively, that the kind of reliability in telling you are aware of something required for knowledge (that you are aware of something) must be a precondition for possessing the concept AWARENESS, a precondition, therefore, for thinking you are aware of something. That explains why those who think they are aware of something know they are. By making reliability of judgment a possession condition for the concept of awareness, we have transformed our epistemological problem into a developmental problem, a problem about how one comes to possess the concept of AWARENESS or, indeed, any concept (like pain) that requires awareness. How do we manage to *think* we are in pain?

This doesn't seem like much progress. If we don't understand how we can make reliable judgments on topic **T**, it doesn't help to be told that reliability is necessary for making judgments about topic **T**.

But though it isn't *much* progress, it is, I think, *some* progress. If nothing else it reminds us that the solution to some of our epistemological problems, problems about how we know that a so-and-so exists, awaits a better understanding of exactly what it is we think when we think a so-and-so exists, a better understanding of what our concept SO-AND-SO is. It reminds us that questions about how we know *p* may sometimes be best approached by asking how we manage to believe *p*.[15] This is especially so when the topic is consciousness and, in particular, pain. Understanding how we know it hurts may require a better understanding of what, exactly, it is we think (and how we manage to think it) when we think it hurts.

Acknowledgments

A version of this essay was first given as a keynote address at the Seventh Annual Inland Northwest Philosophy Conference (INPC) on Knowledge and Skepticism, Washington State University and the University of Idaho, April 30 to May 2, 2004. I am grateful to the audience there for helpful and constructive discussion.

Notes

1. By "things" I mean spatiotemporal particulars. This includes, besides ordinary objects (houses, trees, and armadillos), such things as events (births, deaths, sunsets), processes (digestion, growth), conditions (the mess in his room), and states (e.g., Tom's being married). Events occur at a time and in or at a place (the place is usually the place of the objects to which the event occurs). Likewise for states, conditions, processes, and activities although these are usually said to persist for a time, not to occur at a time. So if one doesn't like talking about pains as objects and prefers to think of them as events (conditions, activities, processes) in the nervous system, that is fine. They are still things in my sense of this word. For more on property-awareness and object-awareness as opposed to fact-awareness, see Dretske 1999.

2. It is for this reason that I cannot accept Shoemaker's (1996) arguments for the "transparency" of pain—the idea that pain is necessarily accompanied by knowledge that one is in pain. Even if it is true (as I'm willing to grant) that pains (at least the pains of which we are aware) necessarily motivate certain aversive behaviors, I think it is an overintellectualization of this fact to *always* explain the pain-feeler's behavior in terms of a desire to be rid of her pain (a desire that, according to Shoemaker, implies a belief that one is in pain). I agree with Siewert (2003, pp. 136–137) that the aversive or motivational aspect of pain needn't be described in terms of conceptually articulated beliefs (that you have it) and desires (to be rid of it). In the case of animals (and young children), it seems to me implausible to give it this gloss. Maybe you and I go to the medicine chest because of what we desire (to lessen the pain) and think (that the pain pills are there), but I doubt whether this is the right way to explain why an animal licks its wound or an infant cries when poked with a pin.

3. Daniel Stoljar and Manuel Garcia-Carpintero (on two separate occasions) have asked me why I think there is anything remaining when I subtract awareness from pain. Why isn't subtracting awareness from pain more like subtracting *oddness* from the number 3 rather than subtracting *married* from a husband? My reason for thinking so is that when we are in pain there is something we are aware of that is ontologically distinct from the pain itself—e.g., the location, duration, and intensity (i.e., the properties) of the pain. These are among the qualities that give pain its distinctive phenomenal character, the qualities that make one pain different from another. They are the qualities that make a splitting headache so different from a throbbing toothache. Take away awareness of these qualities and, unlike the number 3 without oddness, one is left with something—the qualities one was aware of.

4. I introduced crocks as an expository device in Dretske 2003.

5. This way of thinking about pain (and other bodily sensations) is one version of the perceptual model of pain (Armstrong 1961, 1962; Dretske 1995; Lycan 1996; Pitcher 1971; Tye 1995) according to which pain is to be identified with a perceived bodily condition (injury, stress, etc.). Under anesthesia the bodily injury, the object you are aware of when in pain, still exists, but since it is no longer being perceived, it no longer hurts. It isn't pain. I say this is "one version" of a perceptual theory because a perceptual model of pain can identify pain not with the perceived *object* (bodily damage when it is being perceived), but the act of perceiving this object, not the bodily damage of which you are aware, but your awareness of this bodily damage. In the latter case, unlike the former, one does not perceive, one is not actually aware of, pain. When in pain, one is aware of the bodily injury, not the pain itself (which is one's awareness of the bodily injury). I do not here consider theories of this latter sort. As I said at the outset, I am concerned with the epistemology of pain (sensations in general) where these are understood to be things of which one is conscious. If you aren't (or needn't be) aware of pain, there are much greater epistemological problems about pain than the ones I am discussing here.

6. Transparency as here understood should be carefully distinguished from another use of the term in which it refers to the alleged failure (or at least difficulty) in becoming introspectively aware of perceptual experience and its properties. In trying to become aware of the properties of one's perceptual experience, one only seems to be made aware of the properties of the objects

that the experience is an experience of, the things one sees, hears, smells, and tastes. One, as it were, "sees through" the experience (hence, transparency) to what the experience is an experience of.

7. For careful formulations along these lines see Wright 1998, Fricker 1998, and (for "self-intimating") Shoemaker 1996. Chalmers (1996, pp. 196–197) describes awareness as an epistemologically special relation in something like this sense, and Siewert (1998, pp. 19–20, 39, 172) suggests that mere awareness of things (or failure to be aware of things) gives one first person warrant for believing one is (or is not) aware of them. I take it that even animals and children have the warrant. They just don't have the (warranted) belief.

8. Nor *appears* to have these qualities because to suppose that S was aware of something that merely appeared to have these qualities would be to introduce an appearance–reality distinction for mental images. What is it (a part of the brain?) that appears to be a talking white rabbit? This seems like a philosophically disastrous road to follow.

9. If this sounds paradoxical, compare: it can appear to S as though there is a fly in the ointment without there being a fly who appears to be in the ointment.

10. I do not argue for this. I'm not even sure it is true. I use it simply as a more or less plausible example of a relation we bear to objects that has, as a constituent, factual knowledge of that object.

11. This is why functionalism (about the mental) is of no help in explaining why **T** is true. Even if awareness (of an object) is a functional state, one defined by its causal role, its role cannot include the causing of belief that one is aware of something.

12. This echoes a Burgian (Burge 1985, 1988) thesis about belief—that the higher order belief that we believe *p* embodies, as a constituent, the lower order belief (that *p*) that we believe we have. This echo is pretty faint though. The major difference is that awareness of an object is not (like a belief) an intentional state. It is a genuine relation between a conscious being, S, and whatever it is she is aware of. It may be that believing you believe *p* is, among other things, to believe *p*, but why should believing you are aware of something be, among other things, awareness of something? Can you make yourself stand in this relation to something merely by thinking you do? It is for this reason that Bilgrami (1998) thinks that the *constituency thesis* (as he calls it) is only plausible for intentional states like belief (desire, etc.) that have propositional "objects."

13. It isn't clear to me what concept they have—or even whether they have a concept—if they do not have this ability at the requisite (presumably high) level of reliability but are, nonetheless, more often right than wrong in describing something as red. If they don't have the concept RED, what are they saying? What, if anything, are they thinking? Nothing? I take this to be the problem David Chalmers was raising in the discussion at the INPC conference. I ignore the problem here for the sake of seeing how far we can get in the epistemology of pain by requiring a level of reliability (of the sort needed to know) in the capacity to believe.

14. As I understand him, this is basically the same point Gallois (1996) is making against Peacocke's account of why (or, perhaps, how) we (those of us who have the concept of belief)

are justified in believing that we believe the things we do. See, in particular, Gallois 1996, pp. 56–60.

15. In Dretske 1983 I argued that the condition (relating to justification, evidence, or information) required to promote a belief that x is F into knowledge that x is F is also operative in our coming to believe that x is F (in acquiring the concept F). Roughly, if something's being F isn't the sort of thing you can know, it isn't the sort of thing you can believe either. This, of course, is the same conclusion Putnam (1981) reaches by considering brains in a vat.

References

Armstrong, D. 1961. *Perception and the Physical World*. London: Routledge and Kegan Paul.

———. 1962. *Bodily Sensations*. London: Routledge and Kegan Paul.

Bilgrami, A. 1998. "Self-Knowledge and Resentment." Pp. 207–242, in *Knowing Our Own Minds*, ed. Crispin Wright, Barry Smith, and Cynthia Macdonald. Oxford: Clarendon Press.

Burge, T. 1985. "Authoritative Self-Knowledge and Perceptual Individualism." Pp. 86–98, in *Contents of Thought*, ed. R. Grimm and D. Merrill. Tucson: University of Arizona Press.

———. 1988. "Individualism and Self-Knowledge." *Journal of Philosophy* 85: 649–663.

Chalmers, D. 1996. *The Conscious Mind*. New York: Oxford University Press.

Dretske, F. 1983. "The Epistemology of Belief." *Synthese* 55(1): 3–19.

———. 1995. *Naturalizing the Mind*. Cambridge, Mass.: MIT Press.

———. 1999. "The Mind's Awareness of Itself." *Philosophical Studies* 95: 103–124. (Reprinted in F. Dretske, *Perception, Knowledge, and Belief*. Cambridge University Press, 2000, pp. 158–177.)

———. 2003. "How Do You Know You Are Not a Zombie?" Pp. 1–13, in *Privileged Access and First-Person Authority*, ed. Brie Gertler. Aldershot: Ashgate. (Also published in Portuguese, Conference on Mind and Action III, Lisbon, Portugal, 2001, edited by João Sàáguand.)

Evans, G. 1982. *The Varieties of Reference*. Oxford: Oxford University Press.

Fricker, E. 1998. "Self-Knowledge: Special Access versus Artifact of Grammar—A Dichotomy Rejected." Pp. 155–206, in *Knowing Our Own Minds*, ed. Crispin Wright, Barry Smith, and Cynthia Macdonald. Oxford: Clarendon Press.

Gallois, A. 1996. *The World Without, the Mind Within*. Cambridge: Cambridge University Press.

Kripke, S. 1980. *Naming and Necessity*. Cambridge, Mass.: Harvard University Press. (First published in 1972.)

Lycan, W. 1996. *Consciousness and Experience*. Cambridge, Mass.: MIT Press.

Peacocke, C. 1992. *A Study of Concepts*. Cambridge, Mass.: MIT Press.

Pitcher, G. 1971. *A Theory of Perception.* Princeton: Princeton University Press.

Putnam, H. 1981. "Brains in a Vat." Pp. 1–21, in his *Truth and History.* Cambridge: Cambridge University Press.

Reid, T. 1785. *Essays on the Intellectual Powers.* The Edinburgh Edition, ed. K. Haakonssen. Edinburgh: Edinburgh Univ. Press.

Shoemaker, S. 1996. "Self-Knowledge and 'Inner Sense.'" Pp. 224–245, in his *The First-Person Perspective and Other Essays.* Cambridge: Cambridge University Press.

Siewert, C. 1998. *The Significance of Consciousness.* Princeton: Princeton University Press.

———. 2003. "Self-Knowledge and Rationality: Shoemaker on Self-Blindness." Pp. 131–146, in *Privileged Access: Philosophical Accounts of Self-Knowledge,* ed. Brie Gertler. Aldershot: Ashgate.

Tye, M. 1995. *Ten Problems of Consciousness.* Cambridge, Mass.: MIT Press.

Wright, C. 1998. "Self-Knowledge: The Wittgensteinian Legacy." Pp. 13–45, in *Knowing Our Own Minds,* ed. C. Wright, B. C. Smith, and C. Macdonald. Oxford: Oxford University Press.

3 Ow! The Paradox of Pain

Christopher S. Hill

I

It is generally possible to distinguish between the appearance of an empirical phenomenon and the corresponding reality. Moreover, generally speaking, the appearance of an empirical phenomenon is ontologically and nomologically independent of the corresponding reality: it is possible for the phenomenon to exist without its appearing to anyone that it exists, and it is possible for it to appear to exist without its actually existing. It is remarkable, therefore, that our thought and talk about bodily sensations presupposes that the appearance of a bodily sensation is linked indissolubly to the sensation itself. This is true, in particular, of our thought and talk about pain. Thus, we presuppose that the following principles are valid:

(A)   If x is in pain, then it seems to x that x is in pain, in the sense that x has an experiential ground for judging that x is in pain.

(B)   If it seems to x that x is in pain, in the sense that x has an experiential ground for judging that x is in pain, then x really is in pain.

There are alternative ways of expressing these principles. For example, (A) can be expressed by saying that it is impossible for x to be in pain without x's being experientially aware that x is in pain, and (B) can be expressed by saying that it is impossible for x to have an experience of the sort that x has when x is aware of a pain without its being the case that x really is aware of a pain.

(A) appears to hold quite generally—even in cases that are somewhat *outré*. To appreciate this, recall that soldiers and athletes often sustain serious injuries but show no sign of being in pain, continuing to display normal behavior until the end of the battle or the athletic contest, and perhaps even longer. They also deny that they are in pain. We may feel confused in trying to describe situations of this sort, but we realize that we have an obligation to concur with the person who has the injury. If he or she denies that there is pain, and we have no reason to think that the denial is insincere, we will acquiesce in it, and indeed insist on its truth. We feel that it would be deeply absurd to override the testimony of the injured party.

We can appreciate the plausibility of (B) by reflecting on cases of phantom-limb pain. When someone complains of pain in a limb that no longer exists, his or her testimony is taken as fact, provided only that there is no reason to suspect insincerity. To be sure, it will not seem correct, either to the person suffering from the pain or to external observers, to say that the pain is in the part of the body where it appears to be located, for by hypothesis there will be no such part. Because of this, it is true to say that there is a certain discrepancy between appearance and reality in such cases. But this is the only discrepancy. Thus, while we are prepared to say that the victim's perception that the pain is in the right leg is an illusion, we will allow, and in fact insist, that the pain is in all other respects as it appears to the victim. In particular, we will insist that it is a *pain*, thereby manifesting our commitment to principle (B).

It is important that we regard it as *absurd* to say that an agent is in pain in circumstances in which the agent is not aware of a pain, and that we regard it as *absurd* to say that an agent is not in pain in circumstances in which it seems to the agent that he or she is in pain. This suggests that we think of (A) and (B) as necessary truths— that is, as holding either because of deep metaphysical facts about pain, or our awareness of pain, or because of the a priori structure of our concept of pain.

There is one aspect of our thought and talk about pain that seems not to be fully in keeping with (A) and (B). We all know that pain has causal powers with respect to attention and awareness. When we are aware of a very intense pain, it seems to us that it has the power to keep attention focused on itself—indeed, a power to so engage attention as to make it difficult or impossible for us to honor other concerns and interests. Equally, as a pain increases in intensity, it seems to us to increase proportionally in its power to attract attention. Because of these phenomenological facts, we have a tendency to suppose that pain can exert an influence on attention even prior to our becoming aware of it. Thus, we say that pains can wake us up. On one very natural construal of this claim, it presupposes a picture according to which pain can exist prior to and therefore independently of awareness. According to this picture, there are pains that are so "small," so lacking in intensity, that they are incapable of attracting attention, and exist only at a subliminal level. But a pain of this sort can increase in intensity, thereby increasing also in its power to attract attention, and it may finally attain a level of intensity that enables it to open the gates of awareness. When this happens during sleep, we wake up.

Although this picture has a certain appeal, I think that its appeal is due to a recessive strand in our commonsense conception of pain, a strand that pales in comparison to the ones that find expression in (A) and (B). If we were fully committed to the picture, we would be prepared to consider it epistemically possible that an injured soldier actually has a severe pain, despite his professions to the contrary, but that there is something wrong with the mechanisms in his brain that support attention, and that this is preventing the pain from penetrating the threshold of consciousness. When I have asked informants to assess the likelihood of this scenario, however, they have all been inclined to dismiss it as absurd. The fact is that our discourse about pain is subject to divergent pressures. It serves a number of purposes and is responsive to a complex phenomenology. As a result, it is not entirely coherent. We resonate deeply to (A) and (B), but it should not surprise us that we are occasionally led to say things, and to feel the appeal of pictures, that cannot be fully squared with these principles.

At all events, I will go forward on the assumption that (A) and (B) strike us as fundamentally correct, and in fact seem to us to enjoy a kind of necessity. Whether the necessity in question has an ontological ground or is instead due to the a priori structure of our concept of pain is a question that I will defer to section III.

II

The fact that pains do not admit of a substantive appearance–reality distinction is unfortunate, for it presents a serious challenge to an otherwise appealing account of what it is to be aware of a pain. Indeed, we should view the absence of an appearance–reality distinction here as *extremely* unfortunate, for the considerations that favor the account in question actually seem to mandate it. Thus, we have an antinomy. On the one hand, principles (A) and (B) appear not just to be true, but to be partially constitutive of the concept of pain. On the other hand, there are considerations that seem to force us to embrace an account of awareness of pain that is flatly incompatible with (A) and (B). This is the "paradox of pain" that is cited in the title of the present paper.

The account of awareness of pain that I have in mind here is one that likens awareness of pain to such familiar sorts of perceptual awareness as vision, hearing, and touch.

When we consider the familiar forms of perceptual awareness from a phenomenological perspective, we find that they share a number of features, and this impression is reinforced when we view them from the perspective of contemporary cognitive psychology. I will mention several of these common features.[1] It will be clear as we proceed, I believe, that awareness of pain possesses the features as well.

First, perceptual awareness involves subconceptual representations of objects of awareness. Reflection shows that we are able to perceive innumerable properties that we are unable to name or describe. It is often said that we can use demonstrative concepts, such as *that shade of color*, to specify these properties, but even if this observation was fully correct, which is highly questionable, given that animals seem to be capable of highly sophisticated perceptual representations while being very limited conceptually, it would not diminish the case for the view that the properties are represented subconceptually. This is because we must attend to a property perceptually in order to demonstrate it. The representations that are employed by perceptual attention when it provides grounding for demonstrative concepts cannot themselves be conceptual.

Awareness of pain also involves subconceptual representations. To be sure, awareness of pain often takes the form of a judgment to the effect that we are in pain, and judgments put concepts into play. But what we are aware of in being aware of pains can easily transcend the expressive powers of our conceptual repertoire. We have all been aware of pains that have a significant degree of internal complexity. Perhaps it would be possible to put together roughly adequate descriptions of such pains if they were to last long enough, without changing in any way, or if we could remember their particularities for a long enough time after they had disappeared. In actual fact, however, the task of describing them fully is way beyond our powers. What we are aware of is ineffable in practice if not in principle.

Second, all of the familiar forms of perception are associated with automatic attention mechanisms, and also with attention mechanisms that are under voluntary control. Among other things, such mechanisms can increase the resolution of our experience of an object of awareness, and heighten the contrast between an object of awareness and its background. The same is true of awareness of pain. We can attend to pains, and when we do, there is a higher level of resolution and also a more salient contrast between figure and ground.

Third, it is of the essence of perceptual representation to assign locations and other spatial characteristics to its objects. The same is true of awareness of pain.

Fourth, there are a priori norms of good grouping that determine the ways in which perceptual elements are organized into wholes. For example, we group visually presented dots together if they are alike in some respect—that is, if they share a neighborhood in space, or a shape, or a color, or a size, or a common fate. The same is true of groups of pains. Suppose that you have three pains, two in your palm and one on your wrist. The two in your palm will seem to form a unified whole of a certain kind. Equally, two pains that are alike in intensity, or that begin to exist at the same time, will seem to be members of a single "society," even if they lie at some distance from one another spatially.

Fifth, perceptual awareness always represents its objects as having highly determinate forms of the properties that it attributes to them. I can *believe* that an object is blue without believing that it is any determinate shade of blue. But I cannot visually perceive an object to be blue without perceiving it to be navy blue or some other highly determinate shade of blue. Equally, I cannot experience a pain as intense without experiencing it as having a particular level of intensity.

Sixth, perceptual awareness is particularized. I can form a belief that there are three books in a box without being en rapport with any of the particular books that make my belief true. Equally, I can believe that someone has been eating my porridge without having any relevant beliefs about a specific individual. But perceptual awareness is different. If I am perceptually aware of the presence of three books in a box, I must be in some sense perceptually aware of each of the individual books. Equally, if I am aware of the existence of a trio of pains in my arm, I must be aware of each individual member of the trio.

Seventh, perceptual awareness has a certain mereological determinacy. I can form a belief about an object without forming any belief about its parts. But unless an object of perception is atomic, a minimum sensibile of some sort, I am inevitably aware of a range of its parts in being perceptually aware of the object, and also of certain of the structural relationships among the parts. (I mean to be using "part" quite broadly here, so that it applies to temporal constituents of events and qualitative constituents of complex properties as well as to spatial parts of physical substances.) This is also true of awareness of pains. Pains are normally experienced as extended in space, and when they are so experienced, the parts are experienced as well, as are a number of the structural relationships among the parts.

We have been reviewing certain of the common features of paradigmatic forms of perception that are visible from the perspective of phenomenology or the perspective of cognitive psychology. I have maintained that when we reflect, we can see that these features are shared by experiential awareness of pain. I now wish to remind the reader that there are other commonalities that are visible from the perspective of cognitive neuroscience. More specifically, cognitive neuroscience provides a lot of data that suggest that awareness of pain is fundamentally akin to haptic perception, thermal perception, and proprioception. It is widely held that the latter forms of perception owe their character and indeed their very being to the representational functions of certain structures in somatosensory cortex. Insofar as we prescind from the emotional and behavioral phenomena that attend awareness of pain, and think of awareness of pain as awareness of a purely sensory state, we find that awareness of pain also owes its character and its being to representations in somatosensory cortex. Moreover, in addition to this global commonality, there are also many commonalities of detail—commonalities having to do with internal organization, relationship to sensory pickup systems, and connections to higher cognitive centers, such as those that are believed to be involved in attention. In view of these similarities, it would be very uncomfortable to withhold the label "perceptual system" from the structures that subserve awareness of pain while applying it to the structures that subserve, say, haptic perception.[2]

As I see it, then, we are obliged to accept the view that awareness of pain is constituted by representations that are fundamentally perceptual in character, and by the same token, the view that awareness of pain is a form of perceptual awareness. Alas, in accepting these views, we come into conflict with principles (A) and (B), which deny that there is a substantive distinction between its seeming to one that one is in pain and its being the case that one is in pain. The reason for this conflict is that the relationship between perceptual representations and the items they represent is always contingent. Suppose that a perceptual representation R represents a property P. It is always possible for P to be instantiated without being represented by a corresponding token of R, and for R to be tokened without there being a corresponding instance of P. Because of this, if awareness of pain constitutively involves perceptual representations, it must be possible, at least in principle, for there to be pains that one is not aware of, and it must also be possible to be aware of pain, in the sense having an experience just like the ones one has when one is aware of a pain, without one's actually being in pain. That is to say, there must be possible circumstances in which one is in pain without seeming to be in pain, and possible circumstances in which it seems to one that one is in pain without its actually being the case that one is in pain. But (A) and (B) deny that such possibilities exist.

It appears, then, that when we try to combine the folk psychological picture of pain and awareness of pain with various phenomenological and scientific facts, we arrive at an antinomy. I will eventually propose a way of dealing with this conflict, but before I can proceed to do so, it is necessary to take a closer look at its sources. This effort will concern us in the next two sections.

It will be useful to adopt a couple of technical terms. In the future, when I wish to speak of the somatosensory representations that appear to underlie sensory awareness of pain, I will use the term "P-representation." Further, I will use the term "P-state" to stand for the bodily phenomena that P-representations represent. I will eventually recommend a view about the commonsense concept of pain on which it is true that "pain" and "P-state" are coreferential. But this view will also imply that "pain" has a much more complex sense or meaning than "P-state." "P-state" is meant simply to be a name for bodily conditions involving actual or potential damage. "Pain" is different. Its use is governed by (A) and (B), and perhaps by other a priori presuppositions as well. It is not a mere name.

I will assume that P-representations represent P-states as having locations, intensities, and various structural characteristics, such as that of having a central region of high intensity and a surrounding border of low intensity. I will also assume that P-representations assign P-states to such categories as burning, piercing, and throbbing. Further, I will assume that the contents of P-representations are determined by biology, and that the properties that serve as their contents are therefore properties to which evolutionary pressures are sensitive. Thus, for example, while I wish to allow for the possibility that a P-representation can represent a P-state as having the property *being located in the right hand*, I wish to deny that a P-representation can represent a P-state as having the property *being located in a hand that was once shaken by Bill Clinton*. Equally, while I wish to allow that a P-representation can have the content *being of degree of intensity D*, I wish to deny that a P-representation can have the content *being more intense than the pain I felt when I burned my hand last week*.

III

I will now argue that it is possible to explain our commitment to (A) and (B) without making any deep metaphysical assumptions about the nature of pain or our awareness of pain. Rather, (A) and (B) owe their plausibility entirely to the a priori structure of our concept of pain. It follows from this view that there are no factual obstacles to adopting a perceptual theory of awareness of pain. To create logical room for such a theory, we need only make some adjustments in our conceptual scheme.

In developing these views, I will presuppose the familiar doctrine that concepts are defined or individuated by core aspects of the roles that they play in cognition. I will also borrow a term from Christopher Peacocke, who calls the principles that describe such core aspects "possession conditions" (Peacocke 1992). Thus, where C is a concept, the possession conditions for C are a set of propositions that describe core aspects of the cognitive role of C. Each possession condition for C says, in effect, that if one is to possess C, then one must possess a concept whose use has a certain core aspect. Collectively, these individual possession conditions specify a sufficient condition for an agent to possess C.

There are various ways of conceiving of possession conditions. On one of the standard ways, they are concerned with dispositions to accept (or to reject) propositions in various possible situations. It is sometimes supposed, for example, that the possession conditions for a logical concept are concerned with dispositions to infer propositions containing the concept from other propositions, and with dispositions to infer other propositions from propositions containing the concept. (Here, of course, I am conceiving of inference as involving acceptance.) Again, it is sometimes supposed that the possession conditions for the concept of red are concerned, among other things, with dispositions to infer propositions containing the concept from perceptual experiences that represent redness as obtaining in the world. A related view is that the possession conditions for mathematical concepts like the concept of set and the concept of real number are concerned with dispositions to accept the propositions that serve as axioms in the relevant mathematical theories.

The view that possession conditions are concerned with dispositions to accept propositions has considerable intuitive appeal, but it is not inevitable. Many philosophers prefer the view that possession conditions are concerned with *norms* of acceptance. But those who hold this alternative view tend to acknowledge that possession conditions are implicitly or indirectly concerned with dispositions, for they generally maintain that an individual agent must have some tendency to conform to the appropriate norms of acceptance in order to be credited with possessing a concept. Moreover, those who prefer to think of possession conditions as concerned primarily with dispositions generally allow that possession conditions have a normative dimension. They recognize a normative dimension because they believe that the dispositions that figure in possession conditions are held in place, within the internal economies of individual agents, by various cognitive needs and interests. They may also believe that the dispositions that govern concepts are fostered and disciplined by social norms governing the use of words. After all, words express concepts.

In the interests of definiteness, I will assume here that it is at least a rough approximation to the truth to think of possession conditions as explicitly concerned with dispositions to accept and reject propositions. I believe, however, that most of my claims could be reformulated as claims about norms of acceptance.

Returning now at last to the concept of pain, I offer the suggestion that the possession conditions for this concept include (1)–(4):

(1)   In order to possess the concept of pain, one must be disposed to accept the proposition that one is in pain at times when one has a P-representation.

(2)   In order to possess the concept of pain, one must not be disposed to accept the proposition that one is in pain unless one has a P-representation.

(3)   In order to possess the concept of pain, one must be disposed to reject the proposition that one is in pain unless one has a P-representation.

(4) In order to possess the concept of pain, one must be disposed to reject the proposition that one is in pain at times when one does not have a P-representation.

When I use the expression "has a P-representation," I mean that there is currently an active P-representation in the relevant agent's somatosensory cortex, and that this P-representation satisfies whatever further conditions, such as availability to other cognitive modules, are necessary for it to constitute a conscious experience.

The foregoing possession conditions do not provide a full explanation of our commitment to (A) and (B), but they go a long way toward providing such an explanation. In particular, assuming that the condition of seeming to be in pain is constituted by having an active P-representation in somatosensory cortex, they explain why x will not accept a first-person ascription of pain unless it seems to x that x is in pain, and why x will accept a first-person ascription in any circumstances in which it seems to x that x is in pain. In general, (1)–(4) explain why first-person judgments about pain tend to conform to (A) and (B). Now as I see it, first-person judgments about pain are plausibly regarded as more fundamental than third-person judgments. (Thus, among other things, third-person judgments are always trumped by first-person judgments in cases of conflict.) If this view is right, it is reasonable to suppose that (1)–(4) answer the most fundamental questions about conformity to (A) and (B). This is not to say that they answer all such questions. In order to explain why third-person judgments about pain conform to (A) and (B) it is necessary to supplement (1)–(4) with additional conditions. But discussion of the additional conditions lies outside the scope of the present paper.

According to the view I am recommending, the concept of pain has some extremely unusual features. They are shared, I believe, by other concepts that stand for bodily sensations, but most empirical concepts lack them. In the first place, (1)–(4) have the effect of restricting acceptance of propositions of the form *I am in pain* to circumstances in which one is having a perceptual experience of one specific sort—specifically, an experience that involves a P-representation. The possession conditions for empirical concepts normally envision a much broader range of acceptance-promoting circumstances. For example, it is clear that the possession conditions for the concept of red provide for acceptance of *That object is red* in circumstances in which one's experience does not attribute a definite color to the relevant object (say, because it is night), but in which one has been told by a reliable associate that the object is red, or in which one has inferred *That object is red* from other propositions, such as the pair consisting of *All of the objects on the table are red* and *That object is on the table*.

There is another peculiar feature that my proposal attributes to the concept of pain. Possession conditions for perceptually grounded concepts generally make provision for *defeaters*. Thus, for example, it is plausible that there is a possession condition for the concept of red that can be formulated as follows:

In order to possess the concept of red, one must be disposed to accept *That object is red* in circumstances in which (i) one is attending visually to an object *o*, (ii) one's visual experience represents *o* as red, and (iii) one believes no propositions that count as defeaters for *That object is red*.

As is usual, I say that a proposition *P* is a defeater for a proposition *Q*, relative to a believer *x*, if it is the case either (i) that *x* believes *P* and *P* is logically incompatible with *Q*, or (ii) that *x* believes *P* and *P* implies that *x*'s evidence for *Q* is misleading in the context at hand.[3] Thus, where *Q* is the proposition *That object is red*, potential defeaters for *Q* include *That object is not red* and *That object is painted white but it looks red because it is currently illuminated by red light*. Each of these propositions would become a defeater for *Q*, relative to *x*, if *x* were to come to believe it.

It is, I submit, extremely plausible that there is a possession condition for the concept of red that has a component corresponding to (iii). Otherwise it would be appropriate for agents to accept *That object is red* even when they know that the relevant object looks red only because it is illuminated by red light. It is precisely because it is *not* appropriate for agents to accept *That object is red* in such circumstances that the concept of red can be said to stand for the property of being red. The property in question is after all an objective property, and therefore a property whose comings and goings cannot be dictated by human perceptual experience. If the possession conditions for the concept of red authorized acceptance of *That object is red* in circumstances in which it is known that the property of being red is not instantiated, then acceptance of the given proposition would not track the comings and goings of the property, but rather merely the comings and goings of a certain type of human perceptual experience.

Unlike the possession conditions for the concept of red, the possession conditions that I am proposing for the concept of pain do not acknowledge the possibility of defeaters. Thus, condition (1) tells us that if one possesses the concept of pain, then the circumstance of there being an active P-representation in somatosensory cortex is *sufficient* for being disposed to judge that one is in pain. It does not allow that one might have a prior belief that could neutralize the epistemic force and the causal efficacy of a P-representation, thereby modifying or suspending the disposition to accept *I am in pain*. It gives perceptual experience full and final authority over belief.

Since (1)–(4) are unusual in the respects I have indicated, a defense of the view that they succeed in capturing some of the possession conditions for the concept of pain should include an account of why it might be useful for us to have a concept with the features that they imply. What is it about P-representations and P-states that makes it appropriate for us to adopt a concept whose possession conditions make no provision for first-person ascriptions that are not grounded in immediate experience, or for the possibility that such ascriptions might be called into question by other judgments?

I think that the answer has to do with the causal powers of P-representations. P-representations constitute the condition of seeming to be in pain. Because of this they have causal powers that are of great interest to us. They can cause all of the emotions and attitudes that comprise the affective dimension of pain experience, and, depending on such factors as persistence and intensity, they can induce us to engage in all of the usual forms of pain behavior. Because these causal powers are of deep and enduring interest to us, it makes sense for us to have a concept that can be used to keep track of the states in which they inhere.

Now reflection shows that P-representations are not the only states that have the powers in question. P-states possess them as well. It is clearly true that an active bodily disturbance that constitutes actual or potential damage is capable of causing fear, aversion, and all of the other emotions that constitute the affective dimension of pain experience, and also that it is capable of causing such behaviors as nursing and protecting an injured body part, at least if it is sufficiently intense. It is not true, however, that P-states possess these causal powers autonomously. Rather, they possess them only insofar as they are represented by P-representations. Disturbances that are not perceptually represented may occasion reflexive withdrawal, but they have no tendency to trigger a conscious emotional response, and they are powerless to produce intentional pain behavior. By the same token, it is a mistake to say that P-states are proximal causes of emotional responses to pain and of intentional pain behaviors. Rather, they are distal causes: they cause emotional responses and intentional behaviors by causing P-representations. It is P-representations that count as the proximal causes of the emotions and behaviors.

P-states possess the causal powers we have been considering, but they possess them, distally and at second hand, so they cannot be said to be the states in which the powers *most reliably* inhere. It is P-representations that have that distinction.

Since the causal powers in question are of great interest to us, it makes sense for us to have a concept that can be used to keep track of the states in which they most reliably inhere. Hence, it makes sense for us to have a concept that can be used to keep track of P-representations. What I wish to propose is that the concept of pain is the concept that is used to play that role, and that it can play that role precisely because its use is governed by (1)–(4). Those principles imply that our dispositions to accept first-person ascriptions of pain are linked immediately, exclusively, and indefeasibly to the presence of P-representations. Accordingly, insofar as we possess the dispositions that they describe, we possess a concept that is tied to P-representations in the closest possible way.

It may be worth emphasizing, in this connection, that (1)–(4) easily accommodate the fact that injured soldiers and athletes can deny that they are in pain without qualification or acknowledgment that they might be apprised of considerations that would lead them to change their minds, and that (1)–(4) also easily accommodate the fact that victims of phantom-limb pain can affirm that they are in pain with equal confidence. Let x be an injured soldier or athlete of the sort in question; x is in a P-state, but there is no accompanying P-representation. The P-state is not perceived. Accordingly, if x satisfies principle (3), x is disposed to deny categorically that x is in pain. Moreover, this disposition will remain in force until such time as x's P-state comes to be represented by a P-representation. Suppose now that y is a victim of the condition that we call phantom-limb pain: y is not in a P-state, but y does have a P-representation. Hence, if y satisfies principle (1), y is disposed to embrace *I am in pain* unreservedly and without qualification.

I have explained why it is natural and appropriate for us to have a concept that is governed by (1)–(4) in terms of our need to have a concept that can be used to keep track of states that have deeply significant causal powers. Can the explanation be generalized? Presumably our concepts of other bodily sensations, such as the concept of itching, are associated with possessions conditions that are similar in structure and content to (1)–(4). After all, we are not prepared to grant that there can be an itch in one's body without its seeming to one that there is an itch, or that it can seem to one that there is an itch in one's body without there actually being a sensation of that sort. But is it true that when we experience bodily sensations other than pain, we are generally in representational states that have significant causal powers? In most cases, the answer is clearly "yes." This is obviously true of the experience of itching, the experience of bodily pleasure, gustatory experiences, olfactory experiences, and experiences that inform us of higher levels of temperature and pressure. But reflection shows that it is also true, though true to a lesser degree, in other cases as well—for example, in cases in which we experience very gentle sensations of warmth. Even experiences of this sort have a tendency to trigger an emotional response, though of course the response is generally quite mild.

In the present section I have proposed that the concept of pain is governed by conditions (1)–(4). The proposal enjoys considerable plausibility, I feel, in virtue of its ability to explain the most fundamental aspects of our commitment to (A) and (B).

Still, a defense of the proposal must do more than appeal to its fecundity, for its claims about the possession conditions for the concept of pain represent those conditions as fundamentally different in structure and content from the conditions that govern most other perceptually grounded contents. It might be held that the proposal is too *outré* to be acceptable. To counter this feeling, I have tried to show that if our interest in the causal powers of P-representations is taken into account, it is actually reasonable to predict that we have a concept that is governed by possession conditions like (1)–(4).

Assuming that these efforts have been successful, we can conclude that the validity of (A) and (B) is due to a priori features of our concept of pain. We do not need to seek a metaphysical explanation. But this means that we are free to regard the form of awareness that we call awareness of pain as a mode of perception. The problems that arise when we try to combine this view with (A) and (B) can be solved by changing or replacing the concept of pain.

IV

In the previous section we were concerned exclusively with questions about the cognitive role of the concept of pain. I would like to turn now to consider questions about its semantic value or reference. What I believe, and what I will maintain here, is that the concept of pain refers to P-states. In other words, as I see it, although it is built into the cognitive role of the concept of pain that it is used to keep track of the comings and goings of P-representations, it does not refer to P-representations, but rather to the states that P-representations represent.

When it is combined with the foregoing claim about the cognitive role of the concept of pain, this claim about the reference of the concept implies that the concept suffers from a pervasive incoherence. To appreciate this, recall that there are occasions when P-states occur without accompanying P-representations, and occasions when P-representations occur without accompanying P-states. Accordingly, if it is true that the concept of pain is governed by (1)–(4), and true also that it refers to P-states, then there will be occasions when a proposition containing the concept is true but the possession conditions for the concept require that one have an inflexible disposition to reject the proposition, and there will be occasions when the possession conditions require one to be disposed to embrace a proposition containing the concept irrevocably and without qualification even though the proposition is entirely false. This amounts to a lack of cohesiveness between the possession conditions for the concept and its semantic properties, or in other words, a lack of cohesiveness between its cognitive role and its semantic role.

Of course, if it is true that the concept of pain is incoherent, then there is additional motivation for the idea, floated in the last section, that the concept of pain should be changed or replaced.

There are several reasons for thinking that the concept of pain refers to P-states. One such reason comes to the fore when we reflect on what is involved in making demonstrative reference to a particular pain. Suppose that you have been monitoring the progress of two pains and have just noticed that one of them is increasing in intensity. You may be inclined to register this development doxastically. If so, you will probably focus your attention on the increasing pain and make a judgment with the content *That pain is getting worse.* It is clear in this case that the demonstrative concept *that pain* refers to the object of your attention. Now let us consider what is going on here from the perspective of cognitive neuroscience. When we adopt this perspective, I suggest, we see that you are perceptually attending to a P-state. (Or, in other words, we see that you are deploying a perceptual representation with high resolution and fairly narrow focus that represents a P-state.) Since you are attending to a P-state, and the object you are attending to is the referent of *that pain* on the occasion in question, the referent of the concept is a P-state. Further, what holds here must hold generally. It must be true in general that demonstrative concepts involving the concept of pain refer to P-states, for it is true in general that the reference of a demonstrative concept is determined by the attention of the agent who is deploying the concept, and, according to cognitive neuroscience, it is true in general that agents attend to P-states when they attend to their pains.

Moreover, as I see it, anyway, this conclusion about demonstrative concepts containing the concept of pain enjoins acceptance of the view that the concept of pain refers to pains. How could *that pain* refer to a P-state if *pain* refers to something else?

Other reasons for holding that the concept of pain refers to P-states emerge when we reflectively consider the doctrine that is the only serious competitor of this view— the doctrine that it refers to P-representations. I have been assured by a number of philosophers that this doctrine is *obviously* right. As far as I can tell, this boils down to the observation that the doctrine allows us to hold that the cognitive role of the concept of pain coheres fully with its semantic properties, and to the claim that it is desirable to embrace this optimistic perspective. I am in favor of optimism when it is consistent with the facts, but in this case, I believe, it is not a live option.

Observe in the first place that the view is precluded by the semantic intentions that one has when one attempts to describe a pain. In such cases, one attends to the pain and tries to choose descriptive concepts that will do justice to its properties. That is to say, one is concerned to describe the object to which one is attending. But it is clear that one is attending to a P-state in such cases and not a P-representation. Attention is something that we bestow on the objects that are represented by perceptual representations, not on perceptual representations themselves.

The view is also precluded by facts involving the locations and intensities of pains. To be sure, P-representations have locations and intensities: they are located in somatosensory cortex, and they have intensities in virtue of the firing rates of the nociceptive neurons that produce them. But of course, pains can occur almost anywhere in the body *but* in somatosensory cortex. Moreover, while the intensity of a pain may be proportional to the intensity of the P-representation that represents it, the intensity of the pain cannot be the same physical magnitude as the intensity of the P-representation, for the former intensity is associated with a different location than the latter.

It is clear, then, that if the concept of pain refers to P-representations, then it is not literally true that pains have the locations and intensities that figure in our descriptions of them. Amazingly, no one has ever made a true claim about the location or intensity of a pain! This seems unacceptable. It is sometimes maintained, however, that it is possible to bring the present doctrine about the reference of the concept into harmony with our attributive practices by embracing the idea that we use terms in different senses when we mean to be characterizing pains than we do on other occasions. Thus, it is sometimes urged, in effect, that when one says, for example, that a pain is located in one's right hand, what one is actually saying, without being aware of the fact, is that one has a P-representation that represents a P-state as occurring in one's right hand. More generally speaking, the claim is that statements of the form "I have a pain in bodily location L" have truth conditions that can be expressed by statements of the form "I have a P-representation that represents its object as occurring in bodily location L," and that statements of the form "I have a pain of intensity level N" have truth conditions that can be expressed by statements of the form "I have a P-representation that represents its object as having intensity level N." Does this suggestion work? Does it protect the doctrine that the concept of pain refers to P-representations from the objection we are considering?

No. The problems with the suggestion become evident when we consider arguments like the following:

(1)   I have a pain in my right hand.
My right hand is the only part of my body that has been touched by Bill Clinton.
Therefore,

I have a pain in the only part of my body that has been touched by Bill Clinton.

This argument is clearly valid; but if the theory of truth conditions for pain-statements that we are considering was correct, the argument would be fallacious. According to the theory, argument (I) is equivalent to argument (II):

(II)   I have a P-representation that represents its object as occurring in my right hand.
My right hand is the only part of my body that has been touched by Bill Clinton. Therefore,
I have a P-representation that represents its object as occurring in the only part of my body that has been touched by Bill Clinton.

It is easy to imagine a situation in which the premises of argument (II) are true and in which the conclusion is false (especially since it appears to be the case that *everyone* has shaken hands with Clinton at one time or another). This is because the content of P-representations is determined by biology. The selective pressures of evolution are sensitive to properties like *occurring in the agent's right hand*, but are altogether incapable of bestowing contents like *occurring in the only part of the agent's body that has been touched by Bill Clinton* on perceptual representations.

It turns out, then, that the present proposal is incompatible with our intuitions about the logical properties of sentences that ascribe locations to pains. But it also has another flaw: it entails claims about linguistic processing that conflict with the requirements of simplicity.

The present proposal should be understood as making two claims. First, it asserts that all sentences of form (S1) have truth conditions that are captured by sentences of form (S2):

(S1)   I have a pain in bodily location L.

(S2)   I have a P-representation that represents its object as occurring in bodily location L.

And second, it asserts that our language processors compute semantic representations of form (S2) in interpreting sentences of form (S1).

Now as this formulation makes clear, the present proposal is committed to the proposition that our language processors somehow incorporate a complex metaphysical picture of the nature of pain, and are guided by that picture in computing deep structures that represent the truth conditions of such sentences. That is to say, the proposal is committed to the following proposition: under the influence of a meta-physical picture, our language processors compute semantic representations for sentences about pain that depart radically from the apparent logical structures of the sentences themselves. It follows that the proposal is committed to a very large amount of offstage complexity—complexity of representations of truth conditions, and complexity of the processes that compute such representations. Now when a theory postulates hidden complexity, it has a special obligation to marshal evidence from the relevant domain which shows that the postulate is well motivated. Here the relevant domain is discourse about pain, so the evidence should consist of linguistic data. But there is no linguistic evidence that supports the present proposal. On the contrary, all of the motivation for it comes from a philosophical view about the metaphysics of pain. We should conclude, then, that the proposal fails to honor the requirements of simplicity.

As I have stated it here, this second objection is concerned only with the present proposal. But reflection shows that it can be generalized. That is to say, if a proposal resembles the one we are presently considering in that it construes sentences about the locations of pains as equivalent to sentences about the representational contents of P-representations, it will also resemble the present proposal in entailing extravagant and poorly motivated claims about linguistic processing.[4]

It turns out, then, that the present proposal is wrong, and that the same is true of all similar proposals. Now the proposal came to our attention because some such claim is needed to shore up the view that the concept of pain refers to P-representations. Accordingly, the failure of the proposal means that the latter view should be set aside. We are left, then, with the view that I wish to recommend—the view that the concept of pain refers to P-states.

Let us review our findings. Assuming that the foregoing account of the possession conditions for the concept of pain is correct, at least to a first approximation, the cognitive role of the concept is closely bound up with P-representations. On the other hand, we have now found reason to believe that the concept refers to P-states. Putting these views together, we arrive at the conclusion that the state that determines the cognitive role of the concept is altogether different from, and therefore metaphysically independent of, the state that serves as its referent.

But this is only part of the story. In addition to being metaphysically independent, the two states are also nomologically independent. Thus, it sometimes happens that a P-state occurs when there is no corresponding P-representation, and it sometimes happens that a state with the form of a P-representation occurs when there is no state that answers to its content. Because these things can happen, the conditions under which it is appropriate to be indefeasibly committed to a proposition containing the concept of pain can diverge dramatically from the conditions under which the proposition is true. In particular, it can be appropriate for a wounded soldier to deny intransigently that he is in pain even though the proposition that there is a very intense pain in the soldier's right leg is true. Equally, it can be fully appropriate for a victim of phantom-limb pain to commit herself fully and irrevocably to the proposition that she is in intense pain even though the proposition is entirely false.

In view of these considerations, it seems appropriate to conclude that the cognitive role of the concept of pain does not cohere fully with the semantic properties of the concept. The cognitive dimension of the concept and the semantic dimension are nomologically independent, and, moreover, there are occasions on which these dimensions actually come apart.

Is it really credible that one of our most fundamental and frequently used concepts is incoherent? In considering this question, please keep in mind the fact that none of the foregoing remarks either implies or suggests that that the incoherence is readily detectable. If anything, the lines of thought that we have followed suggest that the incoherence should be virtually invisible from the perspective of common sense. Thus, the argument for the existence of an antinomy in section II is heavily indebted to recent work in cognitive neuroscience, and the argument that has been developed in the last two sections borrows heavily both from neuroscience and from contemporary philosophy. I would also remind the reader that there is good reason to suspect other commonsense concepts of incoherence. The semantic paradoxes provide a strong rationale for the view that concepts like truth and reference are governed by inconsistent principles, and it is arguable that Frege's inconsistent theory of sets is actually a successful explication of the commonsense concept of set.

V

Thus far I have attempted to do three things: to describe a paradox involving pain, to explain the paradox by presenting a partial account of the possession conditions that govern the concept of pain, and to maintain that the concept of pain is ultimately incoherent. I would like now to recommend a way of dealing with the paradox. As will be seen, the recommendation also addresses the incoherence.

According to the proposal that I have in mind, we should jettison the commonsense concept of pain, in contexts in which our concerns are philosophical or scientific, and replace it with two new concepts. One of the new concepts should be a concept that refers to P-states, the other a concept that refers to P-representations. In each case, the concept should be endowed with possession conditions that are consistent with its semantic properties. I will not attempt to set out appropriate possession conditions here, for that would be a complex undertaking, and anyway, it is generally possible in practice to acquire dispositions for using a concept that are consistent with its intended semantic value by keeping that value vividly and fixedly in view as one goes about the business of asserting and rejecting propositions that contain it. In other words, in a case in which a reference has been stipulatively assigned to a concept, it is generally possible to establish a practice of using the concept by setting one's sights on accepting true propositions containing the concept and rejecting false ones. It seems to me that this is how theoretical concepts generally acquire a pattern of use in scientific discourse.

At all events, even if I was in a position here and now to formulate two new sets of possession conditions, my proposal would have no practical consequences. That is to say, it is extremely unlikely that we would cease to use the concept of pain in everyday life. The commonsense concept is deeply entrenched, and anyway, the paradox to which it contributes is a theoretical problem—a problem that arises only when one attempts to think systematically about pain, and to consider the picture of it that is embedded in folk psychology in relation to the picture that is emerging in cognitive science. The paradox of pain is not visible from the perspective of common sense. Moreover, the incoherence of the concept does not inconvenience us in everyday life. The point of the present proposal is not to reform our ordinary conceptual scheme, but rather to point to a perspective from which the tasks of constructing an adequate metaphysical account of pain and an adequate scientific account would be more feasible.

VI

I have claimed that P-representations represent bodily disturbances that involve or portend damage. There is a worry about this claim that may have occurred to the reader. I will conclude by saying a few things about this concern.

When one considers the receptive fields of the nociceptive mechanisms in somatosensory cortex, and considers the way in which the mechanisms in question respond to damage in those fields, it can seem that acceptance of my claim about the representational contents of P-representations is obligatory. It should be acknowledged, however, that there is other evidence that somewhat complicates the picture. If a P-representation represents a disturbance in location L, then, one might suppose, the intensity that the representation attributes to its object should correspond more or less closely to the severity of the damage. But much of the time this is not the case. Indeed, it seems that the intensity posited by a representation and the degree of the represented damage are less likely to correspond than to diverge (Price 1999, chapters 4 and 6).

The divergence occurs because signals indicating the amount of damage are very often modified in the spinal cord and in the brain. Upward adjustments in signals can be brought about by several factors. For example, it can happen that the signal is ratcheted up as a result of temporal summation in the spinal cord: the spinal cord keeps track of how long a signal has been coming from location L, and after a temporal threshold has been reached, it magnifies the signal before sending it on. Downward adjustments can also be caused by a variety of factors. Speaking teleologically, such adjustments probably occur in order to prevent the organism from being overwhelmed by pain—that is, from being distracted by it in a way that would preclude flight from danger or quiet recuperation.

In addition to the modulatory adjustments in the spinal cord and brain, there are also adjustments at the sites where damage is located. Damage causes the release of neuroactive chemicals at the site, and these chemicals can accumulate over time, occasioning a commensurate increase in the rate of firing of peripheral nociceptors. In short, there is a kind of peripheral temporal summation. And there is also peripheral spatial summation.

In view of these considerations, if we embrace the view that P-representations represent disturbances involving actual or potential damage, as I have recommended, we will be obliged to say that P-representations systematically misrepresent the disturbances with which they are concerned. In other words, we will have to say that most of our sensory experience of pain is illusory—or so it seems.

At first sight this conclusion can be unsettling, possibly prompting the thought that it cannot be right after all to say that P-representations represent peripheral bodily damage. It might occur to one, for example, that what they actually represent is the level of nociceptive activity in the spinal cord.

On closer examination, however, it becomes clear that there are really no grounds for worry. Thus, reflection shows that the foregoing line of thought has presuppositions that may not be entirely correct. And, what is more important, it shows that there is nothing really wrong with saying that a dimension of experience is systematically deceptive.

In formulating the foregoing line of thought, I have in effect presupposed the following propositions: (i) there is a one-to-one correspondence between the levels of intensity that are posited by P-representations and degrees of damage, and (ii) a P-representation is accurate just in case the level of intensity that it posits corresponds to the current degree of the actual or potential damage that the representation represents. It is natural to assume that these presuppositions are correct, and I believe that they are frequently taken to be correct in the literature. But (ii) might be wrong. It could be that posited intensities are not meant to correspond to concurrently existing degrees of damage, but rather to more complex magnitudes that reflect a variety of factors. Thus, it might be, for example, that a given P-representation was designed to be tokened either when there is concurrent damage of level N, *or* there is damage of lesser level N^* but the mechanisms responsible for temporal summation have determined that the damage is persisting. If so, then the representation would not simply signify the degree of the concurrently existing damage, but would rather have the following disjunctive content: *either there is very recently incurred damage of level N or there is damage of degree N^* that has existed for the interval I.* This suggestion can accommodate the existence of temporal summation without implying that there is misrepresentation in such cases. And it can easily be adapted so as to accommodate the existence of spatial summation. To be sure, philosophers sometimes express reservations about views that assign disjunctive contents to perceptual representations; but there is reason to think that these concerns can be answered (Hill, forthcoming, sec. 6).

Even if this suggestion for accommodating the effects of temporal and spatial summation is right, it is probably true that our experience of pain misrepresents the degree of damage in many cases. But we should not recoil from this observation. Indeed, when we reflect, we come to appreciate that it would be surprising if the representation of damage did not involve a substantial amount of misrepresentation, for misrepresentation is actually quite common in perception, and in cognition generally. There are many familiar perceptual illusions. A number of these occur only in a highly restricted class of contexts, but others pervade a whole dimension of perceptual experience. For example, Dennis Proffitt and his colleagues (Proffitt et al. 1995) have found that we tend to perceive hills as having steeper slants than they actually possess, and that this effect is magnified when we are tired.[5] They have also determined that hills tend to look steeper when viewed from the top than when viewed from the bottom. As Proffitt et al. point out, it is likely that we have evolved with these tendencies because they are beneficial: the first one gives us a preference for avoiding hills, which prevents unnecessary expenditures of energy, and the second one gives us a preference for choosing slopes with lesser slant when we are descending hills, which reduces the risk of accident. (I cite the beneficial character of these illusions to highlight their similarity to illusions concerning the degree of bodily damage. As we saw, the misrepresentation of damage can serve our need to focus our attention elsewhere while escaping from danger, and our need to remain calm and quiet while recuperating from damage.) There is also systematic misrepresentation in other cognitive realms. This is shown, for example, by the work of Kahneman and Tversky on assessments of frequency and probability (Kahneman and Tversky 1974), and also by their work on assessments of utility (Kahneman and Tversky 1981).

VII

We started with the observation that there is no substantive distinction between the appearance of pain and the corresponding reality. We then observed that there is evidence from phenomenology, cognitive psychology, and cognitive neuroscience that appears to mandate the view that awareness of pain is fully perceptual in character. When we combined these observations we arrived at a paradox, for it is of the very essence of perception that there be a distinction between its seeming to a perceiver that such and such is the case and its being true that such and such is the case.

We then saw that it is possible to explain why there is no substantive appearance–reality distinction in the case of pain in terms of features of the cognitive role of the concept of pain. It follows that there are no empirical or metaphysical *facts* that preclude a perceptual model of awareness of pain. It is possible to make the world safe for a perceptual model by making appropriate revisions in our conceptual scheme.

After considering these issues, we went on to notice that there is good reason to believe that the concept of pain refers to bodily disturbances that involve actual or potential damage. Evidently, then, it is true to say that pains *are* bodily disturbances. But bodily disturbances are empirical phenomena, and therefore admit of an appearance–reality distinction. Because of this, there is considerable tension between two aspects of the concept of pain. Those features of its cognitive role that preclude a substantive distinction between appearance and reality are at variance with its referential dimension. This fact provides additional motivation for revising or eliminating the concept of pain.

Finally, we considered an argument that appears to show that our experience of pains systematically misrepresents their quantitative character. After reflection, we concluded that the argument has a questionable presupposition, and also that, despite initial appearances to the contrary, the conclusion of the argument does not pose a threat to the picture of pain and awareness of pain that is offered in the present paper.

Acknowledgments

I have benefited considerably from discussions with Richard Fumerton, Robert M. Harnish, Jack Lyons, Timothy Schroeder, Thomas Senor, and Lee Warren.

Notes

1. Several of the seven features of perceptual awareness that I cite here are borrowed from Hill 2004. Incidentally, the present paper is intended to extend the ideas and arguments presented in that earlier effort. Hill 2004 makes a few observations about the main issue addressed in the present paper, the lack of an appearance–reality distinction for pain, but its treatment of that issue does not address its more difficult aspects. Also, I was unable in that paper to discuss the concept of pain. I hope now to have filled these lacunae in a reasonably satisfactory way.

2. The picture that I have sketched above is presented in much greater detail in Price 1999, chapter 5, and in Basbaum and Jessell 2000. For discussion of the mechanisms that underlie other forms of bodily awareness, see Gardner, Martin, and Jessell 2000, and Gardner and Kandel 2000.

Although the idea that activity in somatosensory cortex is primarily responsible for awareness of pain appears to be widely favored by neuroscientists, there is also support for the view that activity in the insula makes a significant contribution to such awareness. Even if this second view turns out to be correct, however, it will still be true that the mechanisms that support awareness of pain are closely aligned with perceptual mechanisms, for the considerations that count in favor of a role for the insula in awareness of pain have counterparts that suggest that the insula plays a similar role in thermal perception and other forms of bodily awareness.

3. For further discussion of defeaters, see Pollock 1986.

4. It may be useful to illustrate this claim with an example. Consider the proposal that sentences of form (S3):

(S3)   I have a pain in a bodily location that is F.

are equivalent to sentences of the form (S4):

(S4)   (ΣL)(L is a bodily location and I have a P-representation that represents its object as being in L and L is F).

(Here 'ΣL' is a substitutional quantifier. A substitutional quantifier is needed because "I have a P-representation that represents its object as being in L" is an intensional context.) Applied to argument (I) in the text, this new proposal yields the claim that argument (I) is equivalent to argument (II):

(III)   (ΣL)(L is a bodily location and I have a P-representation that represents its object as occurring in L and L is my right hand).
My right hand is the only part of my body that has been touched by Bill Clinton.
Therefore,

(ΣL)(L is a bodily location and I have a P-representation that represents its object as occurring in L and L is the only part of my body that has been touched by Bill Clinton).

Unlike argument (II), argument (III) preserves the validity of argument (I). Thus, this new proposal escapes my first objection to the proposal that is considered in the text. Reflection shows, however, that it is vulnerable to a version of the second objection. It entails claims about linguistic processing that are poorly motivated and at variance with standard principles of simplicity.

5. I am indebted to Michael Pace for bringing Proffitt's work to my attention.

References

Basbaum, Allan J., and Thomas M. Jessell. 2000. "The Perception of Pain." Pp. 472–491, in *Principles of Neural Science*, ed. Eric Kandel, James H. Schwartz, and Thomas M. Jessell. New York: McGraw Hill.

Craig, A. D. 2002. "How Do You Feel? Interoception: The Sense of the Physiological Condition of the Body." *Nature Reviews* 3: 655–666.

Gardner, Esther P., and Eric R. Kandel. 2000. "Touch." Pp. 451–471, in *Principles of Neural Science*, ed. Eric Kandel, James H. Schwartz, and Thomas M. Jessell. New York: McGraw Hill.

Gardner, Esther P., John H. Martin, and Thomas M. Jessell. 2000. "The Bodily Senses." Pp. 430–450, in *Principles of Neural Science*, ed. Eric Kandel, James H. Schwartz, and Thomas M. Jessell. New York: McGraw Hill.

Hill, Christopher S. 2004. "Ouch! An Essay on Pain." Pp. 339–362, in *Higher Order Theories of Consciousness*, ed. Rocco J. Gennaro. Amsterdam: John Benjamins.

———. Forthcoming. "Perceptual Consciousness: How It Opens Directly onto the World, Preferring the World to the Mind." In *Consciousness and Self-Reference*, ed. Uriah Kriegel and Kenneth Williford. New York: Oxford University Press.

Kahneman, Daniel, and Amos Tversky. 1974. "Judgments under Uncertainty: Heuristics and Biases." *Science* 185: 1124–1131.

———. 1981. "The Framing of Decisions and the Psychology of Choice." *Science* 211: 453–458.

Peacocke, Christopher. 1992. *A Study of Concepts*. Cambridge, Mass.: MIT Press.

Pollock, John. 1986. *Contemporary Theories of Knowledge*. Totowa, N.J.: Rowman and Littlefield.

Price, Donald D. 1999. *Psychological Mechanisms of Pain and Analgesia*. Seattle: IASP Press.

Proffitt, Dennis R., Mukul Bhalla, Rich Gossweiler, and Jonathan Midgett. 1995. "Perceiving Geographic Slant." *Psychonomic Bulletin and Review* 2: 409–428.

4 Another Look at Representationalism about Pain

Michael Tye

There is something it is like to feel a sharp pain in the elbow, a migraine headache, a throbbing pain in the leg, a squeezing pain in the heart. These experiences are phenomenologically different, but they also have a common phenomenological core. Do they also have representational content? Some philosophers have thought not. For example, Colin McGinn remarks:

... bodily sensations do not have an intentional object in the way that perceptual experiences do. We distinguish between a visual experience and what it is an experience of; but we do not make this distinction in respect of pains. Or again, visual experiences represent the world as being a certain way, but pains have no such representational content. (McGinn 1982, p. 8)

McGinn's comments on visual experience are uncontroversial. Visual experiences have correctness conditions. For the subject of the experience, the world *seems* a certain way, the way represented by the experience. The way the world is represented is the representational content of the experience. The experience is accurate if the world is that way, inaccurate otherwise.

McGinn's denial that pain experiences have representational content is open to dispute, however. Some philosophers allow that pain experiences have such content as far as the bodily location of pain goes, but they deny that this content, which is shared with other bodily sensations, has anything to do with the characteristic phenomenology of pain experiences (Block 1996). Still other philosophers maintain that pain experiences have a distinctive representational content—for example, they all represent tissue damage at a given bodily location—but they deny that this content captures their phenomenology (Chalmers 1996).

My own view is that pain experiences have a distinctive representational content and that this content *is* their phenomenal character. This view, which is sometimes known as "strong representationalism" for pain, has come under attack from several different quarters in the recent literature. The purpose of this paper is to lay out the strong representationalist position for the case of pain in more detail than has been done before (see Tye 1995a,b, 2000 for earlier discussions) and to take up various criticisms along the way.

1   Some Preliminary Remarks on Pains as Representations

You cannot feel my pains and I cannot feel yours. Even if we are Siamese twins, joined at the hip and stung there by a bee, intuitively there are two pains, yours and mine. Maybe your pain is a bit different from mine. Without further information, that's an open question, which it wouldn't be, of course, if there were just one pain.

Pains cannot exist without owners. There cannot be a pain with no one or no creature around to feel the pain. To be sure, there is the phenomenon of unnoticed pain, but even an unnoticed pain has an owner—the person who fails to notice it.

That pains are necessarily private and necessarily owned is part of our folk conception of pain and it requires explanation. The obvious explanation is that pain is a feeling or an experience of a certain sort. That is certainly how scientists think of pain. Witness the definition of pain by the *International Association for the Study of Pain* (IASP), as "an unpleasant sensory and emotional experience associated with actual or potential tissue damage, or described in terms of such damage" (*Pain* 3[suppl.]: 250, 1986). And it is also part of our commonsense conception.

Since you cannot feel my feelings any more than you can laugh my laughs or scream my screams, and since there cannot be a feeling without a subject for the feeling any more than there can be an unlaughed laugh or an unscreamed scream, pains, conceived of as feelings, are necessarily private and necessarily owned.

Pains, viewed as experiences, intuitively can be misleading or inaccurate. Take the case of phantom limb pains. You have no right leg and yet you experience pain in the leg. How is that possible? Answer: in undergoing the pain experience, you *hallucinate* a right leg.

Referred pains also are nonveridical. You can feel a pain in the left arm, when there is nothing wrong with the arm, the cause of the experience being a disturbance in the heart. Such a pain intuitively is inaccurate or misleading; for without additional information, on the basis of the pain, you would be disposed to nurse the arm, to rub it, to believe that something is awry in the arm itself. The case is one of *illusion.*

If pains are representations, what do they represent? The obvious answer is pain. After all, if I feel a pain in a leg, and I attend to *what* I am feeling, I attend to a quality that *seems* to be tokened in my leg, a quality that I strongly dislike and that, in one ordinary sense of the term 'pain,' is surely pain. A theory that denies this categorically is at odds with what we ordinarily believe.

But how can this claim be accommodated? The answer is to acknowledge that pain is not only a certain sort of experience but also a certain quality or type *insofar as* (and only insofar as) that quality or type is experientially represented.

So, the term 'pain' does double duty (Harman 1990). A pain in a leg (viewed as an experience) represents that a certain quality is tokened in the leg. It is accurate if that quality is tokened there; inaccurate otherwise. The term 'pain,' in one usage, applies to the experience; in another, it applies to the quality represented insofar as (and only insofar as) it is within the content of a pain experience.

Which quality (or type) is represented? Pain experiences normally track tissue damage. So, tissue damage is the obvious naturalistic candidate for the relevant quality. This is not to say that pains always are alike with respect to the kind of tissue damage represented. As I noted in my 1995b, a twinge of pain represents a mild, brief case of damage. A throbbing pain represents a rapidly pulsing disorder. Aches represent regions of damage inside the body rather than on the surface. These regions are represented as having volume, as gradually beginning and ending, as increasing in severity, as slowly fading away.[1] The volumes so represented are not represented as precise or sharply bounded. This is why aches are not felt to have precise locations, unlike pricking pains, for example. A stabbing pain is one that represents sudden damage over a particular well-defined bodily region. This region is represented as having volume (rather than being two-dimensional), as being the shape of something sharp-edged and pointed (like that of a dagger).[2] In the case of a pricking pain, the relevant damage is represented as having a sudden beginning and ending on the surface or just below, and as covering a very tiny area. A racking pain is one that represents that the damage involves the stretching of internal body parts (e.g., muscles).

2   On Exportation and Substitutivity in Pain Contexts

It is well known that in belief contexts, co-referential terms cannot safely be substituted. For example,

The ancients believed that Hesperus = Hesperus

is true, while

The ancients believed that Hesperus = Phosphorus

is false. Exportation also fails. For example,

Many children believe that Santa Claus lives at the North Pole

is true, but

Santa Claus is such that many children believe that he lives at the North Pole is false. Some philosophers have held that these two marks of representation always go together (Quine 1960). But there is no a priori reason why they *should* always go hand in hand; and for pain statements, it is clear that there are cases in which the principle of substitutivity holds but exportation fails. For example,

(1)   Desmond feels a pain in his finger

together with

(2)   Desmond's finger is the body part he recently cut

entails

(3)   Desmond feels a pain in the body part he recently cut.

Here the substitution of "the body part he recently cut" for "his finger" is safe. But, given the possibility of phantom-limb pain, (1) does not entail

(4)   Desmond's finger is such that he feels a pain in it.

Exportation intuitively also fails with respect to the expression "a pain" in (1). To see this, suppose that (1) is true. Still

(5)   A pain is such that Desmond feels it in his finger

may well be false. For Desmond does not feel a *feeling of pain* in his finger; he feels *a pain* in his finger; and if the case is one of phantom-limb pain, it seems clear that the only genuine token pain that exists is Desmond's feeling (or experience) of pain.

Consider now the following statement similar to (1) but in which 'pain' is not part of an indefinite description:

(6)   Samantha feels pain in her toe.

Can we safely substitute "tissue damage" for "pain" here so that if (6) is true then

(7)   Samantha feels tissue damage in her toe

is true also? Well, if the principle of substitutivity holds, we can so substitute *if*

(8)   Pain = tissue damage

is true. But (8) is *not* true, on the view I am proposing. For my claim is that 'pain,' in one sense of the term 'pain,' applies to tissue damage (on the assumption that pain experiences represent tissue damage) *only insofar as* it is within the content of a pain experience. This highly qualified claim does not license the unqualified identity claim made by (8).

The fact that exportation fails in statements such as (1) shows pain experiences are indeed representational; for how else can the failure of exportation be accounted for?

The fact that there is no failure of substitutivity in (1), even though exportation fails, suggests that pain experiences have representational contents different in kind from beliefs. Belief (and thought) contents, in my view, individuate in a more fine-grained way than pain contents. On the natural assumption that conceptual contents generally individuate in the same manner as belief contents, it follows that pain experiences have nonconceptual representational contents. Furthermore, these contents are nonconceptual not just in the sense that the subjects of pain experiences need not have the concepts used to state their correctness conditions; for contents that are nonconceptual in this sense *can* be the contents of beliefs (though not beliefs the subjects of the experiences need be in a position to have themselves). Pain experiences have nonconceptual contents that are *different in kind* from belief contents. I shall return to the topic of nonconceptual content later.

3   The Location of a Pain

We locate pains in bodily parts. We say things like "I have a dull pain in my right leg" and we use descriptions for pain that appeal to bodily parts, for example, "the burning pain in my stomach." The *feeling* of pain isn't in my right leg or my stomach, but the pain, it seems, is. How can this be?

There are four possible strategies for handling pain location. One is to adopt a representational account. More on this shortly. A second is to maintain that pains are real objects of the experience of pain. When I feel a pain in my chest, for example, the pain really is in my chest even if the feeling of pain is not. That, however, cannot be right. The pain, whatever else it is on such a proposal, is a mental object for the feeling. And mental objects cannot exist in chests or legs any more than such objects can exist in walls or tables. Of course, I can feel a pain in a hand (that has recently been amputated) and I can place the stump of my arm against a wall or a table. But the pain I feel isn't *in* the wall or table!

A third response is to say that the word "in" in such contexts has a special causal sense. Pain and the feeling of pain are one. In conceiving of a pain as being in a leg, we are conceiving of it as being caused by damage to the leg, not as being spatially inside the leg. That again seems very implausible. For one thing, it is ad hoc. For another, unless it can be shown to be the only alternative that works, it unnecessarily multiplies word senses. Why insist that 'in' here means something different from what it means when it is used elsewhere?

Those who take this view try to motivate the proposal by noting that inferences like the following are invalid:

(9)   The pain is in my fingertip.

(10)   The fingertip is in my mouth.

Therefore,

(11)   The pain is in my mouth.

How could this be, they ask, if the word "in" is used in (9) and (10) in its ordinary spatial sense (Jackson 1977; Block 1983)? After all, if *A* is spatially inside *B* and *B* is spatially inside *C*, *A* must be spatially inside *C*.

This line of argument would provide some support for the causal proposal, if there were no other explanation of the inference failure consistent with the claim that "in" in such contexts has an ordinary spatial sense.

One alternative explanation is offered by Paul Noordhof (2001). According to Noordhof, the inference from (9) and (10) to (11) displays "precisely the same invalidity" (p. 96) as the inference

(12)   There is a hole in my shoe.

(13)   The shoe is in the box.

Therefore,

(14)   There is a hole in the box.

What such inferences reveal, Noordhof claims, is only that "there is a sense of 'in' that is used to describe the state of an object" (p. 97).

Unfortunately, Noordhof's claim is opaque. One way to read him is as supposing that talk of holes in things is best understood in terms of talk of the things' being in the state of being holed (or perforated). However, quantification over holes is unavoidable in ordinary discourse putatively about holes.[3]

Another way to read Noordhof is as claiming that talk of holes in things is to be understood in terms of talk of the things being in the state of having a hole, where the latter state requires for its tokening that there exist a hole. But then what is the proposed reading for (9) or (11)? That my fingertip (or mouth) is in the state of having a pain? That cannot be right.

Furthermore, although the specimen statements of pain location, (9) and (11), arguably require for their truth that the relevant body parts exist, there are many such statements that lack this requirement. For example,

(15)   I feel a pain in my fingertip

can be true, even if both my hands have been amputated. In this case, there is no bodily part to be in the relevant state. So, the invalidity of the inference from (15) and (10) to

(16)   I feel a pain in my mouth

is left without any satisfactory explanation.

Once the representational approach to pain location is adopted, there is a straight-forward account of this inference failure. The inference from (15) and (10) to (16) fails just as does the inference

(17)   I seem to see a triangle within a circle.

(18)   The circle is within a larger square.

Therefore,

(19)   I seem to see a triangle within a larger square.

But what of the inference failure from (9) and (10) to (11)? Noordhof's claim is that it cannot be explained by appeal to representation; for, on the assumption that (9) and (11) require that my fingertip and mouth exist, the relevant pain representations are now *de re* with respect to these items and thus the inference should go through.

Noordhof is mistaken. There is more than one ordinary spatial sense of 'in.' Where there is a hollow physical object, O, the claim that something X is in O can be under-stood either to assert that X is within the cavity bounded by O or to assert that X is (at least partially) embedded within a portion of the cavity-surround (the top, bottom, and sides of O). For example,

(20)   Tanya has a ring in her nose

uses the second sense of 'in,' whereas

(21)   My car is in the garage

uses the first.

The representationalist explanation for the inference failure from (9) and (10) to (11) is now as follows: the only pain in reality is my experience of pain. According to (9), that pain is accurate if and only if my fingertip is such that there is tissue damage spatially within it (which damage may itself be classified as pain in a second sense of that term insofar as, and only insofar as, it is in the content of my pain experience). (10) asserts that my fingertip is spatially within the cavity bounded by my mouth. From these two claims, it does not follow that my pain experience is accurate if and only if my mouth is such that my pain experience represents that there is tissue damage in a part of its roof, walls, or base. But this is what (11) requires, given that pains represent tissue damage. So, (11), on this proposal, does not follow from (9) and (10).

Perhaps it will be replied that once different spatial senses of "in" are admitted, that is all that is needed to explain the inference failure. The appeal to representational content is otiose. This misses the point. In the case that (9) and (10) are both true, there is in reality no such thing as a pain in my mouth in *either* spatial sense. Moreover, the experience of pain is in my brain, if it is anywhere, not *in* my mouth. To offer an explanation that supposes otherwise is to offer an explanation based on false assumptions. And that is to offer no explanation at all.[4] So, no trouble here for my position and no reason to think that Noordhof has a better explanation in the offing of the invalidity of the pain location inference to which I originally appealed (namely, "I have a pain in my fingertip"; "My fingertip is in my mouth"; therefore, "I have a pain in my mouth.").

4   Complexities in the Content of Pain

The view that pain has distinct sensory and affective–emotional components, subserved by different neural mechanisms, was first proposed by Melzack and Casey in 1968, and in the thirty or so years since then it has been shown to be well motivated by both a wealth of clinical data and neuroscientific evidence.[5]

Normally, in a pain experience, both components are present. But in some cases, the affective component is missing. For example, people who undergo prefrontal leucotomies (operations that sever the neural connections in the deep white matter in the frontal lobes) as a last resort for their intractable, constant, severe pain are typically cheerful and relaxed afterwards. They report still having pains, but they no longer *mind* them.

Similar reports come from people suffering pain who are under hypnotic suggestion or nitrous oxide. Such cases of "reactive disassociation," as Dennett (1978) calls them, are ones in which the distinctive sensory dimension of pain is present but the aversive component is gone. Is pain itself still present? It seems so. The patients *say* that they continue to feel pain. I see no reason not to take these reports at face value. What they show, I suggest, is that pain is not *essentially* an aversive experience.

Pain is essentially a sensory experience, however. Whatever else pain is, at its core, it is a bodily sensation. Take away the characteristic sensory component, and no pain remains. Certain abnormal unpleasant experiences, for example, dysaesthesia, lack any sensory component, and they are not classified by their subjects as pains. Other unpleasant experiences, for example, an irritating itch, are sensory, but they aren't pains since the *distinctive* sensory content of pain is missing.

In any event, a *typical* pain experience has both a sensory and an affective dimension. In the earlier sections, I focused on the sensory side of pain. Let me now turn to the affective dimension of pain.

Pain is normally very unpleasant. People in pain try to get rid of it or to diminish it. Why? The answer surely is because pain *feels* unpleasant or bad, because it is *experienced* as such. But what exactly is experienced as unpleasant? One's attention, when one feels pain, goes to a place different from the one in which the experience of pain is located. The disturbance that is experienced as unpleasant is located in the bodily location to which one attends (in normal circumstances). People whose pains lack the affective dimension undergo purely sensory, nonevaluative representations of tissue damage of one sort or another in a localized bodily region. Those whose pains are normal experience the same sort of disturbance, but now it is experienced by them as unpleasant or bad. It is precisely *because* this is the case that normal subjects have the cognitive reactions to pain they do, reactions such as desiring to stop the pain.

To experience tissue damage as bad is to undergo an experience that represents that damage as bad. Accordingly, in my view, the affective dimension of pain is as much a part of the representational content of pain as the sensory dimension is.

The representational content of pain, as noted earlier, is nonconceptual. Admittedly, my talk above of the unpleasantness of pain, of its experienced badness may sound cognitive. But I do not intend it to be understood in this way. It seems to me that the most plausible view is that we are hardwired to experience pain as bad for us from an extremely early age.

Consider the other side of the coin for a moment. A child as young as two months, upon tasting a little chocolate, typically behaves in a way that signifies that he or she wants more. The child will open and close its lips, push forward toward the chocolate, look happy. Why? The answer is that the chocolate tastes good. *That's* why the child wants more. The child's gustatory experience represents a certain taste and the child experiences that taste as good. The taste is experienced as good by the child in that the child undergoes an overall experience that represents the presence of the taste in the mouth and represents it as good.

Intuitively, this is not a cognitive experience. It does not require concepts. For another example, consider orgasm. Orgasm is a bodily sensation, but it is not only that. The most natural description of an orgasm, and indeed of any pleasant experience is "It feels good." One's orgasm represents a certain change in the region of the genitals as good for one, as something apt to benefit, not to harm one.[6] That isn't a conceptual response. One cannot help but feel the relevant bodily disturbance except as good. One is hardwired by nature to experience it in this way. It is not difficult to fathom why.

Pain is the opposite, as is the experience of smelling vomit, for example. In the latter case, one experiences a certain smell as bad, as something apt to harm. That's part of the representational content of the experience. In the former case, it is tissue damage within one's own body one experiences as bad, as apt to harm.

It is not just the affective dimension of pain that does not require concepts. Intuitively, one does not need concepts to have a pain, period. Given the right stimulus, one feels pain, whatever concepts one has in one's repertoire. This is not to say that one's cognitive assessment of a situation can never influence or affect one's experience of pain. Obviously, it can. The point is simply that the basic experience of pain requires no cognitive sophistication. Humans are hardwired to experience pain when they undergo tissue damage, whatever they think or believe.

5   On the Relationship of Pain Phenomenology and Representational Content

Suppose you have a throbbing pain in your finger. You can certainly have a pain without noticing it, as, for example, when you are distracted for a moment by something else, but if you do notice a pain—if you are introspectively aware of it—then your attention goes to *wherever you feel the pain* (in this case, to your finger). Your attention does not go to where your experience is (that is, to your head, if your experience is a physical thing) or to nowhere at all. In attending to your pain, you are directly and immediately aware of certain qualities, which you experience as being localized in your finger. Among these is a quality you want very strongly to stop experiencing.

The same is true even if you are feeling a pain in a phantom finger. Still, you are directly aware of a quality you strongly dislike, a quality that you *experience* as being in a finger, even though the finger no longer exists. The first point to stress, then, is that the qualities of which we are directly aware in introspecting pain experiences are not qualities *of* the experiences (assuming that there is no massive error), but qualities of bodily disturbances in regions where the pains are felt to be, if they are qualities of anything at all.

Not everyone agrees with these points. Stephen Leeds, for example, writes:

A toothache does not seem to present the world to me in anything like the way that looking at my tooth does: rather, although it seems in some way to present me with properties, these are not presented as properties of my tooth (and still less as properties of my brain or nervous system)—rather they seem to be presented as properties of (or at) point *s* in a somatic field. (2002, p. 6)

I agree with Leeds that there is a big difference phenomenologically between my seeing the tooth (in a mirror, say) while I have a toothache and my experiencing the toothache. But this difference is one that poses no threat to the thesis that when we introspect pain experiences, the qualities of which we are directly aware are not qualities of the experiences; for the properties presented in the two cases are very different (just as they are in the cases of seeing a cube and feeling a cube by running one's fingers over it). In seeing the tooth, I am aware of its color and shape, and I am unaware of the qualities I dislike so much in having the pain experience. In experiencing the toothache, the situation is reversed.[7]

Leeds asserts that in experiencing a toothache, the properties presented to me are not presented as properties of my tooth. I disagree. When I focus upon my toothache, I am aware of an ache *in* my tooth, and in being aware of this ache, I am aware of a quality that I strongly dislike, whatever further account we give of its nature—a quality that is presented to me *as* inside my tooth, as filling the region occupied by my tooth (the tooth to which I direct my dentist when he asks me "Where is the pain?").

Perhaps the thought is that I am introspectively aware of pain in my tooth, but I am aware of this by being aware of something else, a quality or qualities of a region of a somatic field. However, this seems to me no more plausible or well motivated than the corresponding claim that our visual experience of external physical surfaces is mediated by experience of visual sense-data.

In the case of vision, the surfaces of which I am directly aware are experienced by me at varying distances away and orientations. Unless there is some drastic error in visual experience, that is where such surfaces normally are. They are physical surfaces. They can be photographed. We ordinarily assume that others directly see those surfaces too.

In the case of pain, there is, I grant, a body image; and, in one way of talking, the pain in the finger is located on that part of the body image representing the finger. But all this really means is that there is, for each of us, a continuously updating sensory representation of the sort found in general bodily feeling, and that the experience of pain represents the quality or qualities felt as unpleasant as being instantiated at a certain location within the body space represented by the former representation.

Barry Maund (2003) has recently contested these claims. He comments:

. . . the pain one feels in one's leg is a subjective feeling that one 'projects' onto the leg. It is not a real projection. One has a body image which represents the body. The pain is projected on to, is located on, that part of the body image which represents the leg. Likewise with colors. There is a subjective quality which one 'projects' onto an external object, say to the moon, to represent it as yellow.

I confess that I do not understand any of this. If I see the moon, I am not aware of a subjective visual field that represents the moon. I am aware of the moon and perhaps some stars located in distant regions of space before my eyes. Likewise, if I have a pain in my leg, I am not aware of an image that represents my leg. I'm aware of my leg and its condition. To suppose that it is the representation itself—the subjective visual field or the body image—of which I am really (directly) aware in these cases is like supposing that if I desire eternal life, what I really (directly) desire is the idea of eternal life. That, however, is *not* what I desire. The idea of eternal life I already have. What I desire is the real thing. And it does not help, of course, to say that it *must* be the representation of which I am aware, since the case might be one of hallucination—no moon or no leg—for patently, if there is no eternal life, it *still* isn't the idea of such a life that I really desire. If the pain is a phantom one or the visual experience totally delusive, I simply undergo an experience that represents something that isn't there.

It seems to me, then, that the right thing to say is that when I attend to a pain in my finger, I am directly aware of a certain quality or qualities as instantiated in my finger. Moreover, and relatedly, the only particulars of which I am then aware are my finger and things going on in it (for example, its bleeding). My awareness is of my finger and how *it* feels. The qualities I experience are ones the finger or part of the finger or a temporary condition within the finger apparently have. My experience of pain is thus transparent to me. When I try to focus upon it, I "see" right through it, as it were, to the entities it represents. But when I introspect, I am certainly aware of the phenomenal character of my pain experience. On the basis of introspection, I know what it is like for me phenomenally on the given occasion. Via introspection, I am directly aware of certain qualities that I experience as being qualities of my finger or episodes inside it, qualities some of which I react to in a very negative way, *and thereby* I am aware of the phenomenal character of my experience. By being aware of these qualities, I am aware of what it is like for me. This is not to say, of course, that I *infer* the phenomenal character of my experience from my awareness of such qualities. Obviously, no *reasoning* is involved. Still, by attending to what is outside my head,[8] in my body, I know what it is like for me. So, my awareness of phenomenal character is not the direct awareness of a quality of my experience. Relatedly, the phenomenal character itself is not a quality of my experience to which I have direct access.

This conclusion could be resisted if one took the view that there is a massive error in pain experiences, that the qualities represented as instantiated in the body by such experiences do not belong there but are rather qualities of the experiences themselves.

That is highly counterintuitive, however, just as is the analogous position for color experience. And it is far from clear that it is intelligible. Take, for example, a sharp, daggerlike pain. The unpleasant quality one experiences is experienced as distributed throughout, as filling a dagger-shaped region of the body. Could that quality really be a quality of something that isn't extended at all? Is that conceivable? It seems to me conceivable that dualism is true and that my experiences are nonphysical. But is it conceivable that the quality I experience as instantiated throughout a spatial region is really a quality of something that isn't spatial? Whatever we say to this, I know of no convincing arguments that pain experiences involve massive error, and I shall assume that the intuitive view is correct.

Now, as I experience the pain and I attend to it, necessarily, if any of the qualities of which I am *directly* aware changes, then the phenomenal character of my experience changes. If, for example, my finger starts to sting or to ache, the phenomenology of my pain experience changes. Why should this be? Facts like this one are surely not brute. Moreover, they obtain even in the case that I am undergoing a phantom finger pain. An explanation is needed of why the phenomenal character of pain experiences is sensitive in this way to qualities that, if they are qualities of anything at all, are qualities of bodily regions where the pains are experienced to be. The explanation surely is that the phenomenal character *involves* the bodily qualities of which the subject of the visual experience is directly aware, that these qualities at least partly *constitute* phenomenal character.

What, then, is the phenomenal character of pain? One possible hypothesis is that it is a subjective quality (or cluster of qualities) of the relevant bodily region, so that the experience itself inherits its phenomenal character from the relevant subjective quality (or qualities). That hypothesis will not accommodate phantom-limb pain, however; for in that case there is no appropriate bodily region. The best hypothesis, I suggest, is that the phenomenal character of pain is representational content of a certain sort, content into which the experienced qualities enter. This explains why pain phenomenal character is not a quality of an experience to which we have direct access (representational content is not a quality of the thing that has representational content) and why pain phenomenal character necessarily changes with a change in the qualities of which one is directly aware (changing the qualities changes the content). It also explains why the phenomenal character of a pain experience is something the experience *has*, something that can be common to different token experiences, and why pain experiences have phenomenal character even if nothing really has the qualities of which one is directly aware via introspection.

6 Phenomenal Content

The phenomenal character of a given pain, on my view, is one and the same as the phenomenal content it has. The phenomenal content of a pain is a species of representational content. It is content that is *nonconceptual*, not just in the sense that the subject of a pain experience, in undergoing it, need not possess any of the concepts that we, as theorists use, when we state the correctness conditions for that experience but in the full-blooded sense of being content that is different in kind from conceptual content.

Pain experiences represent in something like the way that the height of a mercury column in a thermometer represents. The latter normally tracks a certain temperature in the surrounding air, however that temperature is conceived. Similarly, pain experiences normally track tissue damage in regions of the body, however the tissue damage is conceived. The representational content pain experiences have via such tracking is not content that could be the content of a belief.

The phenomenal content is also *abstract*. That is to say, it is content into which no concrete things enter. Thus a pain in a leg cannot be an experience into which one of the subject's legs enters. The reason is straightforward. One can feel a pain in a leg, while lacking a leg, that is phenomenally indistinguishable from such a pain when the leg is present. The phenomenal character of the two experiences is the same, and thus, on my view, their phenomenal content. But in one of the two cases, there is no leg.

Another condition on the phenomenal content of pain is that it be suitably poised. In particular, it must stand ready and available to make a direct impact on relevant desires the subject has, for example, the desire to get away from the noxious stimulus. Moreover, the qualities in the content must be ones of which the subject of the experience is directly aware via introspection (assuming the subject is sophisticated enough to be able to introspect).

These three conditions on phenomenal content—that it be nonconceptual, abstract, and poised—constitute its essence (Tye 1995a, 2000).

7 Some Remaining Objections

One objection that has been raised to this proposal is that it cannot distinguish between the phenomenology of seeing one's damaged leg and feeling pain there.[9] This, I maintain, is not so. One's visual experience, as one views the leg, nonconceptually represents such features as color, shape, orientation of surface, presence of an edge. It does not nonconceptually represent tissue damage. One's pain does nonconceptually represent tissue damage, but it does not represent the other features.

I concede that, going by the phenomenal look of the leg, one will judge it to be damaged, and thus that it will look *to be* damaged or look *as if it is* damaged. This is a conceptual use of the term 'look,' however. The phenomenal sense of 'looks' is non-conceptual. It is captured by 'looks *F*,' where '*F*' is a term for a quality of which one is directly aware when one introspects the experience.

Maund (2002) has objected that the damaged leg *does* look damaged in the phenomenal sense of 'look.' He comments:

> . . . it seems to me that the phenomenal sense applies to a range of features or cluster of features besides the ones Tye cites: there are rusty looks, wooden looks, jarrah looks, metallic looks, and damaged-tissue looks.

I disagree. If one views something rusty in standard conditions, one is directly aware of a range of color and texture qualities, on the basis of which one judges that it is rusty. The object looks to one *to be* rusty; moreover it looks *like* other rusty things, and indeed, in one ordinary way of speaking, it looks rusty, but it doesn't look rusty in the nonconceptual, phenomenal sense. For rustiness isn't a quality of which one is directly aware when one introspects one's experience any more than is the quality of being feline, when something looks feline to one. Intuitively, felines and twin felines (molecule by molecule duplicates of cats belonging to a different species on a variant of Putnam's planet, twin earth) are phenomenally indistinguishable. In the phenomenal sense, they look alike. But cats look feline to us, whereas twin cats look twin-feline to the inhabitants of the other planet.

What is true here for cats is true *mutatis mutandis* for rusty things, wooden things, and so on. It is true for damaged tissue as well. Imagine in this case that on the twin planet, there is no tissue but an artificial look-alike. Note, incidentally, that this is not to deny that "there is a characteristic experience-type that things which are rusty, wooden, jarrah, metallic and damaged tissue cause in optimal conditions" (Maund 2002). There is such a type; but at the nonconceptual *phenomenal* level, that type is individuated by the cluster of qualities of which the subject is directly aware via introspection (qualities that are also represented by the experience). And those qualities are at the level of shape, color, texture, and so on; they do not include rustiness, woodenness, etc.

A second objection concerns the concepts used in noticing one's own pains. While the experience of pain is nonconceptual, the awareness by the subject *that* he or she is experiencing pain requires concepts. Without concepts, those in pain would be 'blind' to their experiences. But which concepts exactly are the ones needed to notice pain? Murat Aydede (2001) replies: "Certainly, not the concepts of bodily disturbance, tissue damage, disordered states, nociceptive stimulation, etc. All [most people (including small children)] need—and usually have in fact—is the concept of pain. But the concept of pain is not the concept of physical disturbance (tissue damage, etc.)" . . . (p. 55). He continues, with the aim of raising a difficulty for my position:

Consider a mild tickle, itch, tingle, and pain occurring on the very same spot of my back for the same length of time (at different times, of course) . . . They feel different. And if their feel is what they represent, then I seem to be able to discriminate between them, and single out their differences generally, without having the concepts that apply to what physically happens in the relevant part of my body, i.e., without having the concepts for what is represented by these experiences. The concepts I happen to have are none other then the concepts of an itch, tickle, tingle, and pain, and these concepts are not concepts of physical bodily happenings supposed to be represented by them. (Ibid., p. 56)

I agree with Aydede that when I am introspectively aware that I am in pain, the concept I apply is the concept PAIN and not the concept TISSUE DAMAGE. I agree also that the concept PAIN applies to the experience I undergo. But why should this be thought to present a difficulty for my position?

I said earlier that when I experience pain, I am directly aware of the qualities the experience represents. By attending to these qualities, I am aware *that* I am experiencing pain. Introspection of experiences, in my view, is a *reliable* process that takes awareness *of* qualities represented by the experiences as input and yields awareness *that* a certain kind of experience is present as output. It is the reliability of this process that underwrites introspective knowledge of experiences.

Consider the case of introspection of thought contents, which is similar. If I think that water is wet, and I introspect, I become aware *that* I am thinking that water is wet. This awareness is not based upon an inference from other propositional states. Nor is it the result of attention to an internal auditory image of myself saying that water is wet, though such an image may accompany my thought. Intuitively, my introspective access to what I am thinking is direct. It seems plausible to suppose that introspection of thought contents is a reliable process that takes as input the content of the thought and delivers as output a belief or judgment that one is undergoing a state with that content (McLaughlin and Tye 1998).

On this view of introspective knowledge of thought contents, the concept of a thought that *P* is, in its first-person present-tense application, a *recognitional* concept.[10] Those who have mastered the concept can introspectively recognize that an occurrent thought that *P* is present without going through any process of reasoning. In much the same way, we do not have introspective knowledge of our experiences by inferring it from something else. We acquire introspective knowledge that we are undergoing such and such an experience or feeling via a reliable process that triggers the application of a suitable concept. Experiential concepts—the concepts that enable us to form a first-person conception of our experiences via introspection—are, in my view, recognitional concepts of a special sort.

I agree with Aydede, then, that when I introspect a pain, the concept I apply is the concept PAIN or some more specific concept of that type, for example THROBBING PAIN. This concept may be purely phenomenal—and it will be, if all I notice is the phenomenal character of my state—but it may also be broader than that, applying to states that vary in their phenomenal "feel."

Since my awareness is that I am feeling pain and *not* that I am undergoing an experience that represents tissue damage, I do not apply the concept TISSUE DAMAGE via introspection. That tissue damage is the quality paradigmatically represented by pains qua sensory experiences is an empirical hypothesis. It is not a hypothesis, supportable by a priori reflection upon concepts or introspection. So, not only is the concept TISSUE DAMAGE not needed to feel pain, but it is also not needed to be introspectively aware that one is feeling pain.

There is an interesting question here, raised by Aydede, as to why, if the phenomenology of pains is given by their representational content (or some part of it), we have come to develop "an immediate epistemic and utilitarian/practical interest in the experiences in the first place rather than in the objects/conditions represented by them" (2001, p. 58). Aydede elaborates the concern further as follows:

The explanandum is why our primary and immediate interest (epistemic as well as practical) is in the experiences themselves such that the related concept applies primarily to them rather than what they represent. . . . We have no parallel of this phenomenon in standard exteroception, where the whole phenomenology can be accounted (assuming it can) for solely in terms of what is represented, i.e., the perceptual object and its qualities as the representational content of the relevant experiences. (2001, p. 58)

What seems to be worrying Aydede is the thought that in the case of color experience, say, the information flow is such that red causes the experience of red, which then triggers the application of the concept RED to red, the color represented, but in the case of pain, the information flow goes from tissue damage to the experience of pain to the triggering of the concept PAIN, where that concept applies to the experience and not to the damage.

The worry is misplaced. When we introspect our color experiences, we issue reports about the *apparent* colors of things. Here the information flow goes from the awareness of the color red to the application of the concept EXPERIENCE OF RED or the concept LOOKS RED. By being aware of red, I am aware that I am having an experience of red. Correspondingly, when we introspect our pain experiences, we are aware of (among other things) tissue damage and by being aware of those things, we are aware that we are in pain. For these cases, the information flow is of just the same type. Normally, awareness of red is caused by red. Normally awareness of tissue damage is caused by tissue damage. Both awarenesses *of* something external (to the head) bring about an awareness *that* an experience of a certain sort is present.

Of course, if I see something red, I may judge that it is red, and here I apply the concept RED to the object seen. But likewise, if I cut my finger, I may judge that my finger is damaged, and in so doing apply the concept DAMAGE to my finger (or better its condition). The only difference between the two cases that I can discern is that when I introspect my color experience, the concept I exercise (EXPERIENCE OF RED) is a compound concept containing a concept for an entity represented by the experience, whereas in the case of pain, when I introspect, the concept I exercise (PAIN) has no such component.

A final objection I wish to consider is that the painfulness of pains has not been fully accounted for within the above theory.[11] Let me begin my reply by noting first that when we say that something is painful, what we normally mean is that it *causes* the feeling of pain. Thus, we speak of a cut, a bruise, a sore, an operation, a cough, a lashing as painful, that is, as causing pain. This is why, if a person feels a pain in his left arm, and his doctor informs him that the cause of the pain lies in a disturbance in his heart, we allow that what is *really* painful here is the disturbance in his heart. If the doctor has made no mistake, *that* is what is actually hurting the patient.

It is evident that pains themselves are not painful in this sense. Pain does not cause the feeling of pain. So, it is no criticism of my view to say that the painfulness of pains, in this sense of painfulness, is unaccounted for within the representationalist theory.

What, then, has representationalism left out? If what is meant by the painfulness of pains is their phenomenal feel, then *that* hasn't been left out. For a theory has been offered of its nature, a theory that identifies the feel with a certain sort of sensory–affective representational content, the possession of which by pains explains why we normally have such a strong aversive cognitive reaction to them.[12]

So, representationalism seems to me to have much in its favor and no obvious problems. I continue to believe that it provides the best hope for a defensible, naturalistic theory of experience generally.

Notes

1. Cf. Armstrong 1962.

2. I am not suggesting here that one cannot have a stabbing pain unless one has the concept of a dagger.

3. For compelling arguments here, see Casati and Varzi 1995. The reason the inference from (12) and (13) to (14) fails is that the hole in my shoe is an entity occupying a fillable discontinuity in the shoe's surface. It is not thereby an entity occupying a fillable discontinuity in the box's surface, even though the shoe is inside the box.

4. In a later essay written in response to my 2002, Noordhof (2002) presents further inferences that are intended to show that I must proliferate spatial senses of "in" over and above the two I distinguished above. This, he suggests, is an unwelcome result for my view. However, I deny that any such proliferation is necessary.

Let us agree that, where O is a hollow object, to say that X is in O is to say either that X is within the cavity bounded by O (call this 'SENSE 1' of 'in') or that X is embedded (at least partly) in the matter making up O, that is, the cavity–surround (call this 'SENSE 2'). Two of the four inferences Noordhof adduces as creating difficulties for me—namely:

(B1)   There is a flaw in the diamond;

(B2)   the diamond is in the box;

therefore,

(B3)   there is a flaw in the box;

and

(D1)   there is a fault in my computer;

(D2)   the computer is in my office;

therefore,

(D3)   there is a fault in my office

—evidently present no new problem (and it is mystifying as to why Noordhof thinks that they do). In both cases, a hollow object is involved (a box in one and an office in the other) and 'in' in the conclusion has SENSE 2 whereas "in" in the second premise has SENSE 1 (and 'in' in the first premise just means 'spatially within'). So, these arguments commit the fallacy of equivocation.

The same is true for the third inference Noordhof cites:

(C1)   There is a tremor in my hand.

(C2)   My hand is in my pocket.

Therefore,

(C3)   There is a tremor in my pocket.

The only difference is that, unlike a fault or a flaw, a tremor in the hand has a somewhat vague location. It may occur in a single finger or several without there being anything more definite to say about where the tremor is. It may also occur in the hand as a whole. In this case it is located where the hand is. But the tremor is still spatially inside the hand, allowing that a thing is within a location even if it takes up the entire location (just as a thing is a part of itself). Since a pocket is a hollow object, the inference failure is explained in the same way as the previous two.

Noordhof's own account of these cases is to say that in all of them "an object is described as being in a certain state" (2002, p. 154). But Noordhof denies that he has any reductive ambitions. How then does the appeal to states help? Take the second inference. If my computer is in the state of having a fault in it, and my computer is in my office, it surely *does* follow that my office is in the state of having a fault in it, if 'in,' as it occurs within 'state of having an *F* in' means spatially inside. The only solution is either to say, as I do, that 'in' has more than one spatial sense or to introduce further nonspatial senses of 'in.'

Noordhof's remaining inference is

(A1)   There is a poisonous gas in the spacecraft.

(A2)   The spacecraft is in the earth's atmosphere.

Therefore,

(A3)   There is a poisonous gas in the earth's atmosphere.

This case is a little different from the others. The earth's atmosphere surrounds the earth and other objects not too far away from the earth. In this way, the atmosphere is like a multiple cavity–surround. Something in the atmosphere is either spatially inside it, that is, spatially within what the atmosphere surrounds, or spatially embedded in the atmosphere *qua* surround. In (A1), 'in' has SENSE 1. In (A3), however, assuming we agree that it does not follow from the premises, 'in' has SENSE 2. Given that there is a poisonous gas within the spacecraft and that the spacecraft is within the atmosphere, it does not follow that there is a poisonous gas embedded within the atmosphere *qua* surround.

5. For useful summaries here, see Melzack and Wall 1983, Price 1999.

6. The suggestion that the pleasingness of orgasms is part of their representational content is made in Tye 1995c. It is also the view taken by William Seager 2002.

7. For more on the difference between the two cases, see section 7.

8. Ignoring headaches.

9. This is the Leeds's objection for the case of toothache, mentioned in section 1. It was first raised to me in conversation by Ned Block and John Searle; and it has recently been pressed by Barry Maund.

10. For more on recognitional concepts, see Loar 1990.

11. Aydede presses this charge in his 2001.

12. I think it is fair to say that in some of my previous writings, I have not emphasized enough the noncognitive, affective side of pain and its relevance to the overall content of pain experiences. But the importance to phenomenology of the affective dimension of content is clearly noted in Tye 1995c, for example.

References

Armstrong, D. 1962. *Bodily Sensations*. London: Routledge and Kegan Paul.

Aydede, M. 2001. "Naturalism, Introspection, and Direct Realism about Pain." *Consciousness and Emotion* 2(1): 29–73.

Block, N. 1983. "Mental Pictures and Cognitive Science." *Philosophical Review* 92: 499–541.

———. 1996. "Mental Paint and Mental Latex." Pp. 19–49, in *Philosophical Issues* 7, ed. E. Villenueva, Northridge, Calif.: Ridgeview.

Casati, R., and A. C. Varzi. 1995. *Holes and Other Superficialities*. Cambridge, Mass.: MIT Press.

Chalmers, D. 1996. *The Conscious Mind*. Oxford: Oxford University Press.

Dennett, D. 1978. "Why You Can't Make a Computer That Feels Pain." Pp. 190–229, in his *Brainstorms*. Cambridge, Mass.: MIT Press.

Harman, G. 1990. "The Intrinsic Quality of Experience." Pp. 31–52, in *Philosophical Perspectives* 4, ed. J. Tomberlin. Northridge, Calif.: Ridgeview.

Jackson, F. 1977. *Perception*. Cambridge: Cambridge University Press.

Leeds, S. 2002. "Perception, Transparency, and the Language of Thought." *Noûs* 36(1): 104–129.

Loar, B. 1990. "Phenomenal States." Pp. 81–108, in *Philosophical Perspectives* 4, ed. J. Tomberlin. Northridge, Calif.: Ridgeview.

Maund, B. 2002. "Tye on Phenomenal Character and Color." Web symposium on Tye's *Consciousness, Color, and Content* at: http://host.uniroma3.it/progetti/kant/field/tyesymp.htm.

———. 2003. "Tye: Consciousness, Color, and Content." *Philosophical Studies* 113(3): 249–260.

McGinn, C. 1982. *The Character of Mind*. Oxford: Oxford University Press.

McLaughlin, B., and M. Tye. 1998. "Is Content-Externalism Compatible with Privileged Access?" *Philosophical Review* 107: 349–380.

Melzack, R., and K. Casey. 1968. "Sensory, Motivational, and Central Control Determinants of Pain: A New Conceptual Model." Pp. 223–243, in *The Skin Senses*, ed. D. Keshalo. Springfield, Ill.: Charles C. Thomas.

Melzack, R., and P. Wall. 1983. *The Challenge of Pain*. New York: Basic Books.

Noordhof, P. 2001. "In Pain." *Analysis* 61: 95–97.

———. 2002. "More in Pain." *Analysis* 62: 153–154.

Price, D. D. 1999. *Psychological Mechanisms of Pain and Analgesia*. Progress in Pain Research and Management, vol. 15. Seattle: IASP Press.

Quine, W. 1960. *Word and Object*. Cambridge, Mass.: Harvard University Press.

Seager, W. 2002. "Emotional Introspection." *Consciousness and Cognition* 11(4): 666–687.

Tye, M. 1995a. *Ten Problems of Consciousness.* Cambridge, Mass.: MIT Press.

———. 1995b. "A Representational Theory of Pains and Their Phenomenal Character." In *Philosophical Perspectives* 9. Also reprinted with revisions in *The Nature of Consciousness: Philosophical and Scientific Debates*, edited by N. Block, O. Flanagan, and G. Güzeldere, Cambridge, Mass.: MIT Press, 1997.

———. 1995c. "Blindsight, Orgasm, and Representational Overlap." *Behavioral and Brain Sciences* 18: 268–269.

———. 2000. *Consciousness, Color, and Content.* Cambridge, Mass.: MIT Press.

———. 2002. "On the Location of a Pain." *Analysis* 62: 150–153.

Peer Commentary on Michael Tye

5 The Main Difficulty with Pain

Murat Aydede

Consider the following two sentences:

(1) I see a dark discoloration on the back of my hand.

(2) I feel a jabbing pain in the back of my hand.

They seem to have the same surface grammar, and thus prima facie invite the same kind of semantic treatment. Even though a reading of 'see' in (1) where the verb is not treated as a success verb is not out of the question, it is not the ordinary and natural reading. If I am hallucinating a dark discoloration in the back of my hand, then (1) is simply false. For (1) to be true, therefore, I have to stand in the seeing relation to a dark discoloration in the back of my hand, that is, to a certain surface region in the back of my hand marked by a darker shade of the usual color of my skin, a certain region that can be seen by others possibly in the same way in which I see it. Also note that although the truth of (1) doesn't require the possession of any concept by me expressed by the words making up the sentence, my uttering of (1) to *make a report* typically does—if we take such utterances as expressions of one's thoughts. So my seeing would typically induce me to identify something in the back of my hand *as* a dark discoloration. This is a typical case of categorization of something under a perceptual concept induced by perception. Of course, my uttering of (1) does more than attributing a physical property to a bodily region, it also reports that I am seeing *it*.

What has to be the case for (2) to be true? Like (1), (2) invites us to follow its surface grammar and treat it as expressing a perceptual relation between me and something else that has a bodily location, namely, a jabbing pain in the back of my hand. The difficulties with taking this route are familiar, so I won't repeat them here except to say that whatever the true analysis of sentences like (2) turns out to be, one thing is clear: the truth-conditions of (2) put no constraints whatsoever on how things *physically* are with my hand. Anyone who has a sufficient mastery of our ordinary concept of pain has no difficulty whatsoever in understanding how (2) could still be true even though there is nothing physically wrong with my hand. So if the truth of (2) is taken to imply attributing a property to the back of my hand, this property is not a physical property. For many with naturalistic leanings like me, this is one of the main reasons for not taking (2) as primarily making a property attribution to a bodily region. So when I utter (2) to make a report and appear to identify something in the back of my hand as a jabbing pain, whatever else I am doing, I am *not* attributing a property or a condition (physical or otherwise) to the back of my hand and saying that I am feeling *it*.[1]

If this is not what I am doing, what is it that I am doing? What is the proper analysis of sentences such as (2)? On the basis of considerations harnessed so far, one is tempted to argue in the following way. Every genuine case of perception allows reporting an instance of perception in the relevant modality by sentences similar to (1), where the perceptual verb is used dominantly as a success verb. This is for good reason: perception is essentially an activity whereby one gathers information about one's environment in real time (including one's internal bodily environment of course). So it is not surprising that the dominant form of reporting is in the form of a relation between the perceiver and the perceived where the latter are extramental objects or conditions of one's environment. It is also not surprising that perception typically yields conceptual categorization of the perceived object or condition: the typical result of a perceptual process is bringing the perceived object under a perceptual concept. Genuine perception thus puts the premium in the perceived object, not in the perceptual activity itself or in the perceptual mental state whereby one is typically brought into epistemic contact with one's extramental environment. If sentences reporting pain in body parts don't follow the pattern of sentences such as (1), that is, if they are not to be construed as reports of perceptual relations between the perceiver and the perceived, then pain *reports* are prima facie not perceptual reports, reports to the effect that one stands in a perceptual relation to something extramental.

What do sentences such as (2) then report, if not the obtaining of a perceptual relation between the perceiver and a perceived object or condition? The answer is anticlimactic: why, of course, they report pain experiences, that is, mental states or events with certain perhaps complex but characteristic phenomenal character or content. Even though this may be anticlimactic in the sense that it's almost a truism given our ordinary notion of pain, it is not enough to remove the puzzle. Experiences are in the head, and for most physicalists they are in the head by being realized in the brain or central nervous system. If sentences like (2) are reports of the occurrence of certain kinds of experiences, we still need to understand what is going on when we seem to locate pains in body parts. Obviously, if pains are experiences, locating a jabbing pain in the back of my hand is at best confused. How do we reconcile this with the fact that I say something true when I utter (2) (even when, let's assume, there is nothing physically wrong with my hand—say, because I am suffering from a centrally caused chronic condition, which is not uncommon)?[2] In other words, how do we reconcile the commonsense understanding of pain as a subjective experience with the comfort and ease with which the very same commonsense routinely attributes pain to body parts?

There is no easy and comfortable answer to this question. The common answer on the part of the perceptual and strong representational theorists such as Pitcher, Armstrong, and Tye, which I share, is to say that sentences like (2) report the occurrence of experiences that represent a certain condition (or constellation of conditions) in one's body regions. What they represent exactly is a matter of controversy, but Tye's answer is that pain experiences represent tissue damage (or, let's add, an objective physical condition conducive to tissue damage if sustained)—so their location is the location of the actual or potential tissue damage. The experiences represent tissue damage nonconceptually in a way analogous to how our visual experiences represent colors of surfaces. If colors are objective physical conditions of surfaces, the way our visual system represents these conditions is such that we can't necessarily conceptualize them as such. So it is no objection to this view that we don't conceptualize what pain experiences represent as tissue damage. How then do we conceptualize them on the basis of how pain experiences represent tissue damage? The confusion and the error found in the commonsense understanding of the truth conditions of sentences like (2) stems from the fact that we *seem* to use the same term (and perhaps the same concept prereflectively) for the damaged condition of bodily tissues as the term we use to report their experiences, namely pain. It might be helpful to distinguish the former use of 'pain' with a subscript, 'pain$_{td}$', reflecting that it refers to the actual or potential tissue damage, to a physical condition of the tissue which is typically the cause of the pain experience, that is, pain$_e$.[3]

Armstrong summarizes the resulting account of the semantics of sentences like (2) thus:

'I have a pain in my hand' may be rendered somewhat as follows: 'It feels to me that a certain sort of disturbance is occurring in my hand, a perception that evokes in me the peremptory desire that the perception should cease' . . .

The force of the word 'feels' in this formula is no more and no less than the force of the word 'feels' in 'My hand feels hot', where this latter sentence is so used that it neither asserts nor excludes my hand being hot in physical reality. . . . Now while there is no distinction between felt pain and physical pain, there is a distinction between *feeling* that there is a certain sort of disturbance in the hand, and there actually *being* such a difference. Normally, of course, the place where there feels to be such a disturbance *is* a place where there actually is a disturbance, but in unusual cases, such as that of the 'phantom limb', or cases of 'referred pain', there feels to be disturbance in a place where there is no such disturbance.

The problem of the 'location' of physical pains is therefore solved in exactly the same way as that of the location of 'transitive' sensations, such as sensations of pressure. . . . [T]he 'location' of the pain is . . . an intentional location. There feels to be a disturbance of a certain sort in the hand, whether or not there is actually such a disturbance.

This account of the location of pain enables us to resolve a troublesome dilemma. Consider the following two statements: 'The pain is in my hand' and 'The pain is in my mind.' Ordinary usage makes us want to assent to the first, while a moment's philosophical reflection makes us want to assent to the second. Yet they seem to be in conflict with each other. But once we see that the location of the pain in the hand is an intentional location, that is, that it is simply the place where a disturbance feels to be, but need not actually be, it is clear that the two statements are perfectly compatible. (Armstrong 1968, 314–316)

So what sentences like (2) in fact attribute in the first instance is not pain$_{td}$ but rather pain$_e$. What I do when I correctly utter (2) is to self-attribute an experience that then attributes a physical condition to a location in my hand. Intuitively, by uttering (2) I am saying something like "I am undergoing an experience which tells me that something untoward is going on in the back of my hand." If so, the fact that nothing untoward is going on in my hand (nothing physically wrong with my hand) doesn't make (2) false. The fact that I can still correctly point to where it really hurts in my hand after hearing from my doctor that nothing is wrong with my hand is explained by reinterpreting what I say and do with that gesture: I am still undergoing an experience that represents my hand as having something wrong with it.

So according to this perceptualist/representationalist account, when we make claims about *where* it hurts (attribute pain to bodily locations), we in fact rescind from committing ourselves to there being anything physically wrong in those locations—even though we normally *expect* to find some physical disorder in them. In other words, when our attention is drawn to those locations and the felt qualities in those locations, we don't *strictly speaking* commit ourselves to find anything in them. What is primary, then, is the experience itself, that is, that we are undergoing a pain experience in the first instance. Only secondarily is our attention drawn to what those experiences point to, and that only in a noncommittal way. Note that, on the proposed analysis of (2), to say that my pain *experience* attributes a physical condition to a bodily location is not to say that *I* do so attribute it. On the contrary, the fact that I leave open whether the attribution is correct shows that I don't. I simply report an experience that tells me something; whether or not I come to believe what it tells me is a matter of factors not to be read into the analysis of what the truth-conditions of (2) are.

Ironically, this representationalist treatment of pain reports is in tension with the general thrust of any perceptual/representational account of experience, because what it reveals is that in pain our immediate interest (epistemic or practical) is in the experiences themselves in the first place, rather than in what objects or conditions these experiences represent. As we noted above, it is the other way around in standard perception—genuine perception where the concepts we are induced to apply apply in the first instance to the objects and conditions that the inducing perceptual experiences represent. We see an apple and identify it as such. We see a discoloration and identify it as such. In pain, what we identify in the first instance as pain is the experience itself—even though our ordinary reports confusingly point to what these experiences represents (assuming the above account). Indeed, the appropriate classification of access to pain makes it a form of introspection, rather than exteroception or perception of extramental environment. Even when we attend to the qualities in a body region that our pain experience represents, the appropriate classification of our activity is introspection. Indeed, Tye himself, repeatedly uses 'introspection' and talks about introspecting the attended qualities even when he argues for the transparency of pain experiences and identifies these qualities with tissue damage and its features. The term 'introspection' is appropriate given what we have said so far, but it is puzzling to see that it's Tye's choice of word—because whatever else introspection may be, I take it that it is a conceptual truth that introspection is a method of access to mental phenomena and not a method of access, like standard perception, to extramental environment. Whatever else the transparency of genuine perceptual experiences involves it must take us to the extramental world in a *committal* way—or roughly equivalently: sentences like (1) must admit an analysis according to which perception is a relation (at least) between the perceiver and the perceived extramental objects or conditions.[4] But what we access in pain is the experience itself and what it represents without a commitment to its veridicality. If pain experiences are a species of genuine perceptual experience whose phenomenal content is wholly constituted by its representational content as strong representationalism demands, then it's a mystery why pain experiences and the conceptual categorization they induce work the way they do, quite differently from standard perception.

This was the worry I raised in my 2001 to which Tye reacts in his essay in this volume by saying that the worry was misplaced. In the two full paragraphs Tye attempts to explain why that I quote below, I couldn't discern an explanation of why the worry was indeed misplaced. Tye writes:

The worry is misplaced. When we introspect our color experiences, we issue reports about the *apparent* colors of things. Here the information flow goes from the awareness of the color red to the application of the concept EXPERIENCE OF RED or the concept LOOKS RED. By being aware of red, I am aware that I am having an experience of red. Correspondingly, when we introspect our pain experiences, we are aware of (among other things) tissue damage and by being aware of those things, we are aware that we are in pain. For these cases, the information flow is of just the same type. Normally, awareness of red is caused by red. Normally awareness of tissue damage is caused by tissue damage. Both awarenesses *of* something external (to the head) bring about an awareness *that* an experience of a certain sort is present. (This volume, pp. 115–116)

From this, it's easy to get the impression that what happens in perceiving red is exactly the same as perceiving tissue damage by feeling it, and the introspective processes involved in both are thus exactly the same. Note that when I perceive red and identify it as such by the application of RED, I don't do so by first introspecting my visual experience representing red. On the contrary, on Tye's view, introspection of visual experience of red requires that one have the concept of red (and the concept of experience); in other words, one must already have the capacity to apply RED to red things via one's experience of red. As Tye says, "introspection of experiences . . . is a *reliable* process that takes awareness *of* qualities represented by the experiences as input and yields awareness *that* a certain kind of experience is present as output" (ibid.). What kind? The kind that represents those qualities. Since the output is a thought, it requires the concepts that represent what the experience represents. Hence you need the concept RED to introspect your visual experiences of red. Since, according to strong representationalism, the phenomenal content of an experience is completely exhausted by its representational content, introspecting the phenomenal content of an experience is tantamount to having a thought as the output of that reliable process about *what* the experience represents qua experience. If you don't have the concepts of the proprietary range of qualities represented by an experience, you are not only conceptually blind to those qualities, you are also blind to the experiences representing those qualities. This is true in all introspection according to Tye's representationalism. Experiencing red is a matter of perceiving red and conceptually identifying it as such in the first place. Its introspection comes later—if at all.

But with pain, it's the other way around. Tye says "when we introspect our pain experiences, we are aware of (among other things) tissue damage and by being aware of those things, we are aware that we are in pain." I don't think this is correct. When we introspect our pain experiences, we are not aware of tissue damage *in the first place*. For one thing, tissue damage is not the kind of thing that we can *introspect*. For another, when we come to notice pain and report it, we come to notice and report the occurrence of an experience in the first place—as Tye himself says: "when I am introspectively aware that I am in pain, the concept I apply is the concept PAIN and not the concept TISSUE DAMAGE. I agree also that the concept PAIN applies to the experience I undergo" (this volume, p. 114).

This being so, when we introspect our pain experiences, we are aware of our experiences in the first place. The first concepts that we are immediately prompted by the pain experience to apply are concepts that apply to that experience (i.e., to pain), not to what the experience represents, that is, to tissue damage and its features—as indicated by the lack of our commitment to the veridicality of what pain experiences represent (assuming that they represent tissue damage) when we report pain in body parts. It is all too easy to be mislead by our confused common practice of attributing pain to body parts that such practices, all by themselves, support strong representationalism or somehow show that pain experiences are transparent in the strong metaphysical sense that Tye intends it. But they show no such thing—without additional assumptions or arguments.

So if "pain perception" is a matter of introspection in the first place, despite being sustained by the same kind of information flow mechanism underlying other genuine perceptual processes, it is legitimate to express worry and ask why pain processing doesn't conform to the norms of standard *perceptual* processing. In other words, why is pain a matter of introspection of an experience in the first place, and only then a perception of tissue damage—if at all? I am yet to see why this worry is misplaced.

All the early (direct realist) perceptual theorists such as Pitcher (1970, 1971) and Armstrong (1962, 1968), among others, recognized the difficulty as a puzzle for perceptual theories of pain, and acknowledged that this puzzling asymmetry between pain and standard perception needs an adequate explanation consistent with their direct realist perceptualism.[5] And they offered an explanation. Their basic idea was that although pain experiences are indeed genuine perceptual experiences, because of their immediate negative affect, the informational value they have about tissue damage is shadowed by a more "peremptory desire that the perception should cease," a desire evoked by those experiences, and thus forcing the cognitive/conceptual system to adjust its utilities by anchoring the locus of immediate conceptual identification to the pain experiences themselves, rather than to what these experiences represent or carry information about, that is, tissue damage—as reflected in our ordinary concept of pain and pain reports. Whether or not this is an adequate explanation is something I cannot discuss in this essay.[6] The point is that there is a legitimate worry here and an adequate representationalist explanation is needed.

It is important to note that I am *not* objecting to the claim that pain experiences are (or involve) perceptual experiences or that their phenomenal content is constituted by their representational content. None of what I have said above shows that these claims are false. In fact I think that pain experiences are genuinely perceptual (as well as affective-emotional) and their phenomenology (except their affective phenomenology—see below) is exhausted by their representational content. Indeed I favor the same kind of representationalist account of pain attribution sentences as Tye does and other perceptual theorists before him did. But what is needed is an adequate account of why there is a such marked difference between obviously perceptual processes (such as vision, hearing, etc.) and pain (as well as other intransitive bodily sensations like itches, tickles, tingles, orgasms, etc.)—especially when (direct realist) perceptualism and strong representationalism about pain crucially predict otherwise.

Finally, I want to comment briefly on Tye's claim that his strong representationalism can capture the negative affective phenomenology of normal pain experiences. In my 2001, I criticized Tye's brief treatment of pain affect that he had presented in his 1995 and 1997. My criticism was that being a pure qualia representationalist Tye could not account for the negative affect of pain experiences in purely representationalist terms and that he was forced to adopt what I called cognitivism about negative affect, which was also the option adopted by early perceptual theorists such as Armstrong and Pitcher. According to cognitivism, the affective qualities of pain experiences are somehow constituted by one's cognitive conative reactions to them, by pain's evoking in one "the peremptory desire that the perception should cease" (Armstrong 1968, p. 314). In his new essay in this volume, Tye distances himself from any such way out and proposes that the affective qualitative content of pain is entirely representational as his pure qualia representationalism demands:

People in pain try to get rid of it or to diminish it. Why? The answer surely is because pain *feels* unpleasant or bad, because it is *experienced* as such. But what exactly is experienced as unpleasant? One's attention, when one feels pain, goes to a place different from the one in which the experience of pain is located. The disturbance that is experienced as unpleasant is located in the bodily location to which one attends (in normal circumstances). People whose pains lack the affective dimension undergo purely sensory, nonevaluative representations of tissue damage of one sort or another in a localized bodily region. Those whose pains are normal experience the same qualities, but now those qualities are experienced by them as unpleasant or bad. It is precisely *because* this is the case that normal subjects have the cognitive reactions to pain they do, reactions such as desiring to stop the pain.

To experience tissue damage as bad is to undergo an experience that represents that damage as bad. Accordingly, in my view, the affective dimension of pain is as much a part of the representational content of pain as the sensory dimension is. (This volume, p. 107)

So the experience I undergo when I feel a jabbing pain in the back of my hand feels the way it does because it represents tissue damage[7] in my hand and it represents it as bad. In other words, it represents tissue damage as having the quality of being bad.

Of course, normally having one's tissue damaged is bad. We also normally think/judge that it is bad. Not only that, experiencing tissue damage (i.e., feeling pain) is bad. We think/judge so too. But none of these truisms (all by themselves) can support the claim that the experience *represents* the damage *as bad*. And so they should not be confused with this last claim. Indeed what does it mean for one's experience to represent tissue damage as bad? What is the quality of the tissue damage itself that is detected or tracked by the experience so that we can say the experience carries *information* about it? What objective quality of tissue damage could be such that its possession by the tissue damage can be detected by sensory experiences such as pain and turn them also into affective–emotional experiences? Note that Tye is eager to insist that sensory experiences have a proper range of sensory qualities they track. It is the representation of these qualities that gives the experiences their peculiar phenomenal content. (This comes out especially clearly in Tye's response to Maund's objection that things can look rusty, wooden, metallic, etc.) Now we learn that among these proprietary qualities our experiences can detect or track are the badness or goodness of those very same qualities. Recall that Tye presents himself as an informational theorist when it comes to the metaphysics of psychosemantics (indeed this is what the "I" in "PANIC" captures).

I am not sure how to understand Tye's proposal. It seems that whatever the badness or goodness of those qualities properly detected by experiences might come to at the end, these second order qualities are simply not the kind of qualities that can be detected or tracked (in the technical information-theoretic sense). One reason for thinking this is Tye's own test: there could be a molecule-by-molecule identical tissue damage that is not bad (not just experienced as bad, but just not bad—if we abstract away from the regular connotations of "damage" and think of it as composed of whatever physical features and their configurations constitute the damage). There doesn't seem to be any natural property of a tissue damage simple and suitable enough to allow itself to be transduced. Surely the burden of proof here is on Tye to show us that pain experiences normally carry information about a certain quality of tissue damage that constitutes its badness. What natural property is that so it can be detected? If it's not a natural property, what reasons are there to think that it's the kind of property whose instantiations can be detected in the information-theoretic sense? We need to be told.

Perhaps, Tye would want to relax his metaphysics of psychosemantics and allow that an experience can come to represent something as bad not in virtue of some sort of informational relation, but rather in virtue of its causal/functional role.[8] Because these experiences play the role they do in the complex mental and behavioral economy of creatures capable of feeling pain, they come to represent the tissue damage they detect as bad. So the sensory phenomenology of pains gets explained by an informational psychosemantics and their affective-emotional phenomenology by a functional role psychosemantics. I think there is something to this proposal. But note that if Tye opts for this way out, then representationalism about affect becomes otiose. It is a needlessly complex story: the functional role of pain experiences makes for a certain (evaluative) representational content, which is then to be reductively identified with the affective phenomenology of pain experiences. Why not simply say that a certain psychofunctional role that pain experiences play itself *constitutes* their affective phenomenology?[9] Indeed, functionalism or psychofunctionalism in the philosophy of mind has always treated pains as paradigm examples of mental states whose phenomenology can be plausibly captured by functionalist proposals. What made this prima facie plausible was pain's affective phenomenology (as opposed to its sensory phenomenology), which is essentially connected with its being an intrinsically motivational state. But once psychofunctionalism is allowed, we don't need to make a detour through representationalism. In fact, it is hard to see how an experience's representing something as bad can be an intrinsic motivator all by itself. To explain how, a representationalist has to advert to additional (learning?) mechanisms to connect representational content with drive and motivation.

On Tye's view even though we don't need to have any concept to have a pain experience with all its normally awful affective phenomenology, we need to have concepts to introspect it and its affective qualities. Recall that according to his official view, one can introspect phenomenal qualities only if one has the concepts denoting what those qualities represent. So the official story about how we come to introspect our pain experiences and their affective qualities must go something like this. Simplifying quite a bit: we are capable of having pain experiences representing tissue damage and representing it as bad; we are capable of applying concepts like PAIN$_{td}$ and BAD (or their equivalent) to the tissue damage and to its badness. We must also be capable of deploying concepts like EXPERIENCE and REPRESENTS (or their equivalent). When all these capacities are in place, our reliable introspection module can be activated to output an introspective judgment to the effect that we experience pain$_{td}$ as bad and know what it is like to have experiences with that content. I am in sympathy with this kind of view of how introspection works when it comes to how we introspect affectively neutral standard exteroceptive experiences.[10] But it's simply not credible that this is the way we come to introspect pain and its qualities (as well as all the other intransitive bodily sensations). Pain, as already discussed, is a form of introspection from the very beginning, not a form of perception in the first place.

Pain and other intransitive bodily sensations pose serious difficulties for (direct realist) perceptual theorists and for strong representationalists. What those of us with naturalistic and representationalist convictions need to understand is that bodily sensations are special: instead of trying to remove every bit of difficulty in the way of smoothly and completely assimilating them to other obviously perceptual processes, we need to make room for a broader but perhaps messier conception of perception that would involve bodily sensations with all their peculiarities. In other words, we need to pay close attention to how bodily sensations work if we are ever going to succeed in understanding how, despite all their peculiarities, they could after all turn out to be (or at least, involve) genuine perception.

Notes

1. There are different ways of reporting pain in body regions. Many involve subtleties for conveying contextual information. But insofar they are genuine reports of pain in body parts following our ordinary concept of pain, they all follow more or less the same semantic pattern in terms of not involving attribution of physical properties or conditions.

2. Indeed, chronic pain syndromes are not restricted to rare cases like phantom-limb pains and referred pains (although the latter are more common than the former). There are more than 1,500 pain clinics in the United States alone, mostly devoted to treating chronic pains, almost all of which are centrally caused pains felt in bodily locations that are not in any pathological conditions. Forty percent of all Americans suffer from chronic pain at least once and usually late in their lives. Indeed as I write this, I am all too painfully aware of my sciatic pain that I have been suffering for the last two years. I know well that my right leg is just fine as I am told by many medical experts—the cause of the pain is the pressure put by a bone perturbation on the sensory nerve coming out of my lower spine.

3. Tye draws a similar distinction, but he adds a puzzling qualification: "The term 'pain,' in one usage, applies to the experience; in another, it applies to the quality represented *insofar as (and only insofar as) it is within the content of a pain experience*. Which quality (or type) is represented? Pain experiences normally track tissue damage. So, tissue damage is the obvious naturalistic candidate for the relevant quality" (this volume, p. 101, my emphasis). This qualification is puzzling in the light of the identification of this quality with tissue damage. It seems to come down to this: the term 'pain' in the second ordinary usage applies to a relation, to the tissue damage *as being experienced by its "owner."* So the tissue damage is not a pain in any sense, but it's what we might call a protopain (cf. Dretske, this volume), a protopain that becomes a pain, in the second sense, when and only when actually being experienced by someone (i.e., actually being felt or experientially represented by its owner through the appropriate somatosensory modality). If this

is what it comes to, then it seems that it is incorrect to claim that there is an ordinary usage of pain with this sense—if the claim is not simply a terminological stipulation or is meant to be part of a substantive theory of what pains are. For to say that there is such a sense is simply to claim that sentences like (2) has an ordinary reading similar to that of sentences like (1), according to which by uttering (2) we sometimes *ordinarily mean* something that implies that

(3) there is a tissue damage in the back of one's hand such that one is feeling it (by somatosensorily experiencing it).

I doubt there is such an *ordinary* sense of 'pain'—see below.

I am also puzzled about the way Tye describes what a pain experience represents, when he says (in the paragraph preceding the above quotation where he seems to be making the same claim) that pain is "also a certain quality or type *insofar as* (and only insofar as) that quality or type is experientially represented." Assuming representationalism about experiences is right, experiences don't just represent a quality or type, they represent them as *instantiated at a time and place* (even if vaguely). Unlike concepts, that is the only way they represent qualities or types—and that is for good reason again: having a sensory experience is having a state that purports to carry *information* in the technical sense of the term. I can entertain a concept in the abstract without attributing it to anything. But I cannot entertain experiences like that (by the way, sensory imagination may be *like* entertaining a concept, but it is certainly *not* experiencing). Having an experience that purports to carry information that a certain quality is instantiated at a place at a time is having an *illusion* or *hallucination* (if it turns out that there is no such quality actually instantiated there and then), that is, having a (sensory) representation that makes an *incorrect report*.

4. It is a little puzzling why Tye thinks that pain experiences are transparent. He correctly points out where our attention is drawn in feeling pain in a bodily location and emphasizes that our behavior also favors the location we feel pain at. But why would anyone be inclined to think that these show the kind of strong transparency of pain experiences Tye favors? Sense-datum theorists, for instance. would have no difficulty in explaining these facts even though they would reject the strong transparency of pain experiences—see Jackson 1977 among others. When we feel pain in a body location L, we quite naturally expect to find something physically wrong there. Call the general condition TD that we expect to find in L—whatever it may turn out to be. If our feeling pain in L were transparent, if it were simply a matter of perceiving TD in L, we would *not expect* to find TD there, we would *find* TD there. In cases where we don't find TD in L, we don't rescind from reporting pain in L. In such cases, we still *correctly* report pain in L. Compare visual hallucinations. I report a pink butterfly flying ahead of me at three o'clock position. My report is based on a visual hallucination. My report is false. Not only that, my report that I *see* a pink butterfly flying ahead at three o'clock is also false. Upon realization and only then do I revert to introspective mode and report correctly that I visually seem or seemed to see a butterfly. Nothing of this kind happens in reporting pain. If we find that there is no TD in L, we do not make any correction, we still keep reporting the situation, correctly, as exactly in the same way as we did the first time. This reveals that a pain report is an introspective report from the very beginning, hence not a perceptual report *in the first place*.

5. Armstrong writes, for instance: "Now it may well be asked why, in this important field of perception [intransitive bodily sensations], ordinary discourse restricts itself to sense-impression statements, instead of making perceptual claims about what is actually going on in our body. Several reasons may be advanced. In the first place, we have seen that the [conative] reactions characteristically evoked by these bodily impressions are determined almost solely by the impression, whether or not it corresponds to physical reality. Since the reactions are an important part of our concept of almost all the intransitive bodily sensations, it is natural to talk about what feels to us to be the case, whether or not it is the case" (Armstrong 1962, p. 125). Cf. Pitcher 1970, p. 385—among other places.

6. In Aydede 2005 and Aydede and Güzeldere 2005 I argue that this sort of explanation is not adequate—at a minimum incomplete. In the second essay, we offer a fuller account in the context of an information–theoretic account of introspection and discuss why pain and other intransitive bodily sensations are important for understanding the nature of phenomenal concepts and how introspection works that utilizes such concepts.

7. In a PANIC way—see Tye's 1995. Presumably my experience represents a myriad of other features—let's simplify and just focus on tissue damage and its badness in what follows.

8. I haven't seen Tye taking this route before. So my guess in the text is purely speculative. But I hope he would find it plausible and consider it for adoption.

9. I defended this kind of representational-cum-psychofunctional account of affectively non-neutral experiences in my 2000, which also found its way to my 2001. For a more elaborate criticism of pure qualia representationalism along these lines, see my unpublished ms. Clark (this volume) defends a similar view. He proposes that the affective aspect of normal pain experiences is completely reducible to pain sensation's hardwired psychofunctional role. But unlike Tye and myself, he thinks that this aspect of pain is not qualitative in the sense of not being identical to a quale.

10. Indeed, in Aydede and Güzeldere 2005, we propose a detailed empirical (but speculative) theory about how we introspect our exteroceptive experiences that relies on information-theoretic principles laid out in Dretske's seminal (1981) work, which, we believe, is in broad agreement with Tye's views.

References

Armstrong, D. M. 1962. *Bodily Sensations*. London: Routledge and Kegan Paul.

———. 1968. *A Materialist Theory of the Mind*. New York: Humanities Press.

Aydede, M. 2000. "An Analysis of Pleasure vis-à-vis Pain." *Philosophy and Phenomenological Research* 61(3): 537–570.

———. 2001. "Naturalism, Introspection, and Direct Realism about Pain." *Consciousness and Emotion* 2(1): 29–73.

————. 2005. "Pain." In *The Stanford Encyclopedia of Philosophy*, ed. E. N. Zalta. Available at http://web.clas.ufl.edu/users/maydede/pain/pain.sep.pdf.

————. Unpublished. "Against Pure Representationalism about Qualia: The Case of Intransitive Bodily Sensations." University of Florida, Department of Philosophy, 2003.

Aydede, M., and G. Güzeldere. 2005. "Cognitive Architecture, Concepts, and Introspection: An Information-Theoretic Solution to the Problem of Phenomenal Consciousness." *Noûs* 39(2): 197–255.

Dretske, F. 1981. *Knowledge and the Flow of Information*. Cambridge, Mass.: MIT Press.

————. 2003. "How Do You Know You Are Not a Zombie?" Pp. 1–13, in *Privileged Access: Philosophical Accounts of Self-Knowledge*, ed. B. Gertler. Hampshire, U.K.: Ashgate.

Jackson, F. 1977. *Perception: A Representative Theory*. Cambridge: Cambridge University Press.

Pitcher, G. 1970. "Pain Perception." *Philosophical Review* 79(3): 368–393.

————. 1971. *A Theory of Perception*. Princeton: Princeton University Press.

Tye, M. 1995. *Ten Problems of Consciousness: A Representational Theory of the Phenomenal Mind*. Cambridge, Mass.: MIT Press.

————. 1997. "A Representational Theory of Pains and Their Phenomenal Character." In *The Nature of Consciousness: Philosophical Debates*, ed. N. Block, O. Flanagan, and G. Güzeldere. Cambridge, Mass.: MIT Press.

6 Bodily Sensations as an Obstacle for Representationism

Ned Block

Representationism,[1] as I use the term, says that the phenomenal character of an experience just is its representational content, where that representational content can itself be understood and characterized without appeal to phenomenal character. Representationists seem to have a harder time handling pain than visual experience. (I say 'seem' because in my view, representationists cannot actually handle either type of experience successfully, but I will put that claim to one side here.) I will argue that Michael Tye's heroic attempt (this volume) at a representationist theory of pain, although ingenious and enlightening, does not adequately come to terms with the root of this difference.

Representationism is in part an attempt to make an account of phenomenal character comport with G. E. Moore's diaphanousness intuition, the idea of which is that when I try to introspect my experience of the redness of the tomato, I only succeed in attending to the color of the tomato itself, and not to any mental feature of the experience. The representationist thinks we can exploit this intuition to explain phenomenal character in nonphenomenal terms. To understand representationism, we need to know what to make of the phrase 'representational content' as applied to an experience. There is no clear *pretheoretical* notion of representational content as applied to an experience, certainly none that will be of use to the representationist. True, I can speak of seeing *that* and seeing *as*, and more generally of experiencing that and experiencing as. Looking at the gas gauge, I can say that I see that the tank is empty (Dretske 1995). And I can say that I experience my wound as a medical emergency. These (and other) pretheoretical ways of thinking of something that could be called the representational content of experience have little to do with phenomenology or with the kind of properties that the representationist takes the phenomenology to constitutively represent. Thus the representationist thesis involves a partially stipulated notion of representational content. This is not, in itself, a criticism, but as I shall argue, there is a problem about how the stipulation should go in the case of pain.

Thus in my view, the dispute between Tye and Colin McGinn over whether pain even *has* representational content is not a dispute about a matter of fact, but a dispute about how to talk. The same applies to Tye's claim that a referred pain (e.g., a pain in the inside of the left arm caused by malfunction in the heart or a pain in the groin caused by malfunction in the kidney) is nonveridical. Pretheoretically, we might (might!) regard such a pain as misleading but not false or inaccurate or nonveridical. We are willing to allow hallucinations in which it seems to us that there is a colored surface in front of us when there is no colored surface that we are seeing, in front or elsewhere. However, we do not acknowledge pain hallucinations, cases where it seems that I have a pain when in fact there is no pain. Tye does not argue for pain hallucinations in which there seems to be a pain when there is no pain at all, but since he does say that referred pain is nonveridical, he must think that a referred pain in the arm is not actually in the arm. Where, then, is it? In the heart? It is not our practice to assign locations to referred pain in this way, so such a claim is at best stipulative.

In the case of representationism about some aspects of visual phenomenology, there is a fairly natural line of stipulation. My color experience represents colors, or color-like properties. (In speaking of colorlike properties, I am alluding to Sydney Shoemaker's "phenomenal properties" [1994, a,b], "appearance properties" [2001], and Michael Thau's [2002] nameless properties.) According to me, there is no obvious candidate for an objectively assessable property that bears to pain experience the same relation that color bears to color experience. But first, let us ask a *prior* question: what in the domain of pain corresponds to the *tomato*, namely, the thing that is red? Is it the chair leg on which I stub my toe (yet again), which could be said to have a painish or painy quality to it in virtue of its tendency to cause pain–experience in certain circumstances, just as the tomato causes the sensation of red in certain circumstances? Is it the stubbed toe itself, which we experience as aching, just as we experience the tomato as red? Or, given the fact of phantom-limb pain, is it the toeish part of the body image rather than the toe itself? None of these seems obviously better than the others.

Once we have stipulated what we mean by the representational content of pain, it is a substantive and nonstipulative question whether the phenomenal character of pain is that stipulated representational content. The stipulative aspect of the issue is reflected in Tye's presentation by the fact that two-thirds of the way through the paper, he has not yet quite stated what he intends to stipulate. He says "What, then, is the phenomenal character of pain?" and considers the possibility that one might say the representational content of pain is a matter of its representing *subjective qualities* of the bodily region in which the pain occurs. He rejects this proposal on the ground that the phenomenal character of a pain in the leg can be present even when there is no such bodily region (as in phantom-limb pain), suggesting instead that "the phenomenal character of pain is representational content of a certain sort, content into which the *experienced qualities* enter" (this volume, emphasis added). The "certain sort" alludes to his view that the relevant contents are nonconceptual, abstract, and poised.

The problem that is worrying me is what these "subjective qualities" or "experienced qualities" are in terms of which Tye characterizes the representational contents of the phenomenal character of pain. (I will use the former phrase and indicate the problem of the obscurity typographically by talking of Subjective Qualities.) Examples of Subjective Qualities in Tye's sense are what we speak of when we describe a pain as sharp, aching, throbbing, or burning. Here is the problem: why don't these Subjective Qualities bring in the very unreduced phenomenality that the representationist is seeking to avoid?

Let me explain via the comparison with Shoemaker's (1994a,b; 2001) version of representationism mentioned above. Shoemaker honors the diaphanousness intuition without the reductionist aspect of representationism. He holds that when one looks at a red tomato, one's experience has a phenomenal character that represents the tomato as having a certain appearance property *and also* as being red, the latter via the former. Each appearance property of an object can be defined in terms of production by it in certain circumstances of a certain phenomenal character of experience. The view is motivated *in part* by the possibility of an inverted spectrum. If Jack and Jill are spectrum inverted, Jack's experience of the tomato represents it both as red *and* as having appearance property A (the former via the latter). Jill's experience represents the tomato as red and as having appearance property A^*. (Jack's experience represents grass as green and A^*, whereas Jill's experience represents grass as green and A.) What determines that Jack's experience represents appearance property A is that it has phenomenal character PC, and A gets its identity (with respect to Jack) from the production of PC in normal viewing conditions. Red can be identified with the production of PC in Jack, PC^* in Jill, and so on. PC is metaphysically more basic than A since PC is what makes it the case that the experience represents A. But A is epistemically more basic than PC in that in perception of colors one is aware of A rather than PC. And in introspection, one is aware that one's experience represents A. Awareness of PC, by contrast, is at least in part theoretical (which I see as a big problem with Shoemaker's view). Shoemaker's view of the relation between phenomenal character and appearance properties has been in flux, but what I think has been constant is something I can agree with, that PC and A are a pair such that each could be defined in terms of the other taken as basic. Or, if the two are defined in terms of one another as a "package deal," with neither as basic, the definition would not capture the difference between the $\langle PC, A \rangle$ pair and the very different $\langle PC^*, A^* \rangle$ pair. Shoemaker's appearance properties are in that sense of a piece with phenomenal characters.

Shoemaker's (2001) view of pain is that pain experiences are perceptions that represent a part of the body as instantiating an appearance property. Such a view is not problematic for Shoemaker since if he is to be called a representationist, his representationism is nonreductionist: he is not attempting to explain phenomenal character in nonphenomenal terms. But if Tye's Subjective Qualities are appearance properties, then Tye cannot be a representationist in the sense that he at least used to endorse, in which phenomenal character is supposed to be explained in nonphenomenal terms.

Does Tye give us any reason to think that his Subjective Qualities are *not* appearance properties in a sense that undermines his (former?) project? Well, if he said that as a matter of empirical fact, these Subjective Qualities turn out to be (aspects of) tissue damage, then I think they could not be taken by him to be appearance properties. But Tye's view is not that Subjective Qualities are features of tissue damage. Rather, what he says is something importantly different, namely that "pain" applies to tissue damage when it is within the content of a pain experience. And it is good that he does not identify Subjective Qualities with aspects of tissue damage, since that identification would be most implausible given that exactly the same tissue damage in the foot can give rise to a more intense pain—or one that is different in other ways—in me than in you because of differences between my fibers leading from the foot to the brain and yours.

The representationist says that when I try to introspect my experience of the stubbed toe, I only succeed in attending to the Subjective Quality of the toe. My question to Tye has been: why think of the Subjective Quality of the toe as like the redness of the tomato rather than like an appearance property of the tomato? Of course, Tye is a representationist about visual experience as well as about pain, so presumably he will reject the question or regard it as a choice between a correct option (red) and a confused option (an appearance property).

To see the difficulty in such a position, we have to recognize that colors are objective in a way that Subjective Qualities are not. There is an appearance–reality distinction for red but not for a Subjective Quality such as achiness. (Aydede [2005] quotes a characterization of pain from the International Association for the Study of Pain that pretty much makes the point that there is no appearance–reality distinction for pain.) The tomato is red whether or not anyone is looking at it, but the achy toe cannot have its Subjective Quality if no one is having pain. That is, there can be unseen red but not unfelt achiness. Indeed, tomatoes would still be red even if there never had been any people or other creatures who could see them. But in a world without pain-feeling creatures, there would be no Subjective Qualities at all, no burning limbs or achy toes. In the case of color, a physicalist theory has some plausibility. For example, colors may be held to be sets of reflectances. This account fits with the idea that there could be colors in a world with no perceivers, since tomatoes could reflect light even if no one was there to see it. But a physicalist account of Subjective Qualities in terms of tissue damage is not remotely plausible, for the reason given above—the Subjective Qualities of a toe depend not only on the tissue damage but on the connection between tissue damage and the brain. Whether something is red can be an objective matter, but whether my toe aches is something others know about only because of my special privileged relation to it. Finally, as Shoemaker (2001) notes, there is a many–one relation between color–appearance properties and color. Looking around the room, I see all four walls as white, but the color–relevant appearance properties are nonetheless different because of differences in lighting. However, there is no corresponding distinction in the domain of pain. Every slight difference in appearance is a difference in Subjective Quality, indicating that Subjective Qualities are mentalistic in a way that colors are not.

That is why bodily sensations have been a challenge for representationism. If the representationist proposes to explain phenomenal character in nonmentalistic and especially nonphenomenal terms, there must be something for the phenomenal character to (constitutively) represent that is *not itself individuated with respect to phenomenal character.* Color is a better bet to pass this test (even if it does not pass in the end) than are Subjective Qualities.

Of course the view of color that I have been presupposing is itself controversial. It may be said that a physicalistic theory of color ignores the fact that what color something has is relative to the perceiver. Colored objects produce slightly different phenomenal characters in different normal observers in normal circumstances, because the various parts of the eye differ among *normal* perceivers—perceivers who can be assumed to perceive correctly—male versus female, young versus old, black versus white (Block 1999). Perhaps color is not objective after all. So perhaps we should say that in a world without perceivers, nothing has colors. Or perhaps we should say that they have *all* colors—each relative to a different possible but nonactual perceiver. And once we have gone that far, we might say instead that there are no colors even in the actual world, rather merely the projection of phenomenal characters onto objects (Boghossian and Velleman 1989, 1991). But these are all views of color that would deprive representationism of its reductionist point. The challenge to Tye is to manage to assimilate Subjective Qualities to color as an objectivist would see color.[2]

Notes

1. Some say 'representationalism,' but I prefer 'representationism.' "Representationism" is shorter and 'representationalism' is ambiguous, being used also to mean the doctrine in epistemology that seeing is mediated by awareness of a representation, namely, indirect or representative realism. As Aydede (2005) notes, representationism is more akin to direct rather than indirect realism, so the ambiguity is confusing. Since we still have a chance for the more rational use of terms, I hope readers will adopt 'representationism.'

2. I am grateful to Sydney Shoemaker for some comments on an earlier draft.

References

Aydede, Murat. 2005. "Pain." In the *Stanford Encyclopedia of Philosophy*, ed. Ed Zalta. Available at http://plato.stanford.edu/.

Block, Ned. 1999. "Sexism, Racism, Ageism, and the Nature of Consciousness." In the special issue of *Philosophical Topics* 26 (1 and 2) on Sydney Shoemaker's work, edited by Richard Moran, Jennifer Whiting, and Alan Sidelle, 1999.

Boghossian, Paul, and David Velleman. 1989. "Color as a Secondary Quality." *Mind* 98: 81–103. Reprinted in Byrne and Hilbert 1997.

———. 1991. "Physicalist Theories of Color." *Philosophical Review* 100: 67–106. Reprinted in Byrne and Hilbert 1997.

Byrne, Alex, and David Hilbert, eds. 1997. *Readings on Color: The Philosophy of Color,* volume 1. Cambridge, Mass.: MIT Press.

Dretske, F. 1995. *Naturalizing the Mind.* Cambridge, Mass.: MIT Press.

Shoemaker, Sydney. 1994a. "Self-Knowledge and Inner Sense, Lecture III: The Phenomenal Character of Experience." *Philosophy and Phenomenological Research* 54(2): 291–314.

———. 1994b. "Phenomenal Character." *Noûs* 28: 21–38.

———. 2001. "Introspection and Phenomenal Character." *Philosophical Topics* 28(2): 247–273. This paper is reprinted with some omissions in David Chalmers, ed., *Philosophy of Mind: Classical and Contemporary Readings*, New York: Oxford University Press, 2002.

Thau, Michael. 2002. *Consciousness and Cognition.* New York: Oxford University Press.

7   Michael Tye on Pain and Representational Content

Barry Maund

Michael Tye argues for two crucial theses:

(1)   that experiences of pain have representational content (essentially);

(2)   that the representational content can be specified in terms of something like damage in parts of the body. (Different types of pain are connected with different types of damage.)

I reject both of these theses. In my view, experiences of pain carry nonconceptual content, but do not represent essentially. Rather they are apt to represent when the subject attends to them. The experiences carry nonconceptual content not only about tissue damage, but about many other qualities as well, including dispositional qualities.

I appreciate that such a view needs argument, which is not appropriate here. However, I think that there are independent reasons to think that Tye's theses are problematical. It is difficult to see how these theses are consistent both with what we know about pains, and other perceptual experiences, and with the package of claims that are central to Tye's treatment of pain, and more generally of perceptual experiences, whereby phenomenal character is explained in terms of representational content (as presented in Tye 2000).

1   Tye on Content and Phenomenal Character

Tye's two theses have to be understood as embedded in a wider framework that contains the following crucial claims:

1. Experiences of pain have phenomenal character: there is something it is like to have the experience.

2A. Experiences of pain carry representational content (essentially).

2B. The representational content of the experience is nonconceptual content that satisfies certain other constraints. This content is specified in terms of concepts that the subject of the experience need not have.

2C. What makes the nonconceptual content representational content is the fact that the experience has intrinsic features of such a type that they stand in the right kind of causal relations to the properties that are specified in the nonconceptual content.

2D. Experiences of pain, like all perceptual experiences, carry all sorts of nonconceptual content about a variety of different qualities. (The reason for this is that nonconceptual content, in the sense of informational content, is defined in terms of causal relations between the state and other qualities. In general, any perceived quality will be causally related to a whole host of other qualities, especially dispositional qualities.)

3. The phenomenal character of pains (and typical perceptual experiences) is to be explained in terms of the representational content, that is, nonconceptual content of the right type, carried by the experience.

4. The experience of pain is something that one undergoes, and is something that does not involve or require the application of concepts. It is, moreover, something that one can undergo without being aware that one is undergoing it. Even if this situation is rare, it happens often enough, as Tye and Dretske and others point out. There is thus a clear distinction between two kinds of state:

• the experience of having a pain, say in the right leg;

• being aware that one is having a pain.

5. My awareness of the phenomenal character of pain (and other experiences) comes about through introspecting my experience. Such a process of introspection takes as input, the experiences of pain, a nonconceptual state, and has as output a conceptual state: awareness that I have an experience with a certain phenomenal character.

With respect to this last thesis, the claim is not that the subject's awareness is of the form "I have an experience with such and such phenomenal character." It is rather of the form: "I have a pulsating pain in my right leg," "I have a sharp pain behind my right eye." Nevertheless, what Tye says the subject is aware of when he or she introspects is the phenomenal character—even if the subject does not think of it as phenomenal character. However, if it is phenomenal character that the subject is aware of, it is nonconceptual content. But how can that be since the output is expressed in terms of conceptual content. The nonconceptual content is specified, by the theorist, in terms of concepts that the subject need not possess, but if so, they cannot be specific to the content of the output of the process of introspection.

A closely related point is this. Tye says that the process of introspection is one that enables the subject to become aware of what it is like to be in pain (or to have other relevant perceptual experiences). But I cannot see that his theory allows him to explain how this can be. The process of introspection has as input, the experience with its nonconceptual content, but the output is a state of awareness-that: I become aware that I am in pain. But this is not to know what it is like to be in pain. It is to know something else: what it is to be a subject who is aware that he or she is in pain.

It is true that Tye gives an account of phenomenal qualities, in Tye 2000, that is meant to show how we have knowledge of the phenomenal qualities, but it does not seem to me that he is any more successful. "In being cognitively aware of our own phenomenal states," he writes, "we subsume those states under phenomenal concepts." These concepts are simple:

They are also, in part, direct, recognitional concepts. For it is part of their characteristic functional role, qua phenomenal concepts, that they enable us to discriminate phenomenal qualities and states directly on the basis of introspection. In having the phenomenal concept, pain for example, I have a simple way of classifying pain that enables me to recognise it via introspection without the use of any associated reference-fixing intermediaries. (Tye 2000, p. 28)

Introspection of the phenomenal character of pain causally triggers in one the application of the concept of pain (under normal functioning of the introspective mechanism). Thereby via the operation of a reliable process, I know that I am in pain. But once again what I become aware of is that I am in pain. I am said to be able to classify the states that I am in, and to distinguish it from other states, that is, I can tell without inference, that my state is of a certain type, but this is not the same as to know what it is like to be the subject of the experience.

2   Tye on Pain and Representational Content

In the light of these comments, let us look more closely at the details of Tye's treatment of pain. If pains are representations, what do they represent, Tye asks? The obvious answer is 'pain.' "After all, if I feel a pain in my leg and I attend to what I am feeling, I attend to a quality that seems to be tokened in my leg, a quality that I strongly dislike" (this volume, p. 100). This view may present something of a problem since pains are commonly thought of as experiences but the answer to that, Tye holds, is to recognize that the term "pain" does double duty, once as a name for an experience, and once as the name for a certain quality or type "insofar as that quality is experientially represented." A pain in the leg represents that a certain quality is tokened in the leg. This answer raises the further question of which quality it is that is represented. Tye's proposal is that the quality is probably a certain kind of tissue damage, though apparently this would need to be determined by scientific research. (It is plausible, however, that a twinge of pain represents a mild, brief case of damage, and throbbing pain, a rapidly pulsing disorder.)

'Pain,' in one important use, he says here, names a certain quality "insofar as that quality is experientially represented." This may suggest that we can distinguish between the quality—tissue damage—and the same quality as it is *experientially represented*, and accordingly that the subject is aware of the tissue damage, not as tissue damage but "as it is represented as being." But I cannot see that Tye can help himself to such a distinction. All it can mean for a quality to be "experientially represented" is for the quality to be part of the nonconceptual content carried by the experience, albeit suitably poised, and so on. Accordingly, if Tye's theory is right, the tissue damage will be part of the nonconceptual representational content, and it will be part of the specification of how the pain is "experientially represented."

Against what I have written, Tye goes on to explain how it is that pains are experienced as unpleasant. On the face of it, this would seem to be an admission that Tye agrees with the sort of theory I have just described as in conflict with his. It would seem not:

> To experience tissue damage as bad is to undergo an experience which represents that damage as bad. Accordingly, in my view the affective dimension of pain is part of the representational content of pain as the sensory dimension is. (This volume, p. 107)

He reiterates that the representational content of pain, as noted earlier, is nonconceptual:

> Admittedly, my talk above of the unpleasantness of pain, of its experienced badness may sound cognitive. But I do not intend that it be understood in this way. It seems to me that the most plausible view is that we are hardwired to experience pain as bad for us from an extremely early age. (This volume, p. 107)

Tye then goes on to describe the experiences of an infant with chocolate:

> The child's gustatory experience represents a certain taste and the child explains that taste as good. The taste is experienced as good by the child in that the child undergoes an overall experience which represents the presence of the taste in the mouth and represents it as good. Intuitively, this is not a cognitive experience. It does not require concepts. (This volume, p. 107)

The point he seems to be making, with this example as well as the example of orgasm, is that these experiences are experiences of something as good or bad but they are not cognitive responses—on the ground that they are not conceptual responses. "One is hardwired by nature to experience things that way."

There are a number of problems here. First of all, I cannot see why we cannot be hardwired to make conceptual responses, especially if the concepts are of the practical-concept variety, as described by John Campbell (1994, pp. 41–47), nor why all cognitive responses need be conceptual. There is a more important point, however. It seems that Tye is presupposing an account of representational content that goes against his avowed thesis. For him, representational content is a species of nonconceptual content, where each type of content is defined in terms of causal relations between the content carrying state and the state that is represented: "Experience represents various features by causally correlating with, or tracking, those features under certain optimal conditions" (Tye 2000, p. 64). What matters, he further explains, is not actual causal history but the causal correlations that *would obtain*, were optimal or normal conditions operative. Tye qualifies this later by adding a clause:

For state S to represent feature F not only must S causally covary with F under optimal conditions but it must also be the case that if there is some other feature G such that were F to fail to covary with G, the causal covarial link between S and F under optimal conditions would still hold but that between S and G would be broken. (Ibid., pp. 139–140)

Given this account of representational content, it is hard to see how Tye can claim that the affective dimension of pain and pleasure can be part of the representational content.

There is, it seems to me, an alternative way of thinking about representational content that can be thought of as nonconceptual. There are two different ways, that is to say, in which there might be representational content that is nonconceptual. One is the sense presupposed by Tye and similar to that of information content, as described by Dretske. Here the nonconceptual content is defined in terms of those features that the experience is causally correlated with. With this sense, the representational content is nonconceptual content that satisfies certain other conditions.

The second sense is that, according to which, the *representing* is nonconceptual. That is to say, the state represents in a nonconceptual way. In this second sense, "nonconceptual content" is short for "nonconceptual *component or ingredient* of content." E. J. Lowe suggests something along these lines when he says, for example, that he takes perceptual sensations to "constitute a wholly nonconceptual component of perceptual experience" (Lowe 1996, p. 101). A similar idea is expressed in Moreland Perkins (1983, p. 5) and in Barry Maund (2003, pp. 149–164). Perhaps the strongest account of this sense has been provided by Jay Rosenberg (1986, pp. 72–88).

Rosenberg has written of how a person's perceptual awareness that is constitutive of a perceptual experience can contain various elements, one of which is a nonconceptual representation. He begins his discussion with a case in which a man, Bruno, in perceiving a bush, mistakes it for a bear. Bruno sees a bush but does not see it as a bush. Instead he sees it as a bear, that is, he takes it to be a bear. The bush is the occasion of Bruno's experience but the concept of "a bush" does not enter the perceptual experience or awareness. As Rosenberg goes on to argue, however, there is more to the bush's being the occasion of Bruno's perceptual awareness than is captured by the observation that it is the cause of his perceptual experience. We are pressed toward the conclusion that somehow the bush needs to be "present in" that awareness and to be, along with the concept *bear*, an actual "part" or "constituent" of that awareness. Yet if we accept that, there is now an apparent contradiction: the bush is an element of the awareness and yet the concept "bush" is no part of it.

The only way to render these claims consistent is evidently to endorse a certain "Cartesian superstition" (as it has been called): that there is an element of perceptual experience which is *both* representationally significant *and* non-conceptual or non-propositional in form; in short, that the bush which is the occasion of Bruno's perceptual awareness is so by virtue of being *non-conceptually represented* in that experience. (Rosenberg 1986, p. 76)

This is the conclusion that Rosenberg is prepared to endorse. The account that he goes on to develop is one in which perceptual awareness of the bush is a state containing an internal counterpart to the bush. Both the bush and the bear are represented in that awareness. A bear is the content of Bruno's awareness by virtue of being conceptually represented in that awareness; the bush is its occasion by virtue of being nonconceptually represented in that awareness.

There are two important features of this way of thinking of nonconceptual representing. One is that it deals with content, thought of in a Fregean way, as having to do with mode of presentation (representation). The second is that the representational content, at least in the case of perception, obtains at the personal level, and has to do with the use the perceiver can make of the perceptual experiences. Visual experiences, for example, only represent insofar as the perceiver has the capacity to use them. There is a variety of ways in which the perceiver can use the visual experiences: one is to have thoughts about them. Another is to use them to have thoughts about physical objects. A more usual way is to use them to guide one in one's behavior—behavior that is targeted onto objects in the environment. In addition, at least on my way of thinking, visual experiences contain representations with phenomenal qualities and we attribute the qualities of the representation to the objects that we target our actions on: the objects we reach for, grasp, approach, move toward, and so on.

My reason for drawing attention to this alternative view, which of course requires a lot of support, is not to criticize Tye, but to indicate that there is an alternative way of thinking of nonconceptual representations, and to suggest that it is more faithful to the phenomenology of pain-experiences and other perceptual experiences.

References

Campbell, J. 1994. *Past, Space, and Self.* Cambridge, Mass.: MIT Press.

Lowe, E. J. 1996. *Subjects of Experience.* Cambridge: Cambridge University Press.

Maund, B. 2003. *Perception.* Chesham, U.K.: Acumen.

Perkins, M. 1983. *Sensing the World.* Indianapolis, Ind.: Hackett.

Rosenberg, J. 1986. *The Thinking Self.* Philadelphia: Temple University Press.

Tye, M. 2000. *Consciousness, Colour, and Content.* Cambridge: Cambridge University Press.

8 In a State of Pain

Paul Noordhof

Michael Tye and I are both representationalists. Nevertheless, we have managed to disagree about the semantic character of 'in' in 'There is a pain in my fingertip' (see Noordhof 2001, 2002; Tye 2002). The first section of my commentary will focus on this disagreement. I will then turn to issues related to the proper characterization of the content of pain experience. Here, perhaps somewhat surprisingly, there seems to be much more agreement between Tye and me. I restrict myself to three points. First, I argue that Tye has not succeeded in providing a decisive consideration against a related theory that takes pains as representationally unmediated objects of pain experiences. Second, I defend Tye against an objection from Murat Aydede. Third, following on from this, I question whether Tye's characterization of the content of pain experience is correct. The fact that there is so much to discuss is a testament to the richness, interest, and exemplary clarity of Tye's work.

1 In

The nature of the first disagreement concerns the proper account of the invalidity of inferences such as:

(1) The pain is in my fingertip.

(2) The fingertip is in my mouth.

Therefore,

(3) The pain is in my mouth.

Tye argued that this inference is invalid because pain contexts are intensional. If we recognize this, he claims, we will see that the 'in' of 'in my fingertip' is to be understood simply spatially. He cited this, at the time, as an important consideration in favor of the representationalist view (e.g., Tye 1995a, pp. 226–228; Tye 1995b, pp. 111–116). He compared the inference of (1) to (3) with

(4)   I want to be in City Hall.

(5)   City Hall is in the ghetto.

Therefore,

(6)   I want to be in the ghetto (Tye 1995a, pp. 226–228 or pp. 331–332; Tye 1995b, p. 11).

In the latter, we have inference failure due to the intensionality of want contexts.

My initial objection was that the analogy breaks down. Whereas, in the second case, appeal to the intensionality of want contexts preserves the spatial reading of 'in,' in the case of pain, it does not. I tried to bring this out by comparing:

(7a)   There is a pain in my mouth such that I feel pain there (false).

(7b)   There is a place in the ghetto such that I want to be there (true).

The fact that my pained finger is in my mouth does not make the pain there. The fact that City Hall is in the ghetto does make the place I want to be in the ghetto. If the 'in' of 'pain in the forefinger' is simply spatial, then (7a) and, for that matter, the following should be true.

(8a)   A pain is in my mouth.

In fact, it is false. Whereas, by comparison (7b) and

(8b)   City Hall is in the ghetto

are true. So, my thought was, there's no *pure* spatial sense of the 'in' in 'pain in the mouth' to preserve.

I made a proposal of my own. When I say I have a pain in my mouth or a pain in my finger, I am describing a state of my mouth or the finger, that of being painful or hurting. Because my mouth and finger have spatial locations, the pain has a spatial location too. Nevertheless it has a spatial location in virtue of being a state of something that does. The 'in' in such formulations is not simply spatial. It has a state-attributing character with spatial implications in the case of spatial things. I compared the case of pain with the following.

(9)   There is a hole in my shoe.

(10)   The shoe is in the box.

Therefore,

(11)   There is a hole in the box.

The claim that there is a hole in my shoe describes a state of the shoe. The box is not in that state just because a shoe is and the shoe is in the box. Therefore, one cannot conclude from (9) and (10), (11).

I am quite prepared to allow that the failure of certain inferences relating to pain may be due to the intensionality of pain experience, for instance:

(12)   I feel that my left foot is painful.

(13)   My left foot is the foot with harder soles.

Therefore,

(14)   I feel that the foot with harder soles is painful.

It is quite conceivable that there may be inference failures because of my proposal and the fact that pain experiences are intensional. One example would be when (1) and (3) are explicitly written as the content of our pain experience. My point is that, even if we allow that pain is intensional, it does not follow that the 'in' of 'pain in the foot' should be taken to be simply spatial and, in fact, it is not.[1]

Tye's attempted rescue of the spatial thesis is to distinguish three spatial senses of 'in.' I list them.

in_1: spatially within

and, in the case of hollow objects,

in_2: spatially within cavity of hollow object

in_3: embedded within cavity-surround of hollow object

Thus, he explains why we cannot conclude that there is a pain in the mouth by saying that the context-determined use of 'in' is 'embedded within the cavity-surround of a hollow object'—in this case the mouth—and by having a finger in one's mouth the pain is not embedded in the surround.

In response, I provided a series of further invalid inferences that sought to question Tye's diagnosis of the situation (Noordhof 2002). In so doing, I had two aims. First, I wanted to supply cases with which his approach might have difficulty. Second, I wanted to underline the relative smoothness of the application of my own approach.

I think I was unsuccessful under the first heading with the diamond case I actually described so I'll set that one aside. Although there are different criteria regarding what constitutes a flaw for different kinds of object, Tye needn't be sensitive to this in the diamond case I gave because his approach can just appeal to the first and third senses of 'in' as he notes (this volume, p. 117). However, I think he does have difficulty with the following.

Suppose a wealthy nobleman is so displeased with the discovery of a flaw in a diamond that he decides to use it as building material for a wall. The following inference would be invalid.

(15)   There is a flaw in the diamond.

(16)   The diamond is in the wall.

Therefore,

(17)   There is a flaw in the wall.

There may be nothing wrong with the wall. The flawed character of the diamond does not make it poor building material and it may contribute exactly as it should to the proper structure of the wall. I take it that (15) would involve sense one of 'in' and that the same sense is at work in both (16) and (17). I'm not quite sure how Tye would develop his spatial proposal further to deal with this type of case. My own proposal deals with it perfectly easily. At a rough first pass, in attributing a flaw to an object, we are attributing a state that explains why the object does not meet with some ideal standard. The state in question can have spatial location in virtue of the spatial location of the object. But the use of 'in' in attributing it is not purely spatial. Of course, if I'm right about this, then Tye's treatment of the original diamond case is incorrect.

Let me now turn to the others I cited. Consider:

(18)   There is a fault in my computer.

(19)   The computer is in my office.

Therefore,

(20)   There is a fault in my office.

Here I was supposing that 'fault' referred to an error *in* software functioning, for instance, in some of the files needed to boot and reboot one's computer. Now I have no doubt that, in fact, an error in functioning can be given a spatial location given that the computer is a spatially extended object and its programs load from a spatially locatable memory. Nevertheless, the 'in' of an error 'in' functioning refers to a particular breakdown in the temporal sequence of the computer's operations. There is little doubt that, in English at least, 'in' can be used in a purely temporal sense, for example, 'He arrived just in time,' 'In the moments after he first heard of her death . . . ,' 'In the middle of the performance, the leading actor lost his wig.' Talk of errors in functioning has some similarities to and may well be an extension of, this usage. If this is what we mean by a fault in the computer, then 'in' does not have a purely spatial meaning. Instead, the spatial location *derives* from the location of the object with the error in functioning.

It is conceivable that Tye can insist that there is a nonderivative spatial sense of 'in' in which a fault can be *in* an object just because it is a feature of its workings. But rather than proliferate another spatial sense of 'in,' we might instead note that the failure of the inference stems from the fact that just because there is an error in functioning in the computer, it doesn't follow that there is an error in functioning of my office. Our attribution of this kind of (malfunctioning) state to the computer does not imply that it should be attributed to the office.

I had a related idea in mind with regard to the following inference.

(21)   There is a tremor in my hand.

(22)   My hand is in my pocket.

Therefore,

(23)   There is a tremor in my pocket.

A tremor is a feature of the activity of a hand. Again, I have no doubt that because a hand is a spatially extended object, it will be possible to provide a location for the tremor even if (as Tye envisages) this might end up being the whole hand. Nevertheless, when we say that there is a tremor in the hand, we don't think of a particular spatial location in the hand whether vague or precise but rather attribute a feature to its active and passive state: to its workings. This is brought out by the fact that the various options for spatial location appear forced. We don't think of a tremor as either vaguely located—for example, as a mountain may be—or as taking up the whole hand. Rather its location seems perfectly precise—the hand—while at the same time not taking it all up but only characterizing properties of its active or inactive state. Other illustrations of the same usage are 'There is a limp in his step' or 'There is a quickness in his hand.' For that reason, I don't think that Tye happily construes the 'in' of (21) as just the first sense: spatially within. It is true that if he insists upon it, then he has an account of the invalidity that appeals to a simply spatial sense of 'in.' But the tremor is not an isolated case of description of activity.

Here is another inference that fails. It illustrates the same point but lends itself even less to Tye's treatment.

(24)   There is a jauntiness in their playing of a piece of music.

(25)   Their playing of a piece of music is in the house.

Therefore,

(26)   There is a jauntiness in the house.

I should make clear that I'm envisaging that (24) describes *how* they are making the music sound and not the fact that they are moving their limbs in a jaunty manner.

So understood, the 'in' of (24) is characterizing a state of musical activity. It is not happily interpreted as saying that the jauntiness is spatially within the played music even if this is understood to mean taking up the entire location of the music (whatever that might mean). It is partly because 'in' has this sense in (24) and (26) that the inference does not follow. Again it is possible for Tye to appeal to a spatial sense of 'in' in which something can be spatially 'in' something else by being a characterization of its workings or activity but again it is unclear why we must proliferate spatial senses of 'in' rather than recognize the point I have been stressing about the characterization of states of spatially located objects.

Perhaps Tye will insist that the failure of (24) to (26) reflects the peculiar character of music. There is no need to appeal to a spatial 'in' deriving from the characterization of an object's workings or activity. He might insist on a similar response in the case of the computer. His point would then be that these uses of 'in' just should not be understood in the same way as the use of 'in' in 'a pain in the foot.'

The following inference failure suggests that this line of response will be unfruitful. It concerns the fact that the bottom step of a staircase is made of springy wood.

(27)   There is a spring in the step.

(28)   The step is in the house.

Therefore,

(29)   There is a spring in the house.

I take it that Tye would say that (27) involves sense one of 'in.' (28) involves the same sense. The house is not a cavity within which a step is present. Part of the structure of the house is the step just like part of the wall was the diamond (if you don't like this, think of a spring in the wall instead). Therefore, (29) should also be a case of sense one. Yet the inference does not follow. It might be right to call an element of a structure springy without calling the structure springy. The explanation of this is (again) that it is not appropriate to attribute a state to an object—as 'in' can do—just because a part of it is in that state. This case does not appeal to some consideration about what would be an ideal state of an object—as the revised diamond case did— which Tye might hold importantly differentiates the other cases from the cases in which he is interested.

My last example was:

(29)   There is a poisonous gas in the spacecraft.

(30)   The spacecraft is in the earth's atmosphere.

Therefore,

(31)   There is a poisonous gas in the earth's atmosphere.

Although I'm not entirely clear, I think that Tye's treatment of (29) to (31) is this. The poisonous gas in the spacecraft is within the cavity surrounded by the atmosphere. Hence sense two of 'in' is in play. Whereas in (31), the third sense of 'embedded in the cavity-surround' is in play. There are three problems with this suggestion. First, Tye is assuming that part of the earth's atmosphere is not in the spacecraft. But at that stage in the flight, this need not be right. Air from the earth may not yet have been replaced by the stores of oxygen and so forth on the spacecraft, in which case, it is not true that the poisonous gas is not 'embedded' in (part of) the earth's atmosphere. Second, and in many ways more important, gases don't embed in other gases. Rather they mingle with other gases. There is a sense of 'in' where x is in y because it is mingled with elements of y. Consider a shoal of fish, shoal A, and another shoal of fish made up of shoal A and shoal B, call it shoal C. It is appropriate to say shoal A is in shoal C. Nevertheless, it seems inappropriate to say shoal A is embedded in shoal C. Embedding conveys the idea that something is in something else without being dispersed. Shoals and gases are not like that. Third, it is not clear why having the poisonous gas in the spacecraft fails to count as being embedded in the atmosphere unless embedding is tacitly being understood as mingling in this case. For instance, consider

(32)   Tanya has a ring in her nose.

(33)   A lock of her lover's hair is in the ring.

Therefore,

(34)   Tanya has a lock of her lover's hair in her nose.

It seems to me that, in the circumstances, the latter inference is valid. I'm not convinced that there are contextual factors that legitimately settle that embedding can mean mingling in the sense required to explain the inference failure of the spacecraft case.

By contrast, my own proposal can explain the inference failure. In saying that the poisonous gas is in the spacecraft we are characterizing a state of the interior of the spacecraft. Just because the spacecraft is in the atmosphere, it does not follow that it is legitimate to characterize a state of the atmosphere in this way. Once more, characterizing the state of a part does not imply the same characterization of the state of the whole.

So let me summarize my first line of objection. I have explained the failure of certain inferences relating to the location of pain in terms that still give pain a spatial location and that appeal to a perfectly standard use of 'in' not restricted to mental contexts—the state-attributing sense. Tye's insistence on a purely spatial reading of 'in' has trouble with other inferences involving the state-attributing sense of 'in' leading to a proliferation of special *purely* spatial senses of 'in.' I suggest that at some point he is going to have to recognize the state-attributing sense. When he does, he will find that he has a neater explanation of both these cases and his original inference involving pain. As I have remarked, this is compatible with allowing that the intensionality of pain experience will have a role to play in explaining inference failure. The point is just that Tye cannot cite as a consideration in favor of representationalism that it enables us to see that the proper reading of 'in' in 'pain in a foot' is purely spatial. It is not.

2   The Content of Pain Experience

According to Tye, "Pain" has a twofold use. It describes a particular mental state— what I shall call the experience of pain—and it describes a particular quality we experience in the body. Tye claims that the obvious naturalistic candidate for this quality is bodily or tissue damage. The following passage sums up his view: "If pains are representations, what do they represent? The obvious answer is pain. . . . Which quality (or type) is represented? Pain experiences normally track tissue damage. So, tissue damage is the obvious naturalistic candidate for the relevant quality" (this volume, pp. 100–101). He holds that our experiences of pain represent pain in something like the same way as the height of the mercury column in a thermometer represents atmospheric pressure (this volume, p. 112). From his previous work, we know that the theory of representation he has in mind is, roughly, causal covariance in optimal circumstances (e.g., Tye 1995b, pp. 100–105).

   Given that this is his position, it is a little surprising to read him remarking about another theory in which pains are real objects of mental states: "When I feel a pain in my chest, for example, the pain really is in my chest even if the feeling of pain is not. That, however, cannot be right" (this volume, p. 103). If pain in the second sense is bodily or tissue damage, then pain may be located in the chest. Moreover, he is quite explicit here that he is considering what should be said about pain and not the feeling of pain. He continues: "The pain, whatever else it is on such a proposal, is a mental object for the feeling. And mental objects cannot exist in chests or legs any more than such objects can exist in walls or tables. Of course, I can feel a pain in a hand (that has recently been amputated) and I can place the stump of my arm against a wall or table. But the pain I feel isn't *in* the wall or table" (ibid., p. 103). I have a couple of comments about this passage.

First, although pains can't be located in walls or tables it seems very natural to locate them in the body, indeed, that's where our experience represents them to be. If mental objects cannot be located in the body, then we should not think of pains as *mental* objects. So characterizing them is not essential to the theory at hand and foisting this upon them is not a sympathetic development of the theory under consideration. Second, if we do allow that pains may be located in bodies, that does not mean that we can't rule out their being located in walls or tables. Once we recognize that the 'in' in 'pain in a phantom limb' has a state–attributing character, no amount of jiggery pokery will impel us to conclude that pain is located in something that cannot hurt. In the absence of a limb, the pain will not be located in the wall. The pain is not represented to have a pure spatial location that, because we have pressed the stump against the wall, lies within the wall. Instead, our experience of pain represents the pain to be a state of the limb and if the limb doesn't exist, the default position is that we have an illusion.

Of course, the default position is not realized if we have a case of referred pain. How might opponents of representationalism think of this? It is compatible with their position that pain experiences are partly representational, namely that part relating to the representation of the location of the pain, in which case the content of pain experiences would be purely predicational. There would be a gap for an instance of the property of pain. If the subject stands in the right relation to this instance, then it will be tied to a predicational element representing the location of the pain. Thus we might analyse 'John has a pain in his left shoulder' as

(35)   R(John, pain, $\langle \ldots$, in the left shoulder$\rangle$).

Here 'pain' stands for an instance of the property of pain, 'R' the appropriate relation between John, the instance of pain and a certain predicational content—$\langle \ldots$, in the left shoulder$\rangle$. This leaves open the possibility that there can be misrepresentations of location. That is what opponents of representationalism might choose to say in cases of referred pain generally and, specifically, for those phantom-limb cases in which the verdict of referred pain is plausible. As far as I can see, Tye has not provided any grounds for rejecting such a theory.

Let me now turn to Tye's characterization of the content of pain experience as concerning tissue or bodily damage and its relationship to the intuitive thought that, standardly, pain hurts.

According to Tye, a bit of bodily or tissue damage hurts if it gives rise to a certain affective-emotional response in us. Murat Aydede questions whether such an approach can be adequate. As Aydede rightly says: "One would think that it is *because* the damage feels so painful, hurts so much, that it causes dislike in me, not the other way round" (Aydede 2001, p. 33). Aydede's concern is that Tye can't capture the *causal* connotation of the 'because.' Tye, in effect, concedes to Aydede that this is correct but questions whether he needs to respect any such intuition (this volume, p. 107).

It seems to me that a proponent of Tye's theory can do rather better than either party allows. First, according to Tye, the affective-emotional response is responsible for representing a particular case of bodily damage as hurting. So we can still capture the intuition that it is something represented as hurting to which we are responding affectively and emotionally. This is not enough by itself to capture the causal intuition but it does serve to explain some of its force. Second, as yet no theory of being painful has been provided. One option is to take it as the disposition to give rise to the affective-emotional response. Different theories of dispositions yield different verdicts as to whether dispositions cause their distinctive manifestations. I think it is by no means clear that, if he were to understand painfulness in this way, he would have to conclude it was inefficacious (see Noordhof 1997 for relevant considerations put in terms of functional properties).

Nevertheless, introducing the disposition of painfulness into the characterization of the phenomenal content of pain experience has its cost. As Barry Maund notes, representationalists need an answer to the objection that a visual experience of bodily damage is phenomenally different to a pain experience of bodily damage. If phenomenal differences are just representational differences, then these two states should have the same phenomenal content. There are various moves that representationalists may make. In his response to Maund, Tye elects to deny that our visual experiences do have bodily damage as part of their nonconceptual representational content.

The problem is that the introduction of painfulness in the content of our pain experience seems to threaten the argumentative force of Tye's response to Maund. Maund suggests that it is appropriate to attribute quite rich phenomenal contents to visual experience, including the fact that objects look rusty, wooden, or metallic. He sees these as of a piece with saying that a particular part of our body may look damaged. Tye denies that objects look to have these properties in the nonconceptual phenomenal sense. His grounds for this are that, intuitively, there may be things that look phenomenally just like these items and yet lack these properties (this volume, p. 113). Of course, this does not by itself suggest that looking rusty, wooden, or metallic are not part of the nonconceptual phenomenal content of our visual experiences. An essential step to this conclusion is that the existence of something that looks just like a rusty object, but is not rusty, reveals a more general phenomenal property that both of these objects have. Otherwise, we would have to stick to a phenomenal content involving rustiness and just conclude that one object was correctly phenomenally represented to be rusty and the other was not.

The difficulty for Tye is that it seems just as conceivable that there may be features of body parts that give rise to an experience of painfulness in the absence of any genuine bodily or tissue damage. So if it is legitimate to rule out bodily or tissue damage from the proper characterization of the nonconceptual phenomenal content of our visual experience, it seems just as plausible to rule it out from the nonconceptual phenomenal content of our pain experience on the same grounds.

Perhaps Tye will reply that it is not conceivable that features of body parts that don't involve damage will be correlated with experiences of pain in optimal circumstances. To this thought, there are two things to say. First, it is quite conceivable that natural selection should have so arranged matters that some false positives in optimal circumstances take place. There will be some kinds of bodily activity that, although they are not bodily or tissue damage, are sufficiently similar to bodily or tissue damage that it is better that the creature whose body it is experiences pain. Second, Tye is prepared to divorce the characterization of "optimal circumstances" from biology to capture intuitions about the consciousness of swamp people. In this case, let us consider a possible world populated by swamp people who have experiences of pain that are not correlated with bodily or tissue damage at all. They may not live long but their brief lives are sufficient to establish that bodily or tissue damage should not figure in the nonconceptual phenomenal content of our experiences of pain if it should not figure in the nonconceptual phenomenal content of our visual experiences of bodily or tissue damage.

So the problem for Tye is that once you allow painfulness to figure in the characterization of the phenomenology of pain experience, it threatens the centrality of bodily or tissue damage.

Note

1. In his response, Michael Tye takes me as supposing that he is defending the thesis that the intensionality of pain entails that the 'in' is spatial. I regret giving this impression, which was certainly not my intention. My point was rather that recognizing that pain is intensional does not enable us to retain a thesis that Tye takes to be antecedently plausible, namely that the 'in' of pain is purely spatial.

References

Aydede, Murat. 2001. "Naturalism, Introspection, and Direct Realism about Pain." *Consciousness and Emotion* 2(1): 29–73.

Maund, Barry. 2002. "Tye on Phenomenal Character and Colour." Web symposium on *Tye*: *Consciousness, Colour, and Content*. Available at http://host.uniroma3.it/progetti/kant/field/tyesymp.htm.

Noordhof, Paul. 1997. "Making the Change, the Functionalist's Way." *British Journal for the Philosophy of Science* 48: 233–250.

———. 2001. "In Pain." *Analysis* 61(2): 95–97.

———. 2002. "More in Pain . . ." *Analysis* 62(2): 153–154.

Tye, Michael. 1995a. "Representationalist Theory of Pains and Their Phenomenal Character." Pp. 223–239, in *Philosophical Perspectives, 9, AI, Connectionism and Philosophical Psychology*, ed. James E. Tomberlin. Reprinted in Ned Block, Owen Flanagan, and Güven Güzeldere, eds., *The Nature of Consciousness*, Cambridge, Mass.: MIT Press, 1997, pp. 329–340.

———. 1995b. *Ten Problems of Consciousness*. Cambridge, Mass.: MIT Press.

———. 2002. "On the Location of a Pain." *Analysis* 72: 150–153.

9   In Defense of Representationalism: Reply to Commentaries

Michael Tye

I am very grateful to Murat Aydede, Ned Block, Barry Maund, and Paul Noordhof for their extensive and insightful comments. Let me begin my reply with some general remarks about the transparency of experience.

Consider first the case in which I am having a visual experience of a red, round surface, as when I am viewing a ripe tomato. If you tell me to attend to aspects or qualities of my experience, I find myself attending to qualities that are not qualities of my experience at all—qualities such as redness and roundness. These are qualities that belong to the object of my experience (assuming that I am not hallucinating so that there really is such an object). In this way, my visual experience is transparent to me. My introspective gaze seems to go right through it to qualities it does not possess. By being aware *of* these qualities, I am aware *that* I am having a visual experience of a red, round surface, but I am not aware *of* the experience itself or its qualities.

This much the sense-datum theorists would have accepted. Their mistake (by my lights) was to suppose that the qualities of which I am aware when I introspect are qualities, not of an ordinary physical surface (if they are qualities of anything), but of an immaterial surface presented to me by the experience (a sense datum common to both veridical perception and cases of illusion and hallucination).

Turning now to the case of pain, I claim that the situation is analogous. If I am feeling a pain in a thumb and you tell me to attend to aspects or qualities of my experience, I find myself attending to my thumb and to qualities that I experience as being instantiated in my thumb, among them a quality that I strongly dislike. This quality is one of which I am aware, and which, by being aware of it, I am aware that I am feeling pain. But it is not a quality of my experience of pain any more than redness is a quality of my experience of red. After all, my experience is not in my thumb; so the quality I experience cannot be a quality of my experience unless my pain experience involves a massive error and somehow projects onto my thumb a quality that does not belong there but instead belongs to the experience.

Maund takes me to hold that whenever I introspect a pain, I am aware of the phenomenal character of my pain experience. He then objects that if it is phenomenal character I am aware of, then, on my view, I am aware of nonconceptual content of a certain sort; and this conflicts with my supposition that introspection of a pain experience is a process that yields as output a *conceptual* state: awareness *that* I have an experience with a certain phenomenal character.

This gets me wrong in two ways. Take the case of belief. If I introspect a belief, in one way of speaking, I am aware of the content of my belief. But this awareness-of is really awareness-that. It is awareness that I have a belief with so-and-so content—likewise, I maintain, with the phenomenal character of experience. For phenomenal character, on the representationalist view, as Maund notes, is a certain sort of content. Second, while my awareness that I am undergoing a pain experience with a certain phenomenal character is itself a higher-level conceptual state, this does not imply that the pain experience is itself a conceptual state any more than my (conceptual) awareness that my body temperature is 98 degrees, say, implies that having a body temperature of 98 degrees is a conceptual state.

Maund says that my being aware via introspection that I am feeling pain does not suffice to know what it is like to feel pain. Rather, it suffices only to know what it is to be a subject who is aware that she or he is in pain. But here we just differ. According to the analysis of knowing what it is like I have elaborated and defended elsewhere (see Tye 2000, ch. 2), knowing what it is like is precisely either being in such a state of awareness-that via introspection or, if such a state is not present, having a cluster of relevant abilities (to remember the experience, to recognize it when it comes again, to imaginatively recreate it to some degree). On this view, once the first disjunct is satisfied, the disjunction as a whole is met and the subject *does* know what it's like to undergo the given experience.

In taking the above position on pain, I am not denying that when I introspect a pain experience, I am aware *of* pain. For in my view, it is perfectly proper to call the quality I dislike and that I experience as being in a certain body part 'pain.' What complicates things is that we also call the experience itself 'pain.' Let us call the second sort of pain 'pain$_E$' (for the experience of pain) and the first sort 'pain$_O$' (for pain as an object or quality experienced in having pain$_E$). Via introspection, then, I maintain, I am aware of pain$_O$ and by being aware of it I am aware that I am in pain$_E$, but I am not aware of pain$_E$.

So, in my view, and contra Aydede, introspection of pain experience really is like introspection of perceptual experiences. We do not "access" the experience itself, as Aydede puts it, when we introspect; we "access" a quality (or qualities) we experience as belonging to a part of the body where we feel pain. But what exactly is the status of these qualities? In particular, what exactly is the status of the quality I am calling 'pain$_O$.' Pain$_E$ is certainly mental, as are all experiences. But what about pain$_O$?

Since pain experiences normally track tissue damage, the obvious naturalistic candidate for the referent of 'pain$_O$' is tissue damage. So, I claim, the quality I dislike, of which I am aware, in feeling pain in a given body part, is tissue damage.[1] And this is the quality of which I am aware when I introspect my experience of pain. To this, Aydede remarks, "When we introspect our pain experiences, we are not aware of tissue damage *in the first place*. For one thing, tissue damage is not the kind of thing we can *introspect*" (this volume, p. 128). I disagree.

Consider again, the case of introspecting the experience of red. The quality, red, of which we are aware, in my view, is a complicated objective, reflectance property.[2] Its nature, of course, is not given to us in introspection or perception. We are not aware of red as being such-and-such a reflectance when we introspect or perceive. Indeed, we are not aware of red under any mode of presentation at all. We are aware of it directly, not via anything else. In this respect, our awareness of the color red is *not* like our awareness of a red surface. In the latter case, in being aware of the surface, we are aware of it *as* having various qualities; but not so in the former. To suppose otherwise is to take the first step down the slippery path that leads to the thesis of revelation (the thesis that the natures of the qualities we experience are fully revealed to us in our experience of them). And the thesis of revelation is a philosophical thesis, not a thesis of commonsense. The intuition we all have in the redness case is that the red surface does not look to have so-and-so reflectance. But it does not follow from this that it looks not to have so-and-so reflectance. Our visual experience simply does not comment on the nature of the color, red. It certainly does not tell us that red is a simple, sensous quality with no hidden nature at all.

The same goes for pain$_O$. Furthermore, just as in the visual case, when we classify our visual experience as one of something red, we bring to bear the concept *red* and not the concept *reflectance R* with respect to what we experience, so too when we classify our experience as one of pain, we bring to bear the concept *pain$_O$* with respect to what we experience and not the concept *tissue damage*. Even so, it can certainly turn out that the nature of pain$_O$ is tissue damage, just as it has turned out (in my view) that the nature of red is a certain kind of reflectance pattern.

Noordhof objects that "it seems conceivable that there may be features of body parts that give rise to an experience of painfulness in the absence of any genuine bodily or tissue damage" (this volume, p. 161) and further that if we are not to suppose that when things look rusty to us, rustiness is part of the representational content that gives the visual experience its phenomenal character (as I say elsewhere); we should not say that tissue damage is part of the representational content of the experience of pain that gives pain its phenomenal character.

To this, I reply again that the concept *pain*$_O$ is not the same as the concept *tissue damage*, so, of course, it is conceivable that pain occurs without tissue damage (indeed, there are actual cases). But that is consistent with the two concepts having the same referent in all metaphysically possible worlds—just as it is conceivable that I am not Michael Tye even though in actual fact the concept *I*, as I use it, and the concept *Michael Tye* pick out the same individual (and do so moreover in all metaphysically possible worlds). The salient difference between the case of rustiness and that of pain$_O$ is that rustiness is not a quality of which I am directly aware when something looks rusty to me. The qualities of which I am directly aware are certain shape, texture, and color qualities that together lead me to believe that the surface is rusty. But rustiness is not itself such a quality. In the case of pain$_O$, the quality of which I am directly aware when I feel pain *is* pain$_O$. That much should be uncontroversial. Pain$_O$ has as its nature tissue damage, I claim, but I am not aware of it as tissue damage; indeed I am not aware of it in itself as anything at all. Parallel points apply to redness. The quality of which I am directly aware in having a visual experience of red is red. But I am not aware of red as so-and-so reflectance. I am not aware of it in itself as anything at all. Since, on my view, the qualities of which I am directly aware in experience are the ones that contribute to phenomenal character—these are all qualities that are nonconceptually represented and suitably poised to bring about cognitive responses—redness and pain$_O$ are phenomenally relevant but rustiness is not.

I must now address a complication I have ignored so far. I have written above as if pain$_O$ is identical with tissue damage. But, as I note in my original paper, this unqualified identity claim is false, on my view. Pain$_O$ is tissue damage *only insofar as* it is represented by pain$_E$. This claim puzzles Aydede, Block, and Maund, and I should have said more to explain what I was getting at.

As Block notes, in a world without experiencers, there can be no pain but there certainly can be tissue damage. I agree with Block that this shows that we should not say that pain$_O$ is tissue damage simpliciter. Instead, we can accommodate Block's point by holding that we apply the term 'pain' to tissue damage only in a certain context—the context provided by tissue damage being represented by a token of pain$_E$. In my view, pain$_O$ is really pain$_O$ for person P at time T; and pain$_O$ for P at T is tissue damage represented by a token pain$_E$ of P at T. So, in a world without experiencers there is no tissue damage represented by a token pain$_E$ and correspondingly no pain. Pain$_O$, thus, is not purely objective. It has the status of an "intentional inexistent."

Relatedly, on my view, if there appears to be pain, there is pain, just as Block asserts. For there appears to be pain just in case there is a pain$_E$ that represents tissue damage. Tissue damage so represented is pain; thus, as soon as there appears to be pain, there is pain. No acknowledgment of special, irreducible phenomenal qualities is needed. I shall return to this point later.

One important claim I made in my essay was that the affective component of pain is an aspect of its representational content, an aspect we capture pretheoretically when we say that pain feels bad. Aydede wonders what it can mean to assert that one's experience represents tissue damage as bad. He asks, "What objective quality of tissue damage could be such that its possession by the tissue damage can be detected by sensory experiences such as pain and turn them into affective–emotional experiences?" (this volume, p. 131). Relatedly, Maund claims that it is hard to see how, on the causal covariational account of representational content, the affective dimension of pain can be part of its content.

As to the general point that badness or aptness to harm is part of the representational content of a (normal) pain experience, we need only look at the parallel between our use of 'feels bad' and our use of such a predicate as 'looks square.' Something can look square without really being square. This is possible since an object looks square via its causing an experience that represents it as square, and misrepresentation is always possible. Likewise, a bodily disturbance can feel bad without really being bad for one. Suppose, for example, that some diseased tissue is damaged with the result that the virus it is harboring dies and the tissue is no longer apt to harm. Here tissue damage (let us suppose) is a disturbance that feels bad (that is, it triggers an experience that represents it as bad) when in fact that disturbance is apt to benefit the organism. Alternatively, consider a taste that tastes good but has negative effects, as with certain poisonous mushrooms. Just as squareness is part of the representational content of the visual experience, so badness and goodness are within the representational contents of the relevant bodily and gustatory experiences.

Badness, that is, aptness to harm, is certainly an objective quality. But how is it represented? The picture I have here likens the case of pain, as far as the representation of badness goes, to the bodily aspect of an emotional experience such as depression. When one feels depressed, one's overall bodily state is not the one present when one is operating in a normal, balanced way without experiencing any mood. There is a shift in the "body landscape," as Damasio (1994) puts it, with the result that functional equilibrium no longer obtains. One senses this departure from functional equilibrium, and in so doing, one feels depressed.[3]

Something like this happens in the case of a normal experience of pain. Damaged tissue releases organic acids called 'prostaglandins.' The release of these prostaglandins has nonlocal effects. For example, they effect blood pressure in a negative way. The ensuing shift in the body landscape, occurring as pain is felt, is not good for the subject. It is a departure for the worse from functional equilibrium. And this the subject experiences. In this way, pain is usually an emotional experience as well as a sensory one.

So far as I can see, a picture of this sort presents no obvious difficulty for a normal tracking or causal covariational account of the representational content of pain. Consider again the case of feeling pain in a thumb. One undergoes an experience that represents that a disturbance is occurring in a thumb, a disturbance that involves damaged tissue and that is bad for one. There are thus three represented aspects: bodily location, tissue damage, and badness (or aptness to harm). An experience of this complex sort is no more problematic for a normal tracking account of representational content than is a visual experience that represents that there is something red and round and situated to the left of the subject.

The relevant sort of representational content for the experience of pain, in my view, is nonconceptual. It is nonconceptual not just in the sense that the subject of the experience need not possess the concepts required to state the correctness conditions for the experience but in the stronger sense that it is content of a kind that *could* not be the content of a thought or a belief. Maund prefers a view of the content of (perceptual) sensations under which there is no difference in kind for the content itself from the case of thought or belief. The difference is rather that perceptual sensations are nonconceptual states whereas thoughts and beliefs are not. On this view, sensations have Fregean contents involving modes of presentation.

I can only say here that I reject this approach. On my view, the relevant nonconceptual contents for pain and for perceptual experiences involves no modes of presentation at all.[4] They are thus coarse-grained in a way that thought contents are not. To suppose otherwise, while still adopting a nonconceptualist line, is to make experiences nonconceptual states having conceptual contents. And that, I have argued elsewhere, is an unhappy combination.[5]

To my comment that I do not intend my talk of the experienced badness of pain or the experienced goodness of certain tastes to be understood cognitively or conceptually, and that we are hardwired by nature to experience some things as good or bad for us, Maund objects that we could be hardwired to make conceptual responses. This is true, of course, but my point was that we do not need to *possess* concepts in order for pain to feel bad to us or chocolate to taste good. One does not undermine this point by noting that we may be disposed by our wiring to make certain conceptual responses. In general, the disposition to apply a concept C in certain circumstances does not suffice to possess *C*. If you have any doubts about this, consider the case of Frank Jackson's Mary. Mary, while in her black and white room, does not possess *phenomenal* concepts of a sort the rest of us exercise in our introspective awareness of experiences of the various hues, since she does not know what it is like to experience the hues. And not knowing this, she does not have any understanding of the relevant phenomenal concepts. So, she is not capable of thinking thoughts into which such phenomenal concepts enter. But Mary in her room *does* have the disposition to exercise those concepts in classifications she makes of how objects appear to her if and when she sees objects with the various hues.

Block's complaint about my representationalist view is broader than Maund's. Block asserts that "there is no clear pretheoretical notion of representational content as applied to an experience, certainly none that will be of use to the representationist" (this volume, p. 137).[6] I find this comment puzzling. Surely, for the subject of an experience, there is a way things subjectively seem to be (internally or externally) just as for the subject of a belief, there is a way things are believed to be. The way things are believed to be is the representational content of the belief. If things are the way they are believed to be, the belief is true or accurate; otherwise it is false. Likewise, the way things subjectively seem to be is the representational content of the experience. If things are the way they seem, the experience is accurate; otherwise it is inaccurate.

Blocks tries to bolster his initial complaint by asserting that "we do not acknowledge pain hallucinations, cases where it seems that I have a pain but in fact there is no pain." He adds that since I say that referred pains are nonveridical, I must think "that a referred pain in the arm is not actually in the arm. Where, then, is it? In the heart?" (this volume, p. 138).

I disagree with Block about pain hallucinations. Consider the case of having a visual experience of red. This experience can certainly be hallucinatory, as Block would grant. But even if the subject is hallucinating, redness itself exists. The subject is aware of the quality, redness, and undergoes an experience that (mis)represents that something in the vicinity of the viewer has it. This is what happens in the case of experiencing a red afterimage, for example. There is no filmy, red spot floating in space, but to the subject it seems that there is. Here the experienced quality, red, has no bearer.

Likewise in the case of phantom-limb pain. Suppose I feel a pain in a leg that no longer exists. It seems to me that I have a pain, and I am conscious of a quality it is correct to call 'pain' ($pain_0$ from earlier). But that quality has no bearer. My experience represents that there is a tissue damage in a leg, or so I say. But there is no tissue damage and no leg. My pain experience is a hallucination even though I am certainly conscious of pain.

In the above example of phantom-limb pain, exportation of the expression 'a leg' with respect to the report "I feel a pain in a leg" is not safe and neither is exportation of 'a pain.' For in this case, it is not true that a pain is such that I feel it in a leg any more than it is true that a leg is such that I feel a pain in it. To suppose otherwise is to take a position that leads to some embarrassing questions: for example, could that very pain I am feeling in a leg be felt by someone else? Could that very pain exist without anyone to feel it? Could I feel that very pain in another part of my body? I take it that these questions are pseudo-questions. And the representational account of pain reveals nicely just why and how they are.

The situation in the case of phantom-limb pain is rather like that of my having a pictorial representation of a horse (though no horse in particular). Given that I have such a picture, it does not follow that a horse is such that I have a picture of it. Nor is it sensible to ask whether someone else could have a picture of that very horse or whether it—that horse—could exist without any picture of it. Likewise with the phantom-limb pain.

Block asks whether a referred pain in the arm is really in the heart. My answer is that referred pains are not objects that their subjects feel and so the question of where they are located does not arise. Let me explain. An experience of arm pain, I claim, is an experience with representational content. What is the content? My answer is that it is content that is existential in character as follows: it represents that *there is* tissue damage in an arm. This content does not involve a token of tissue damage. It is not object-involving in this way. So, in the case of referred pain, it is not that the experience represents a tissue damage token (a token of pain$_O$) that is really in the heart *as* being in an arm. The experience does not represent a tissue damage token at all. It is inaccurate since it carries the (mis)information that there is some tissue damage in the arm, and there is no tissue damage there. The trouble lies elsewhere.

Block claims that it is implausible to hold that the subjective qualities we experience in undergoing pains of various sorts can be identified with aspects of tissue damage, since, he says, the same tissue damage can give rise to pains that feel different to different people in virtue of differences in the fibers leading from the bodily regions in which the pains are felt to the brain. I have already denied that the qualities we experience in feeling pains should be identified with aspects of tissue damage without any qualification. But I do not find Block's reason here for rejecting such identities convincing. After all, the same reflectance pattern can give rise to different experiences of color in different perceivers in virtue of differences in the retina and the optic nerve in different subjects. But this does not show that colors are not reflectance patterns (or so I would say). What it shows is simply that where the color experiences differ with respect to the color of the same surface, not all of them are accurate. This raises the further question admittedly of how we can tell who gets the color right. But there is no deep problem for color objectivism here.

To see this, imagine that in a country in Latin America after a revolution, car tires are replaced as they wear out in a haphazard way and that at some given time in this country there are 47 cars traveling on the road, all with replaced tires, and all going exactly 30 mph. The speedometers in these cars read different things: some read 30, some read 31, some read 37, some read 28, and so on. Is there any way of telling which of the 47 speedometers is representing accurately actual car speed just by looking at the speedometers? Of course not. The speedometers were designed for use with tires of a certain size. They give an accurate readout of the speed just in case they are operating as they were designed to operate with tires of the appropriate size.

Similarly, in the case of color. The color experience system in the brain was designed by evolution for use with an optic nerve and retina meeting certain further conditions. There is no way of telling who, among present-day actual color perceivers, has the retinal apparatus and optic nerve meeting the historical design specifications (Dretske 1995). But whoever does gets the colors of things right in experience. That people of different ages and skin color and gender should be slightly more or less accurate in their perceptions of the real colors of things in virtue of retinal differences seems to me no more problematic for an objectivist view of color than is the fact that people of different ages and skin color and gender are slightly better or worse at depth perception for the supposition that distance-away is objective.

In my view, then, there is a greater similarity between pain experience and color experience than Block acknowledges. But, let me stress, there is an important difference. Colors are identical with certain reflectance patterns *period*. Pain is not identical with tissue damage *period*. However, this difference does not require us to suppose that in the pain case, there are special, subjective qualities of a sort not found in an objectivist story about color.

Block makes two further points in connection with the disanalogy he sees between the color case and that of pain. He claims that "whether something is red can be an objective matter, but whether my toe aches is something others know about only because of my special privileged relation to it" (this volume, p. 141). He also claims that there is a many–one relation between color-appearance properties and color, and he cites the example of looking around a room, seeing all four walls as white, and yet experiencing different color-experience properties because of differences in lighting. According to Block, there is no corresponding distinction in the domain of pain. He comments, "Every slight difference in appearance is a difference in Subjective Quality, indicating that Subjective Qualities are mentalistic in a way that colors are not" (this volume, p. 141).

I have no firm grasp on the supposed 'color appearance properties' to which Block (and Shoemaker) advert. That is, I cannot find any such properties in my experience. This is a big topic, of course, and one I cannot pursue here. Suffice it to say that I believe that inverted spectra can be handled without reference to such qualities[7] and inverted spectrum cases are the main motivation for their introduction. As for the example of the room whose walls look white and yet still look different, my account of this is straightforward. The walls all look white to me but they look illuminated to different degrees. Color experiences represent not only color but also level of illumination, and level of illumination is objective just as color is.

In the pain case, felt differences among pains require represented differences (as in the color case). So, we must suppose that pain experiences are all alike in representing tissue damage but different in representing different aspects of tissue damage or other qualities associated with such damage as well as different bodily locations. In my original paper, I gave various examples of these representational differences.

Block is right to say that whether my toe aches is subjective in a way that something's being red is not. But this can easily be explained by my view. The achiness of a toe is partly a matter of tissue damage and partly a matter of locational indeterminacy but only insofar as these are represented by a pain experience. So, for example, my toes aches just in case I undergo a pain experience that represents that there is gradually beginning tissue damage (or some closely allied disturbance) in a toe that increases in severity and slowly fades away, extending across a volume of the toe (a volume that is not represented as precise or sharply bounded). So, for others to know that my toe aches, they have to know that I am having a pain experience with the specific representational features just mentioned; and that requires them to know about my psychological state. This makes achiness subjective in a way that redness is not; but the subjectivity is no threat to the representationalist view. If Block now responds that a pain experience just is an experience with his special subjective qualities, so I am sneaking in phenomenality through the back door, I reply that, according to my theory, experiences have no such qualities. Pain experiences are states having poised, abstract, nonconceptual representational contents of an existential sort into which the type, tissue damage, enters. What makes such states experiences of pain just is their having such a content.[8]

I come finally to the issues raised by Noordhof about pain location. Noordhof summarizes his main contention by saying that even if pain contexts are intensional, "it does not follow that the 'in' of 'pain in the foot' should be taken to be simply spatial and, in fact, it is not" (this volume, p. 153). Here I agree with the first part of Noord-hof's assertion. My contention in my earlier exchange with him was that the inva-lidity of the inference from

I feel a pain in my finger

and

My finger is in my mouth

to

I feel a pain in my mouth

provides no good reason to deny that the 'in' in these statements is *not* spatial. In par-ticular the invalidity of this inference does not require us to introduce a special, causal sense for the term 'in' in pain contexts. We can account for the invalidity of the infer-ence (a) by acknowledging that there is more than one standard spatial sense of the term 'in,' (b) by noting that the spatial sense of 'in' operative in the first premise is different from that operative in the second, and (c) by claiming that the contexts operative in the first premise and the conclusion are intensional. Step (c) is needed, since without it a spatial reading of 'in' in pain statements prima facie requires the admission that necessarily whenever I feel a pain in a finger, there is a pain spatially in my finger, and that, I held (and continue to hold), is absurd.

Later in his commentary, Noordhof remarks, "The point is just that Tye cannot cite as a consideration in favor of representationalism that it enables us to see that a proper reading of 'in' in 'pain in a foot' is purely spatial. It is not" (this volume, p. 158). But this was not *my* contention. I wanted to insist that we can rest content with an ordi-nary spatial reading of 'in' in this context, if we adopt representationalism. I did not claim that representationalism *shows* that 'in' in this context is spatial. However, I continue to think that it is, and it seems to me that Noordhof has supplied no good reason for supposing otherwise.

Admittedly, Noordhof gives a variety of examples designed to demonstrate that 'in,' as used elsewhere, is sometimes not spatial. But this seems to me to carry no weight unless it can be made plausible first that a nonspatial sense of 'in' is applicable in the case of the specimen inference above and second that it accounts for the invalidity of that inference without involving us in a plainly unacceptable metaphysics. Some of Noordhof's examples, such as the one involving jauntiness in the playing of a piece of music and the one involving a spring in the step, seem to me clearly irrelevant, since whatever else is true of pains, they can be counted as when, for example, one can have two (or more) pains in a leg. But the same is not true for jauntiness and springiness. There cannot be two jauntinesses in the playing of a piece of music or two springs in the step.

Noordhof's view is that 'in' in pain contexts has a state-attributing character. When we say, for example, that John feels a pain in a toe, we are asserting that John's toe tokens the state of pain. But then what about phantom-limb pain? Here Noordhof says that "the experience of pain represents the pain to be a state of the limb" (this volume, p. 159). So in this case we have a representational proposal. In the case of referred pain, a third proposal seems to be floated, namely that a three-place relation R obtains between John, an instance of pain, and a predicational locational content.

This appears to be a radically disjunctivist position and for that reason alone I find it unacceptable. But leaving this to one side, let me conclude by saying that, in my view, it is simply false that (in general) if I feel a pain in a leg, my leg tokens the state of pain, even if I have a leg and my pain experience is accurate. Pain here is not a state of my leg. If it is a state at all, it is a state of a spatial part of my leg, namely the damaged part. Accordingly, even on a state analysis, 'in' in 'pain in a leg' is spatial. Furthermore, the only plausible candidate, it seems to me, for the relation R in the case of referred pain is a representational one. But it will not be a representational relation connecting John to a concrete pain object, an instance of pain. For that then leaves open the possibility that someone else could feel that very instance of pain or that that very instance of pain could exist even if there were no one around to feel it. Instead, the representational relation will relate John to a content that is existential in nature, as indicated earlier.

Notes

1. I shall qualify this shortly, but for the moment the qualification does not matter.

2. For more here, see Bradley and Tye 2002; Byrne and Hilbert 2003.

3. This is to oversimplify, since there is an external aspect to the experience of depression. For present purposes, that can be ignored. See my forthcoming.

4. I do not wish to deny that perceptual experiences often have conceptual contents as well as nonconceptual ones. As far as perceptual content goes, I am a pluralist.

5. For an extended discussion of the nature of the nonconceptual content of experience, see my 2005.

6. I prefer to use the term 'representationalism' rather than 'representationism,' as Block does, largely because the latter term is an abomination. As Block notes, 'representationalism' is also used for another very different position in the history of philosophy; but it is hard to believe

that anyone interested in the current debate about the nature of consciousness could confuse the two positions.

7. See my 2000.

8. So, on my view, it is a necessary truth that pain experiences represent tissue damage. This truth, however, is not a priori.

References

Bradley, P., and M. Tye. 2002. "Of Colors, Kestrels, Caterpillars, and Leaves." *Journal of Philosophy* 98: 469–487.

Byrne, A., and D. Hilbert. 2003. "Color Realism and Color Science." *Behavioral and Brain Sciences* 26: 3–21.

Damasio, A. 1994. *Descartes' Error*. New York: G. P. Putnam's Sons.

Dretske, F. 1995. *Naturalizing the Mind*. Cambridge, Mass.: MIT Press.

Tye, M. 2000. *Consciousness, Color, and Content*. Cambridge, Mass.: MIT Press.

———. 2005. "Nonconceptual Content, Richness, and Fineness of Grain." In *Perceptual Experience*, ed. T. Gendler and J. Hawthorne. Oxford: Oxford University Press.

———. Forthcoming. "The Experience of Emotion: An Intentionalist Theory."

10 Painfulness Is Not a Quale

Austen Clark

When you suffer a pain are you suffering a sensation? An emotion? An aversion? Pain typically has all three components, and others too. There is indeed a distinct sensory system devoted to pain, with its own nociceptors and pathways. As a species of somesthesis, pain has a distinctive sensory organization and its own special sensory qualities. I think it is fair to call it a distinct sensory modality, devoted to nociceptive somesthetic discrimination. But the typical pain kicks off other processes too. For one it can grab your attention in a distinctive way, alerting you to its presence and sometimes obliging you to focus attention on the damaged member. Intense pain can eliminate your ability to think about anything else. Pain typically has direct and immediate motivational consequences: one wants it to stop, has an incentive to do whatever one can to reduce it, and is gratified by its termination. As these desires and motives collide with neural reality, emotional components of mental anguish, anxiety, and dread arise. The suffering involved in suffering from pain has multiple strands: it is not just the painfulness of the sensation, or the frustration of the desire that it end, but also the anguish over the possibility that it will never end, and the impossibility, if the pain is sufficiently intense, of focusing one's attention on anything else.

The ordinary word 'pain', then, points dimly to a process that has multiple components. What is more, these components can be dissociated from one another. For example, nociceptive input is poorly correlated with the sensation of pain: phantom pain, causalgia, and neuralgia yield intense pain to minor or nonexistent stimuli; while stress-induced analgesia can block pain sensation in a soldier or athlete who has had major trauma. Further, the sensory aspect of pain can be divorced from its typical motivational consequences. There are well known reports of patients who have had frontal lobotomy or who have been given various opiate analgesics and who say they still feel pain, but that it no longer bothers them. It is no longer hurtful or distressing. (Sever the cingulum—a fiber bundle from the cingulate cortex to the limbic system, about the thickness of a telephone cord—and one can get the same effect.)

Philosophers have drawn various conclusions from the premise that pain has multiple components that can be dissociated from one another. The most famous is that our ordinary term 'pain' has no reference; nothing out there satisfies the conditions necessary for there to exist something you could accurately call 'pain'. To ask "does that hurt?" is, according to one such proponent, to ask an ill-formed question (Hardcastle 1999, p. 146), but this is precisely the question on which I wish to focus. If pain has multiple dissociable components, what sense if any can we make of the commonsense notion that sometimes these episodes hurt? They can be awful, painful, bad. They typically have an aversive, nasty, to-be-avoided quality. If pain fractionates into multiple dissociable components, what makes the congeries bad? Wherein lies the painfulness of pain?

On this more localized question philosophers have drawn a variety of conclusions. Some describe the hurtfulness of pain as one of its intrinsic qualities, sitting firmly on the sensory side of things. It has been identified as the essential property of pain, and the one by which the reference of the term is fixed. Some moral realists have argued what makes pain bad is intrinsic to pain: the hurtful, painful, to-be-avoided character of pain is an intrinsic property of the mental state, and it provides a reason to avoid the state. I will argue against all of these positions. The argument has two steps. First I will describe what painfulness could not be. Then I will describe something that it could be.

I

In order to show that painfulness is not a quale, it is vital first to accept some stipulations as to what it is to *be* a quale. I am going to accept, for the sake of argument, two stipulations that are common to all of those who think qualia pose a problem for materialist or functionalist theories of mind. That is, for the sake of argument, I am going to *accept* these stipulations as providing partial definitions for what we mean by 'quale.' (It is only then that I can show that painfulness does *not* satisfy them.) They are:

1. qualia are properties that are (somehow) instantiated in various sensory episodes; and

2. that in virtue of which two sensory episodes instantiate the same particular quale cannot be defined in any functional or behavioral terms.

The first can be interpreted in various ways whose differences are not critical to this argument.[1] The second is commonly expressed by saying that "qualia cannot be functionally defined" or that "qualia are distinct from any functional properties or behavioral dispositions." More precisely, there is no specification of functional roles or behavioral dispositions whose satisfaction by two episodes guarantees that both episodes are instances of the same particular quale, or whose satisfaction is required in order for two instances to be instances of the same quale.[2] That same quale might, on other occasions, serve different roles. For example, we currently have the disposition to stop at red lights at intersections, but we cannot define the qualitative character of the sensation we have at such junctures in terms of its role in any such dispositions, since it is readily conceivable that those same dispositions could be engaged (under a different traffic authority) by sensations whose qualitative character corresponds to that of seeing something green. The quale can have various causes and effects, but (according to our stipulation) that quale cannot successfully be identified in terms of such causes and effects.

Stipulation (2) is a powerful one, since it rules out the possibility of successfully identifying qualia using *any* functional or behavioral terms. We need somehow to comprehend the compass of what is meant by "any" functional term. For the sake of argument this should be read as broadly as possible: take any theory offered as an explanation for some psychological phenomenon. The theory can be formulated in any idiom one pleases: the theoretical terms can be biological, psychophysical, neuropsychological, cognitive, computational, representational, commonsense, or what have you. Apply the Ramsey–Lewis technique for defining the theoretical terms of that theory, and one can derive what are called "Ramsey functional correlates" for those theoretical terms, relating them to one another and directly or indirectly to observations (see Block 1980). We can call the Ramsey functional correlate "behavioral" if it includes any terms that can be satisfied by observations of behavior. Then stipulation (2) implies the following claim. Take any set of Ramsey functional correlates (including behavioral correlates) from any psychological theory: two sensory episodes can equally satisfy all such correlates yet fail to instantiate the same particular quale. So sameness of qualitative character cannot be thus defined.

An implication is that there can be no relational specification whose satisfaction makes the difference between a particular quale's being present and its being absent. Put another way, if the presence or absence of some property Q depends essentially upon its putative instances satisfying or failing to satisfy some set of relations, then Q is not the sort of thing that our stipulations stipulate qualia to be. The argument above uses 'functional term' to include any Ramsey correlate for *any* term that might help to explain psychological phenomena. Ramsey correlates provide "definitions" by specifying the relations that the terms have to one another, to stimulus inputs, and to behavioral outputs. If we had a relational specification for Q, whatever it is, it would be grist for the mill of the Ramsey–Lewis technique. The relata will be named with terms that are either theoretical or observational, so any such specification would eventually be disgorged, as a Ramsey correlate. Hence it follows from stipulation (2) that relational characterizations must fail to identify that in virtue of which two sensory episodes are episodes of the same quale. Qualia elude such characterization.

Here's an intuitive image for this idea. Let us construct the finest possible sieve we can, using all the psychological distinctions available to us from any vocabulary that might be of use. Observational distinctions of course are all allowed, but any theoretical term connected in some way to some observations might contribute to the project, adding some distinction somewhere. Under this criterion we would find ourselves adding vast chunks of neurobiology, biochemistry, physics, and other disciplines as well. When the sieve is finished we run two sensory episodes through it. The idea is that they might both drop through in exactly the same place—the sieve fails to differentiate them—even though they manifest different qualia. The qualitative difference between them is not captured *anywhere* in that vast structure; they sail right through, completely untouched.[3]

Stipulation (2) drives us into a rather "narrow" reading for what constitutes "a" quale, and for later developments we need to spell out this implication. If there are qualia, then there must be such a thing as "a" quale. What then constitutes "a" quale? How are we to determine whether we have one of them or more than one?

If we conjoin our two stipulations, we get a quick consequence: one quale can encompass no more than a fully determinate sensible quality. A fully determinate sensible quality is one that is as distinguished as completely as it possibly can be from all other sensible qualities. For suppose we allow that "a" quale is not a fully determinate sensible quality, but instead has some determinable components. Then that quale would have some componential structure that could be analyzed in relational terms. Discrimination data could tease apart the relations between the sensible qualities that compose the quale. As this contradicts the second stipulation, banning such relational differentiation, we must suppose that a quale encompasses nothing more than a single fully determinate sensible quality.

By 'fully determinate' I mean that no possible discrimination data could distinguish different instances of the supposed quale. Start with the entirety of all possible pairwise discrimination tests: stimuli that present the same quale are entirely indiscriminable from one another. More strongly, one of them can be discriminable from some third stimulus if and only if the other is as well. So stimuli presenting the same quale must not only match one another; the sets of stimuli each matches must also be identical. If they pass this test, they are what one can call "globally" indiscriminable: no distinction between them can be found anywhere in the realm of discriminability, including matches and relative similarities with *other* stimuli.

A paradigm example of a quale is, then, a fully determinate shade of color, where a "fully determinate shade" is one all of whose instances are globally indiscriminable from one another, yet each of which is discriminable from instances of any other shade. These are *points* in quality space. There is no qualitative difference, at all, between their instances.[4]

Now pains are very often taken to be another paradigm example—perhaps *the* other paradigm example—of states that manifest qualia. I have no wish to deny that episodes of pain typically have some sort of sensory character. But I do want to deny that what makes these episodes *painful* is a quale. Painfulness is not, and could not be, a quale. Pains have a sensory character, but it is not their sensory character—specifically, not their qualitative character—that makes them so awful. So as paradigm examples of qualia they leave something to be desired.

II

If we abide by the stipulations above, we can identify the qualitative character of pain in a fashion that is sufficiently precise to distinguish it from other aspects that are *not* qualitative.

It helps to note, first, that pain seems to be subserved by neuroanatomical machinery that is in many respects similar to that of other sensory modalities. It has its own receptors, dedicated pathways, and central loci. We have a fast, epicritic system, that uses myelinated fibers; and a slower, protopathic system that uses the unmyelinated and well-known C-fibers. The epicritic system ascends the spinal cord in at least three lateral pathways, and like many other sensory systems has its first central synapses in thalamic nuclei, which project to somatosensory cortical areas. The slow pain system proceeds up the spinal cord in at least three different medial pathways. A fact that will become critical later on is that several of these tracts synapse first not in the thalamus, but in various limbic structures in the midbrain and pons.[5]

As interesting as the neural machinery is the phenomenology it supports. Consider all the different ways in which painful bodily conditions can be perceived to be similar or different. Across occasions we sense differences in apparent locations and differences in qualities that appear at those locations. Here there is a dull cramp; there, a sharp stabbing pain. Or: here there is a burning, tearing feeling; later, at the same place, one senses only a dull ache. The variety of features that appear in painful episodes is rather astounding; here is an ordering provided by Melzack and Wall (1983, pp. 58–59):

temporal: flickering, quivering, pulsing, throbbing, beating, pounding

spatial: jumping, flashing, shooting

incisive pressure: pricking, boring, drilling, stabbing, lancinating, sharp, cutting, lacerating, splitting

constrictive pressure: tender, pinching, pressing, gnawing, cramping, crushing

traction pressure: tugging, pulling, rasping, taut, wrenching, tearing

thermal: hot, burning, scalding, searing

brightness: tingling, itchy, smarting, stinging

dullness: dull, sore, aching, heavy

All these terms characterize some particular aspect of the appearance of a portion of one's body during a painful episode. Each must have an apparent location; a pricking pain that does not seem to be anywhere would not be a pricking pain. We can treat them in a relatively straightforward way as features that characterize the appearance of a portion of one's body. "Appearance of a portion of one's body" must be read opaquely, so that it applies even to cases of phantom pain, in which in fact there *is* no portion of one's body at the place where the pain appears to be located. So these are phenomenal properties, or, to use the equivalent term in neuropsychology, sensory features. They appear in, appear to characterize, are attributed to, what appear to be regions or volumes of one's own body. Such locations are, obviously enough, phenomenal locations. Usually the apparent location suffices to pick out a real one (the wrenching feeling apparently in your abdomen directs your attention to your real abdomen), but sometimes they do not.

To detail the full sensory content of such an episode, one can proceed as follows. We need to canvas the content of all the similarities and differences one can possibly sense among episodes of pain. This may or may not be a tractable task. If we are lucky the features listed above arrange themselves in incompatibility groups, or "contrary ranges": axes along which just one such feature can appear to characterize a minimally discriminable location at a minimally discriminable time. There will be a number of distinct but independent contrary ranges. Specifying a value on one says nothing at all about the other, and so specifying the full sensory content requires a specification of a value on each of the independent axes along which different pains can be sensed to resemble or differ. But, again if we are lucky, some of the contrary ranges will turn out not to be independent of the others, but instead to redescribe content already captured. Such axes may be linguistically distinct, but are phenomenologically redundant. Their values are some function of values already specified.

Once you have specified a value on all the independent axes of variation of sensed similarities and differences, these all fall out as freebies.

The qualia of painful episodes are, like all sensory qualia, fully determinate sensible qualities. So the qualia of painful episodes will correspond to points in this somesthetic quality space. The contrast between a "flickering" and a "throbbing" pain is a prototypical qualitative contrast, though doubtless the English words pick out vast swaths of discriminably different somesthetic qualities. A quale can contain no qualitative variations within itself; qualia are the *points* that make up this space.

So much for the phenomenal properties; we need to apply a similar strategy to their sensed locations. Here too one can imagine cataloging the capacity of all possible discriminations among felt locations. Presumably this space is three dimensional (recall William James and his voluminous pains), but its metric is defined not by physical space but by the sensitivity of spatial discrimination. To what extent can one sense a difference between a pricking pain here and a pricking pain there? The distance needed between "here" and "there" before our hapless subject can feel two pricks *as* two varies enormously at different places on the body. A gap of a millimeter suffices on the fingertips, but one needs fifteen times that distance on the back. We want to plumb the capacities to sense differences in apparent location, and for this we need a measuring line marked in units of discriminability. These do not map in an isotropic way onto millimeters.

So to detail the sensory content of pain we need to describe the sensory content of various somesthetic qualities, and the sensory capacity to identify the different portions of one's body that those qualities appear to characterize. Both tasks will push us to the limits of what it is possible for our subject to discriminate. This catalog will need to extend to the "global indiscriminability" of two sensory qualities. That is, the difference between phenomenal properties P and Q may not be revealed in any direct comparison between them. But if P is routinely judged to match some quality that Q does not, then P and Q cannot be qualitatively identical. They stand in slightly different places within the entire corpus of possible judgments of similarity and difference. And if their places can be differentiated in any way at all, then they are not identical.

III

The catalog of possible somesthetic sensory contents is, then, rather compendious. But now I want to argue that the property earlier identified as the painfulness of pain— what makes pain hurtful, bad, aversive, awful—is not going to be found in that catalog. It could not be.

Here is a warm-up argument. What quality must two episodes share (or appear to share) if those two episodes are felt to be equally painful? With your typical sensory quality, this question has a straightforward answer. These two tastes are equally salty because they both present the same quality of saltiness. These two bananas match in hue because each has the color that the other does. Two episodes match on that dimension of variability because, we say, they both present the same quality. But painfulness is a different beast.

One episode can hurt as much as another, can be equally awful, even though their sensory character differs. Consider a range of examples: the ischemic pain one gets from *very* cold hands or feet; the diffuse pain of sunburn after an unprotected day on the beach; a heavy dull abdominal pain; or the burning tearing pain of a wrenched muscle. These can in their various ways be equally painful. If you had to choose between them on the basis of how much they hurt, you might be indifferent. As punishment any one of them will do, thanks.[6] But what is the sensory resemblance between the intense freezing pain of an almost frozen foot and the diffuse hot pain of an sunburned back? The point is exactly parallel to one that Sidgwick made about the notion of pleasure:

[F]or my own part, when I reflect on the notion of pleasure—using the term in the comprehensive sense which I have adopted, to include the most refined and subtle intellectual and emotional gratifications, no less that the coarser and more definite sensual enjoyments—the only common quality that I can find in the feelings so designated seems to be that expressed by the general term "good" or "desirable" which we have before examined. Hence, while I cannot define Pleasure—at least when we are considering its "strict value" for purposes of quantitative comparison—as the kind of feeling which we actually desire and aim at, I still recognize as its essential quality some relation to desire or volition. I propose therefore to define it as feeling which, when experienced by intelligent beings, is at least implicitly apprehended as desirable or—in cases of comparison—preferable. (Sidgwick 1893, p. 128)

For my part, when I reflect on these episodes of pain, the only common quality I can find in the feelings so designated seems to be that expressed by the general term 'bad' or 'aversive'. A cramp and a pinprick might be equally painful, even if no sensory quality is common to the two episodes. They are in their various ways equally undesirable. Perhaps the essential property in virtue of which two episodes are both painful is not any sensory quality at all, but simply the undesirability of whatever collection of such qualities they happen to manifest. Sidgwick held this of pleasure; I think it is true of pain as well.

But this is just a warm-up argument. Someone in the audience can and will insist that, no, when I peer into *my* consciousness I do sense a common quality, call it painfulness, that characterizes both episodes. I have a beetle in *my* beetle box. Alas, we cannot prove this person wrong. Best to agree politely—and a very nice beetle it is, too. We need a more systematic argument.

We need to specify as an essential quality some "relation to desire or volition" but in the case of pain the exact relation is of some delicacy. One typically wants an episode of painful bodily sensation to stop, and typically one would have liked to avoid it altogether, but there is no necessity in those connections. As Shaffer (1976) notes, the masochist wants some pains to continue; someone worried about peripheral nerve damage might *want* to feel pain; a proponent of natural childbirth might be gratified when the pain begins. So apart from its overinclusiveness one cannot define painful sensations as those one wants to stop. More broadly, what is often called the "motivational–affective" component of pain has an array of possible targets:

desire: to avoid the pain, reduce it, or have it stop

drive: the urgency to do something about it; the degree to which the motivations aroused by pain override all others

interest: the degree to which pain grabs and holds attention, and prevents one from attending to other projects or plans; the degree to which one can be distracted from the pain

preference: the extent to which presence of pain changes preferences among alternative states of affairs

incentive: the degree to which reduction of pain provides a reward for other behaviors

reinforcer: the degree to which pain decreases the probability of some behaviors (aversive conditioning and avoidance learning)

Pains might exert varying influences on these different levers, cogs, and pulleys within the motivational machinery. One might not want the pain to stop, exactly, even though it is true that one would prefer, if it were possible, to live through the same episode without the pain. Or perhaps even this preference is missing—the presence of the pain adds a certain piquant vivacity to the adventure, which one would scarce do without—but perhaps even then the pain has done its job as a negative reinforcer, and certain muscle twitches one would otherwise make have been unconsciously suppressed. Or perhaps one finds no evidence of such negative reinforcement, but nevertheless our hero feels a warm sense of gratification when the adventure is over, and this gratification prompts him to repeat the adventure again at some later date. Pain might worm its way into our motives in any of these fashions, and so one cannot pick any one of them as its essential relation to desire or volition.

But a trick similar to that used to get "global indiscriminability" can work here too. A direct comparison between two alternatives might leave an agent indifferent between them. Take this as an analog for pairwise discriminability of stimuli, and then generalize it to get a notion of "global indifference." Think of a vast array of measurements: of *all* your motives, desires, inclinations, interests, preferences, reinforcers, incentives, and drives. We are homunculi, down in the basement of the nuclear power plant, scanning hundreds of measuring instruments. A state of affairs is a matter of global motivational indifference if it can be registered without causing a single twitch on any of those meters. Adding it to or subtracting it from any other state of affairs does not cause a single needle to budge, at all, anywhere, ever. *That* is what you might call utter indifference. Alone or in combination with others, the state of affairs in question makes no difference at all to any of one's desires, drives, preferences, reinforcers, inclinations, interests, or motives. The notion is like global indiscriminability in that there is no way for a difference to manifest itself, no matter what the permutations or combinations.

Now there is no essential direct connection between any particular pain and any particular desire. That S is painful cannot be tied directly to a particular desire that S cease. Nevertheless, I think there is an essential global connection. If state S is in fact painful, if it hurts, then to its bearer it cannot be a matter of global motivational indifference. One might not want it to stop, exactly, but in one way or another it will not be a matter of utter indifference. The same is true of something that is pleasant. Pleasures and pains push one from the neutral point of utter indifference in what are, intuitively, opposite directions. Typically, something pleasant is desirable, attractive, and gratifying. It seems good. Something painful is typically undesirable, aversive, and punishing. It seems bad. These opposite poles, or the opposition between these poles, cannot be understood in purely sensory terms. But perhaps they can be understood as opposing motivations.

One can summarize the effect of something painful on one's global motivational state by saying: it arouses an *aversion*. This is a change in the disposition of one's motivations. As a disposition, it might not manifest itself in any behavior. The disposition might remain unexercised. Furthermore, the motivational change might show up in any of the different ways I listed above: as a change in the content of desires, preferences, inclinations, incentives, reinforcers, and so on. The aversion might show up in any of these ways; no single one of them is essential.[7] What is essential to painfulness is that it arouse one or another of them. Another way to put this: if S is painful, then S arouses some disposition to avoid. Avoidance, like aversion, is a catch-all term. It indicates the general direction of one's motivations, it gives their general drift, so to speak, without spelling out the precise details of heading, wind speed, tiller angle, tack angle, water currents, and sails aloft. Desires, inclinations, reinforcers, drives, interests, preferences, are all heading, roughly, that-a-way.

If S is painful, then, it cannot be a matter of global motivational indifference. It must be aversive. This doesn't say much, but it says enough to show that painfulness could not be a quale. For to say S is aversive implies that one must have some disposition to avoid it. Aversiveness is an essential but relational property of those states we call "painful." But then the properties that such relational characterizations describe are not qualia. For according to our earlier stipulations, there can be no relational characterization whose satisfaction makes the difference between quale Q's being present and quale Q's being absent. However, unless a sensory episode stands in the appropriate relations to one's motivational states, it is not aversive, and hence not painful. Hence painfulness is not a quale. It is a conative or motivational property, not a sensory one.

Suppose, for example, that the aversiveness of a particular sensory episode S consists in your desire that S cease. But then S is aversive only given a constellation of surrounding desires. Surround the same sensory state with a different constellation of desires, and the aversiveness of S could change. The same quale Q could come to seem less painful. The same holds for the other ways in which the aversiveness of S might be manifested. Perhaps one has some tendency to try to avoid it or to end it; or some preference for other states of affairs that lack the painful aspect; or some tendency to stop any actions that stimulate the pain; or some drive to indulge in behaviors that reduce the pain, and so on. No matter which of these strands one picks, that strand is some proclivity, inclination, or preference toward some outcomes and away from others. So aversiveness is a relational property: it characterizes that sensory state in terms of the relations in which it stands to one's preferences, inclinations, and desires.

It follows that one could have two instances of mental states that are qualitatively identical, that share all the same sensory qualia, yet which are not equally aversive. Surround that same sensory state with a different constellation of preferences, and this second instance of the same state may not be equally painful. So painfulness is not a quale. It is at best a motivational disposition occasioned by a quale. To paraphrase Wittgenstein and Anscombe (Anscombe 1957, p. 77): no immediate phenomenological quality could be an aversion, because it cannot have the consequences of aversion.

IV

Pain straddles the divide between sensation and desire, and it also slops into neighboring compartments of motive, emotion, and mood. It is useful to point out that pain is not alone in this regard; other of our primitive "feeling" states seem also to be demanding sensations or sensible desires: states that uneasily combine sensation and volition. Consider, for example, thirst. We speak of "sensations of thirst," but on the other hand someone who is very thirsty must, it seems, want to drink. How can there be a *sensation* of wanting to drink? The same argument as applied to pain will show that thirst *cannot* be a mere sensation. No nonrelational specification of sensation could have the consequences that follow when we call it "thirst."

The sensations of dry mouth, parched throat, weakness, and perhaps dizziness must somehow engage the appropriate desire before we can label the result "thirst." Is it logically possible to have sensations exactly like those you have when you are thirsty, but not desire to drink water? Obviously so, Socrates. Sensations are one thing, desires another. Desires color our awareness of the things we sense, but they come from a logically distinct part of the zoo. But if so, it is likewise logically possible to have sensations exactly like those you have when you have pain, but have no desire to have those sensations cease. It feels just like pain, but it is not awful; it does not bother you; its continuation or cessation is a matter of indifference. Would that state be pain? Well, Socrates, yes and no.

Some states of affairs satisfy your thirst. Others frustrate it. To identify a particular desire, one must identify what it is a desire *for*, what its "satisfaction conditions" are. Notice that the behaviors produced by thirst could be produced by various distinct desires. Thirst could be the desire to put liquid in one's mouth; the desire to swallow liquid; the desire to put liquid into one's stomach; the desire to absorb liquid; the desire to end the symptoms of dehydration. Which is thirst? We will know only when we know the satisfaction conditions for the desire.

But this raises the problem. Can you *sense* that you are thirsty?[8] If to identify the state as thirst one must identify that which would satisfy it, then the answer must be "no." The satisfaction conditions for the desire describe an intentional object. If the desire is currently raging, unsatisfied, then its intentional object, the state of affairs that would satisfy it, is currently nowhere present. It may not exist anywhere. And whatever else one might say about sensing, it is very clear that one cannot sense a *merely* intentional object. You might seem to, perhaps, but you cannot literally do so. So you cannot sense what would—if only it were present—satisfy your currently raging thirst. One can represent such states of affairs in other ways, but we do not reckon such representation to be sensory.

Similarly with pain and pleasure. Some states of affairs relieve the pain; others aggravate it. To say "*S* is painful" implies certain things about that creature's motivations. And per hypothesis the mere presence of a quale cannot imply anything about the creature's motivations. It follows that the mere presence of a quale cannot guarantee any particular effect on one's motivations. So no phenomenal quale is *per se* painful; no matter what one tries to pack into the list of "immediate phenomenological qualities," one cannot guarantee that when that congeries is sensed, it will be found to be aversive.

Kripke says the reference of "pain" is fixed by its "immediate phenomenological quality," by being "felt as pain" (Kripke 1980, pp. 151–152). Consider "thirst" as an analogy. Suppose being thirsty entails having particular motivational states—wanting to drink, for example. Two consequences follow. First, wanting to drink does not have any intrinsic phenomenological character. Any phenomenal character it happens to have is inessential to its being a desire to drink. Second, whatever phenomenological character you pick as essential to thirst, if that character is not connected with a desire, it is not connected to thirst. At best it is the "immediate phenomenological character" of a *sensation* of thirst. But a sensation is not a sensation of thirst in virtue of having one sort of phenomenological character instead of another. The essential property, which is relational, is the connection to desire. Similarly there is no phenomenological character specific to painfulness. At best there is a phenomenological character of sensations that are painful. Again the essential feature in virtue of which those sensations are painful is the relational and contingent connection to dispositions to avoid.

V

Einstein advised that our theories of nature be as simple as possible, but no simpler. Painfulness could not be a sensory quale. What could it be? The next simplest hypothesis is that it is not simply a sensory state, nor simply a motivational state, but a tandem product, a state of affairs constituted by the two of them standing in a certain relation.

A painful quale is one identified by its effect on motivations. The typical bodily pain engages such motivations—it is painful—but it also presents a distinctive sensory character: some range of qualities within those discriminable by nociceptive somesthetic discrimination, as already described. The qualities sensed and the painfulness thereof are not always distinguished, but they are logically distinct. Call the former $pain_{sensory}$ ($pain_s$, in short) and the latter $pain_{motivational}$ ($pain_m$). If painfulness is not a quale then no characterization of $pain_s$ implies a characterization of $pain_m$. The simplest alternative that would explain the facts is that there exists a very strong but nevertheless contingent connection between them. The ordinary term 'pain' refers to a state of affairs constituted by both, standing in a certain relation to one another.[9] Suppose that we are built so that having certain kinds of somesthetic qualitative states attracts one's attention and causes an immediate, overriding, and compelling desire that such sensing cease. The link here is immediate, direct, and very strong. Suppose that sensing that kind of quality almost invariably arouses a strong aversion to that quality. The causal connection is so strong and exceptions are so rare that ordinary language can label the conjunction with just one word, and almost never go wrong. The two can be treated as a sum because they almost always occur together. But there is no logical necessity binding them, and in odd cases they can diverge. I will call this the "tandem" model. It is in the odd cases that we see the need to note the logical distinctions between the components of what ordinary language identifies, indiscriminately, as pain. The patient on opiates says he still feels the same burning pain, but it no longer bothers him. Is he in pain or not? To Socrates we said yes and no; more helpfully, perhaps the patient does have pain$_s$—all the same sensory qualities are present—but does not have pain$_m$—their normal motivational effects have been blunted.

On this hypothesis the painful, hurtful, or to-be-avoided character of pain lies firmly on the side of pain$_m$. If those same sensory qualities could lose their aversive character, could cease to arouse any disposition to avoid, and if the patient became utterly indifferent to their occurrence or nonoccurrence, then they would lose any claim to being painful. That same burning feeling no longer hurts. This is difficult to imagine because in our experience that kind of burning sensation invariably arouses a strong and overriding desire that it cease, and so it is difficult to imagine that sensory quality occurring without its being aversive. Per hypothesis there is a very strong causal connection between the two. But it is not a logical connection. We are just built that way.[10]

If the connection between a somesthetic quality and its aversiveness is contingent, then we can explain various oddball cases, as well as what makes them odd. So, for example, it is possible that the very same quality that in us sets off alarms, avoidance, and evasion in the masochist stimulates some other motivational disposition. Does the masochist feel pain? Yes and no. The masochist might have the same pain$_s$ states, but different pain$_m$ states.[11] It is likewise easy to accommodate the "madman" of David Lewis (1980): the madman has the same pain$_s$ as we do, but it is connected to a different motivational state. (It makes him snap his fingers and think about mathematics.) "Must pain be disliked?" is an *ambiguous* question: the sensory qualities themselves cannot necessitate any particular like or dislike, but the aversions they almost invariably arouse must constitute dislike of something.

Philosophically the tandem model may seem ad hoc, but neuroanatomically there is a very interesting regularity that supports it, and in fact, suggests it. Most of our sensory modalities conduct the bulk of their traffic on myelinated pathways that proceed from receptors to the thalamus and thence to specialized areas of sensory cortex. Pain is very different. Along with the fast spinothalamic tracts, it includes at least four unmyelinated spinal pathways that synapse first in limbic structures in the midbrain and pons. It has direct connections into the reticular activating system of the midbrain, which controls alertness, arousal, and sleep; and it has an enormous number of projections and offshoots into a wide variety of nuclei in the limbic system. The latter is an ancient portion of our nervous system that we share with the earliest reptiles. It controls the sympathetic and parasympathetic nervous system, mediating the fight or flight responses; it has the nuclei that seem to associate positive and negative reinforcers with stimuli; it maintains homeostasis, and controls appetite and mood; it contains the famous pleasure centers, others, less famous, that yield frightful pain; it has nuclei that release endogenous opiates and can block out pain altogether. Addictions to illegal substances are all housed here. Various parts are overactive or underresponsive when one is anxious or depressed. In motivational and emotional terms, the limbic system is a potent place. Pain wraps its tendrils around all parts of it. Unlike most other sensory systems, pain has direct access into the innards of our preference functions.

Only a few systems do this: pain, smell, and taste.[12] Hunger, thirst, and other appetites are likewise limbic phenomena. These are arguably the most primitive of all our mental states. Trace your family history back far enough, and you will finally reach a primitive segmented worm, who was the common ancestor of insects, mollusca, *c. elegans*, and us (see Katz 2000). A worm is basically a moving alimentary canal, and all its mental states are imbued with the task of managing the alimentary canal. The first necessity is to wiggle away from places that cause bodily destruction, and lo, there is pain. An obvious design improvement is to avoid swallowing materials in one's mouth that are not fit for the alimentary canal, and lo, there is taste. Even better, before putting something in one's mouth, get a sense of whether it would be good to do so. Lo, there is smell. We are thereby fitted with primitive approach–avoid systems: wiggle toward the places that smell good, and away from those that smell bad, or cause pain. All three of these systems have initial synapses in the midbrain or lower, and all three have very strong and direct connections to parts of the limbic system. These comprise what I called the primitive feeling states. All three have direct and strong connections to motivational states. The tandem model works for all of them.

In these primitive feeling states the sensory and motivational components really do work in tandem: they do not arise entirely independently of one another, nor is one entirely antecedent to the other. Neuroanatomically, there is considerable interaction between neurons in the medial and ventral tracts from the dorsal horn of the spine upwards (see Millan 1999, sec. 4), and many interconnections between the central nuclei to which they ascend. Furthermore, the sensation and the aversion have not only a common ancestor but a common object. That is, the sensory state presents certain somesthetic features as located in parts of one's body. One feels something tearing and burning here, or sharp and piercing there. The simplest hypothesis is that the "objects" to which one is averse are those very same features. One would like to get away from that tearing burning feeling, but unfortunately one cannot. The dislike is directed at the appearance of a portion of one's body. The burning and tearing feeling is what is painful; it hurts here, we say, pointing to the place that seems to be characterized by those qualities. "It hurts here" means, on the tandem model: "I sense various somesthetic features F as occurring here, and I am averse to those features." For somesthetic feature F, plug in any of the sensory qualities of pain listed above. Those features are the objects of both the sensory state and the aversion. *Pace* Hardcastle, "it hurts here" is not only well formed and meaningful; it is sometimes true.

The tandem model though is true of all our primitive feeling states, not just pain. Compare "it stinks there" or "that tastes awful." An appearance is attributed to a locale and there is a disposition to avoid those features in that locale. Many of our words for bad tastes and bad smells have a striking logical similarity to those we use to characterize pains. Taste aversion in particular provides an excellent analogy for pain. Suppose you come home from vacation and, without thinking about it, take some milk from the refrigerator. Suppose that milk is not only sour but spoiled—in fact, *very* spoiled. You make acquaintance with a taste that is sour, stale, rank, loathsome, disgusting, repulsive. It arouses a strong and immediate desire to eject the liquid. If you are lucky you can act upon the desire immediately and then rinse out your mouth. Now just imagine that you cannot eject the liquid; you must keep it in your mouth. This is a good model for the aversive character of pain. There are certain distasteful features that are sensed to characterize a particular locale, and which you'd desperately like to avoid, but which you cannot avoid. The distasteful features are the joint objects of both sensory and motivational states: they are both sensed and disliked. Having a loathsome taste in your mouth is logically analogous to having a pain in your foot; "it hurts there" is logically analogous to "that taste is repulsive."

The primitive worm at our functional core did not have the need to distinguish between its sensations and its desires. In tastes, smells, and pains we can see its ancestry in us still.

VI

You might ask why we should have various sensory modalities coupled so intimately with motivational powers. *Why* should such and such a feeling be negatively reinforcing? Why? A good way to probe the motivational component of pain is to consider what happens to those who cannot feel pain. These are the ones in whom the job of pain, whatever it is, is unperformed. Then ask: how would you design a creature so that it would avoid such a fate? So consider the fate of someone who does not feel pain:

Tanya was a four year old patient with dark, flashing eyes, curly hair, and an impish smile. . . . Testing her swollen ankle, I found that the foot rotated freely, the sign of a fully dislocated ankle. I winced at the unnatural movement, but Tanya did not. . . . When I unwrapped the last bandage, I found grossly infected ulcers on the soles of both feet. Ever so gently I probed the wounds, glancing at Tanya's face for some reaction. She showed none. The probe pushed easily through soft, necrotic tissue, and I could even see the white gleam of bare bone. Still no reaction from Tanya.

It seems apparent that Tanya suffered from a rare genetic defect known informally as "congenital indifference to pain." She was healthy in every respect but one: she did not feel pain. Nerves in her hands and feet transmitted messages about changes in pressure and temperature—she felt a kind of tingling when she burned herself or bit her finger—but these carried no hint of unpleasantness. Tanya lacked any mental construct of pain. She rather enjoyed the tingling sensations, especially when they produced such dramatic reactions in others. . . .

Seven years later I received a telephone call from Tanya's mother. . . . Tanya, now eleven, was living a pathetic existence in an institution. She had lost both legs to amputation: she had refused to wear proper shoes and that, coupled with her failure to limp or shift weight when standing (because she felt no discomfort), had eventually put intolerable pressure on her joints. Tanya had also lost most of her fingers. Her elbows were constantly dislocated. She suffered the effects of chronic sepsis on her hands and amputation stumps. Her tongue was lacerated and badly scarred from her nervous habit of chewing it. (Brand and Yancey 1993, pp. 3–5)

So suppose you are God up in genomic heaven, designing survival machines to transmit and propagate your genes. Clearly one would want to lower the odds that Tanya will continue to injure her body. What is the simplest and cheapest mechanism one can imagine that would serve this end?

We would like (for example) to lower the odds that she will stand with all her weight on a fully dislocated ankle. For this we need some mechanism to register the difference between the dislocated ankle and the ankle in normal working condition; some means of making Tanya notice the difference (to get her attention, and shift it to the damaged member); some signal that makes it less likely that she will continue to stand on the fully dislocated ankle—ideally one that would make her stop immediately (provided there are no other overriding survival needs, like running away from a tiger); and (if possible) some more enduring state that would motivate Tanya to repair her injuries, or at least allow them to heal. One would not want these signals to be easy to ignore, and one would not want the recipient to be able to habituate to them, since then they would lose their motivating power. In these ways the signals must be different from other sensory signals.

In effect we need a sensory modality that also has its foot on the brakes. It has to yell "STOP" at Tanya, and yell in a way that cannot be ignored. Avoiding further damage is probably the first thing to get right. This needs a negative reinforcer: something that decreases the odds of any action that is followed by signals of increased bodily damage. That sort of signal, I submit, is the simplest precursor for the states that could become our pains.

We can imagine elaborations. It would be nice if our signal would prompt Tanya to shift weight off the damaged leg, and then freeze. Imagine little green letters appearing on the heads-up display in Tanya's cockpit: "Damage to left ankle. Standing on it is ill advised. Shift rightwards."[13] We have to make Tanya pay attention to such information, and ideally make her obey. Little green letters on the screen won't do. We need whips and lashes, pleasure and pain.

In fact people have tried to develop prosthetic pain systems, and have run up against precisely this barrier. The nerve damage caused by leprosy leaves the patient insensitive to pain in feet and hands. Paul Brand showed that flesh does not rot in leprosy, nor do fingers or toes fall off; instead all the damage is self inflicted, and arises because the patient does not feel any pain. With colleagues he put in an NIH proposal to develop a prosthetic pain system, using pressure sensors in gloves or socks and a warning signal to alert the patient if some activity was damaging. Despite years of effort the system failed, and the reasons for its failure are quite instructive. Brand says:

Patients who perceived "pain" only in the abstract could not be persuaded to trust the artificial sensors. Or they became bored with the signals and ignored them. The sobering realization dawned on us that unless we built in a quality of compulsion, our substitute system would never work. Being alerted to the danger was not enough; our patients had to be forced to respond. Professor Tims of LSU said to me, almost in despair, "Paul, it's no use. We'll never be able to protect these limbs unless the signal really hurts. Surely there must be some way to hurt your patients enough to make them pay attention." (Ibid., p. 194)

So they decided to make the signal painful: a high voltage, low current electric shock to the armpit, a place where most leprosy patients could still feel pain. But even that didn't work very well; he says most patients saw the shocks as punishment for breaking rules, rather than signals of danger one would naturally want to avoid. He says:

In the end we had to abandon the entire scheme. . . . Most important, we found no way around the fundamental weakness in our system: it remained under the patient's control. If the patient did not want to heed the warnings from our sensors, he could always find a way to bypass the whole system. . . . Why must pain be unpleasant? Why must pain persist? Our system failed for the precise reason that we could not effectively duplicate those two qualities of pain. Their mysterious power of the human brain can force a person to STOP!—something I could never accomplish with my substitute system. And "natural" pain will persist as long as danger threatens, whether we want it to or not; unlike my substitute system, it cannot be switched off. (Ibid., pp. 195–196)

As a selfish God up in genomic heaven, we must write in some aversive subroutines, or all our creatures will meet Tanya's fate. None of the genes will propagate. Pain, I suggest, can be read as a message from the genes. It says "Stop that, you stupid organism; you're hurting my chances!" And the only way to make sure that the organism will obey is to wire the perception of these sensory qualities directly into the creature's preference functions. Sensing in that fashion causes an immediate and overriding aversion. The organism wants to get on with its life, so it wants the pain to stop. It obeys. As Proust said, "To knowledge we make promises only; pain we obey."

Notes

1. Some take qualia to be properties of mental states, while others take qualia to be phenomenal properties, or characterizations of the appearance of one's surroundings. On the latter reading, but not the former, qualia can be ascribed to regions of space *outside* the body of the sentient organism, and it is proper to speak of sensing them: so that one can, for example, *see* visual qualia, and hear auditory ones.

2. This leaves open the possibility that the general relation of qualitative sameness does have functional consequences and can be functionally defined (as argued in Shoemaker 1975). That is, perhaps we could give a functional specification identifying those conditions under which two episodes happen to be qualitatively identical to one another. The problem remains of giving a functional specification which suffices to identify those conditions under any episode presents (say) a *particular* shade of blue.

3. Perhaps this is the source of Block's suggestion (in "Troubles with Functionalism") that it is hard to see how psychology "in its current incarnation" *could* define qualia; the latter properties seem not only untouched but completely untouchable (see Block 1980, p. 289). I do not share these intuitions, but I remind the reader that here I am accepting stipulation (2) for the sake of argument. Dispute about the *truth* of these stipulations is a dispute that must take place elsewhere.

4. If you believe qualia are nonrelational then even global indiscriminability does not suffice to define qualitative identity. Such a philosopher could admit that the two stimuli are globally

indiscriminable because they both present the same quale, but if sameness of quale is nonrelational then global indiscriminability does *not* (and cannot) suffice to guarantee qualitative identity.

5. Specifically, the spinomesencephalic tract synapses at the periaqueductal gray area (PAG); two "spinoparabrachio-" tracts synapse first in the parabrachial nucleus (PBN), one of which proceeds to the amygdala and the other to the hypothalamus; and the spinohypothalamic-spinotelencephalic tract has its first synapses in the hypothalamus and thalamus, and from there projects to the pons, amygdala, and striatum (Millan 1999, p. 31, table 4).

6. Interestingly, you might have strong preferences based on the nonsensory representation of these various contingencies. You might realize, for example, that ischemic pain goes away as soon as circulation is restored, with no lasting ill-effect; but bad sunburn can increase your risk of developing skin cancer. The wrenched muscle might prevent you from skiing again tomorrow, while the abdominal pain might just be indigestion, which will go away tonight. On the other hand, it could be stomach cancer. The sensory system cannot represent any of these future or counterfactual contingencies; from its point of view all the episodes are, simply, equally painful.

7. Suppose I am averse to strawberries. I might positively dislike them, or I might not. Perhaps I simply want to avoid them. But I might lack that desire too. Perhaps I simply prefer dishes without them. Perhaps I simply fail to repeat orders for dishes that contain strawberries at the same rate at which I repeat orders for dishes that do not contain strawberries, even though I am not consciously aware of the presence or absence of strawberries, or of this behavior of mine. Aversion is a catch-all term, covering this entire range.

8. Descartes seems to have noticed this problem. He identifies pain, hunger, and thirst as "confused modes of thought" that arise from the union of mind and body. He goes on: "When we say, then, with respect to the body suffering from dropsy, that it has a disordered nature because it has a dry throat and yet does not need drink, the term 'nature' is here used merely as an extraneous label. However, with respect to the composite, that is, the mind united with this body, what is involved is not a mere label, but a true error of nature, namely that it is thirsty at a time when drink is going to cause it harm" (Descartes 1984, p. 59).

9. This hypothesis is meant to apply to that species of bodily sensation we typically call "pain." It is *not* meant to encompass all the states of affairs that fall under one or another sense of the word 'painful'—as in painful emotions, painful encounters, a painful loss, and so on. Similarly, there are many conditions that people find aversive (glaring light, sufficiently loud noise, bad smells, and so on) but only those that involve sensation in the nociceptive somesthetic modality are typically called "pains."

10. Descartes, *Meditation 6* (6th para.): "When I inquired, why, from some, I know not what, painful sensation, there follows sadness of mind, and from the pleasurable sensation there arises joy, or why this mysterious pinching of the stomach which I call hunger causes me to desire to eat, and dryness of throat causes a desire to drink, and so on, I could give no reason excepting that nature taught me so; for there is certainly no affinity (that I at least can understand) between

the craving of the stomach and the desire to eat, any more than between the perception of whatever causes pain and the thought of sadness which arises from this perception."

11. It is also possible, and perhaps more likely, that the pains of the masochist arouse the normal dispositions-to-avoid, but those dispositions are overridden by other desires.

12. Other modalities send collaterals into the reticular activating system (so that a flash of bright light or a loud bang can be startling, grab attention, and increase alertness), but they do not have the wealth of relatively direct connections found in pain, smell, or taste.

13. The old allusion is to Descartes's pilot, who (in *Meditation 6*) perceives damage occurring to his vessel. Pain, hunger, and thirst demonstrate that we are *not* housed in our bodies the way the pilot is in his vessel: they "arise from the union, and as it were, intermingling, of the mind with the body" (Descartes 1984, p. 56). The more recent allusion is to Arnold Schwarzenegger as the Terminator. Asked whether it hurt to have bullets dug out of his back, he replied: "I receive data that could be interpreted as pain." Perhaps "does it hurt?" is ill formed when asked of the Terminator.

References

Anscombe, G. E. M. 1957. *Intention*. Oxford: Basil Blackwell.

Block, Ned. 1980. "Troubles with Functionalism." Pp. 268–306, in *Readings in Philosophy of Psychology*, vol. 1, ed. Ned Block. Cambridge, Mass.: Harvard University Press.

Brand, P., and P. Yancey. 1993. *Pain: The Gift Nobody Wants*. New York: Harper Collins.

Descartes, René. 1984. *The Philosophical Writings of Descartes*, vol. 2. Translated by John Cottingham, Robert Stoothoff, and Dugald Murdoch. Cambridge: Cambridge University Press.

Hardcastle, Valerie. 1999. *The Myth of Pain*. Cambridge, Mass.: MIT Press.

Katz, L. J. 2000. "True Pleasure, from Plato to Neuroscience." Revised text of a talk at Tufts University.

Kripke, Saul. 1980. *Naming and Necessity*. Cambridge, Mass.: Harvard University Press.

Lewis, David. 1980. "Mad Pain and Martian pain." Pp. 216–222, in *Readings in Philosophy of Psychology*, vol. 1, ed. Ned Block. Cambridge, Mass.: Harvard University Press.

Melzack, Ronald, and Patrick D. Wall. 1983. *The Challenge of Pain*. New York: Basic Books.

Millan, Mark J. 1999. "The Induction of Pain: An Integrative Review." *Progress in Neurobiology* 57: 1–164.

Shaffer, Jerome A. 1976. "Pain and Suffering." Pp. 221–233, in *Philosophical Dimensions of the Neuro-Medical Sciences*, ed. S. F. Spicker, and H. T. Engelhardt, Jr. Dordrecht: R. Reidel.

Shoemaker, Sydney. 1975. "Functionalism and Qualia." *Philosophical Studies* 27: 271–315.

Sidgwick, Henry. 1893. *The Methods of Ethics*. 5th ed. London: Macmillan.

11 An Indirectly Realistic, Representational Account of Pain(ed) Perception

Moreland Perkins

1 Indirect Realism Exhibited in Perception of a Toothache

A hockey player with a newly loose tooth may be distracted by it. For a time his attention is held by this new condition of his tooth. The form his attention to his tooth takes is feeling it. With his tongue he feels his tooth move; and he feels this motion with the gum in which his tooth is embedded. He becomes tactually conscious of his tooth's motion.

A tooth can hold our sensory attention through other modalities than touch; for example, a tooth's aching may hold it. When the athlete's loose tooth holds his attention through its motion, his attention takes the form of sensory consciousness of his tooth caused by the action of sensory receptors within his gum and tongue. When my tooth holds my attention through its aching, my attention takes the form of sensory consciousness of my tooth caused by the action of sensory receptors within the tooth itself. To what feature of the tooth is attention given when the tooth holds one's attention in virtue of its aching? To the tooth's aching. The form of this attention, too, is called "feeling": we feel the tooth's presence by feeling it ache; we become in a sensory way conscious of its aching. The tooth's aching here plays a role like the role the tooth's moving plays when a hockey player feels his loosened tooth.

This account of feeling a toothache begins to sound like an account of sense perception of a tooth.

On the other hand, the fact that within the experience of feeling a toothache there is experience we can think of as merely feeling pain can lead us, upon reflection, to think of feeling toothache as not a form of perception of the body. This is because considerations that make us think of feeling pain as not perception may lead us to think of feeling toothache the same way. Why do we think that merely feeling pain does not qualify as sense perception of pain?

Here is a first step along one route to a plausible answer. To perceive through the senses a sensible property or state of an object is to gain knowledge of the property or state's instantiation in virtue of the property or state's contribution to the object's acting causally upon one's sense organs so as to bring about the perception. Thus a tooth's motion counts as a state of the tooth that the athlete perceives by the sense of touch because the tooth's motion before the organs of touch embedded within his tongue and gum causes him through its action upon these organs to feel that motion. (Although in a strict sense, because not orientable to external stimulation, these sensory receptors do not count as "organs," they function sufficiently like organs in the strict sense for it to be convenient for me so to call them.)

For a second step suppose that something we attentively feel is inseparable from our feeling it, hence dependent for its existence upon our being conscious of it. What we feel in this way cannot be situated externally to our sense organs, since objects and events (including bodily tissues and their states) located before our sense organs exist independently of our being conscious of them. Hence what we feel that is inseparable from our feeling it cannot (in the proper way) act upon our sense organs; so it cannot by so acting cause us to feel it; therefore we cannot be said to perceive it through use of sense organs; hence feeling it is not sense-perceiving it.

Third and last step: pain is an example of something that is inseparable from our being conscious of it; pain exists only when we feel it. Our intuition that this is so is most vivid for us when an anesthetic "stops our pain"—as we invariably report when we speak spontaneously. Intuitively, we do not believe that the pain goes on existing in the tooth when the novocaine causes us to cease to feel pain. Upon reflection, therefore, we judge that pain belongs to our consciousness of it and exists only where a conscious state exists, which is most assuredly not within a tooth, hence not where the pain could act from without upon the receptor cells embedded in the tooth. Consequently, when we think of feeling a toothache as an instance of merely feeling pain we find ourselves thinking of feeling toothache as not an instance of perception by sense.

A fatal error in this way of thinking of toothache is evident. What I feel in feeling a toothache is not pain alone, but also my tooth. It is because in feeling a toothache I feel my tooth's presence in my mouth that my feeling a toothache is an instance of sense perception of my body, specifically of my tooth. In feeling his loose tooth move, the hockey player perceives his tooth by perceiving a state of his tooth, its motion. In feeling a tooth ache I also perceive my tooth by perceiving a state of my tooth, its aching.

However, the feature of a tooth that we perceive in feeling a toothache offers difficulties. A tooth's aching includes in its definition—in its nature—that it causes its owner to feel something (pain), whereas a tooth's moving includes within its nature nothing subjective. Therefore an exact account of how a tooth's aching can be perceived by means of sense organs presents a special challenge that must be met. For now I rely upon my reader's good sense to tolerate for the nonce the working hypothesis that in attentively feeling a toothache we perceive by the sense of pain a tooth in our mouth.

Before meeting the challenge cited two sentences back, I will first develop a thesis implicit in the reasoning of three paragraphs back: that the pain we feel in feeling a toothache is *both* something we are conscious *of* and a state of our consciousness. I exhibited intuitive support for this contention in my remarks about the inseparability of pain from feeling it. Since pain exists only in so far as felt, when pain occurs it qualifies—belongs to—our feeling. When we feel a painful tooth we have a pained feeling of our tooth. When an athlete feels the moving state of his tooth his feeling is not itself in motion. However, when I feel what we call the painful state of my tooth, my feeling is itself painful.

Connections between feeling, sensory attention, and consciousness now need explicit notice. Suppose a tooth's aching suddenly commands my attention. In what does this attention consist? In part it consists in my attention's suddenly being held by a pain. What form does this attention to my pain take? It consists in my becoming conscious of the pain. What sort of consciousness is this? It is not mere thought—I do not suddenly begin merely to think of pain. It is sensory consciousness; specifically, it is *feeling* the pain. Attention to my pain, sensory consciousness of this pain, and feeling the pain: these are the same state of affairs described in three ways. From pain's way of being inseparable from our feeling it I have drawn the conclusion that pain qualifies our feeling when we feel pain. Therefore pain qualifies our consciousness and our attention when these are directed upon pain. Our sensory consciousness of an aching tooth is filled with pain—it is a painful consciousness; our attention to this tooth is a pained attention.

I have now drawn this preliminary sketch of attentively feeling a toothache: it is attentive sense perception of an aching condition of a tooth—hence it is sensory consciousness of a tooth. This perception includes within itself sensory consciousness of pain that makes this consciousness of the tooth a painful consciousness of it. Since the pain we are aware of is not before the sense organs and our awareness of this pain is not mediated by awareness of something else in virtue of which we are aware of the pain, we are *directly* conscious of the pain. In special circumstances we might have become conscious of such a pain without being conscious of a tooth as its source, hence without consciousness of a tooth; but we can never in a sensory way be conscious of a tooth's aching without being conscious of pain. So when our tooth aches we are conscious of the tooth before our sense organs in virtue of being directly conscious of pain that is not before them: hence we are indirectly conscious of the tooth in our mouth in virtue of being directly conscious of pain that belongs to our consciousness. Feeling a toothache is indirectly realistic perception of a tooth, hence of the tooth's states, hence of its aching.

Several challenges to explaining and justifying this proposition remain to be confronted.[1]

2   Two Problems for the Thesis That We Perceive Toothaches; One Solution

That we perceive a toothache by use of our sense organs is, upon reflection, none too easy to believe: the subjectivity of the toothache's pain is the obstacle. It makes difficulty in at least two ways. I can begin to elicit one way by asking a question. To our sense of pain, what does an ache in a tooth appear to be?

For comparison, to the sense of touch what does the movement of a loose tooth appear to be when using the tongue to feel it? It appears to be a movement—what else? As to what else: one feels pressure on the tongue, and one feels a displacement of the source of the pressure as one's tongue pushes the tooth. That's how the motion of the tooth appears to touch. It turns out that the tactual appearance of a moving tooth offers no obstacle to realization of this constraint upon perception of the tooth's motion: that with respect to motion, the perception must know how it is with the tooth. The physical motion of the tooth is in fact what it appears to touch to be: it *is* a displacement away from the tongue of a locus of pressure responding to the push of the tongue.

What appearance does an aching in a tooth present to the sense of pain? It presents itself as a pained condition of the tooth—as pain in the tooth. Putting this matter the other way round: we feel toothache by feeling pain as if it belongs to the tooth. However, pain does not belong to the tooth. So what we feel as if it *is* the aching condition of the tooth cannot be that. Our feeling a toothache seems to fail to meet this condition upon perception of a toothache: that with respect to aching, the perception must know how it is with the tooth. There we have a first problem for the contention that feeling a toothache is sense-perception of the toothache.

To solve this problem, perhaps we need first to decide what an aching of a tooth *is*, then inquire whether we can know its presence on the basis of a deceptive appearance.

However, a second difficulty arises when we focus on what a tooth's aching is. An aching condition of a tooth consists in the tooth's causing us to feel pain. So in order attentively to perceive—be conscious of—the aching of one's tooth, one must be conscious of the tooth's causing one to feel—to be conscious of—pain. Hence attentive perception (consciousness) of a toothache must consist in our being in a sensory way *conscious* of the tooth's causing us to be *conscious* of pain. Perceiving a toothache requires us to be conscious of being conscious. But occurrence of iterated consciousness whenever we feel a toothache seems improbable. This is a second problem for the thesis that feeling a tooth ache is perceiving it.

There are, however, considerations that help solve the problem of iterated consciousness. Paradigmatic pain is intense pain, which is to say, dangerous pain. Feeling dangerous pain can be expected to set off an alarm system in the entire organism. Because of the threat to survival that some injured or diseased tissue presents to us, we not only become aware of the danger signal the injury sends—we not only feel pain—we also achieve a higher level awareness, we go into a state of "general alarm" about the fact that we're discerning this signal: we become conscious of being conscious of intense pain. Doing so makes immediately available to us *all* the resources we have for dealing with a threatening injury. If this is the way intense pain works, then the fact of iterated consciousness does not seem improbable. We can then plausibly infer that *some* degree of higher-level consciousness of our being conscious of pain is aroused even by less intense pain in virtue of both the physiological affinity between its cause and intense pain's cause, and in virtue of a (related) indeterminacy about what constitutes a dangerous injury. This solves the second problem for the thesis that feeling toothache is perceiving it.

I shall return later to confront the problem posed by the fact that we always perceive a tooth's aching condition as pain in the tooth although pain is not in the tooth. It will help first to make sure we can make coherent to ourselves an aspect of the indirectness of our perception of toothache.

3   A Toothache's Pain as the Content of a Sensuous Attribution of Pain to a Tooth

I have argued that our being conscious of pain in feeling toothache means that our conscious state is itself painful. How can we instructively analyze this painful conscious state? We can describe functions that painful states of consciousness perform within a larger conscious structure. Consider two functions.

Because it is part of our nature to suffer an instantaneous aversion to pain, it acts as a motivator. In virtue of feeling an aversion to pain, one is moved to do something to eliminate the pain—perhaps go to a dentist. The motivating role of pain exhibits one way it functions as a conscious state.

I will later conjecture that integration of our felt aversion to pain into perception of toothache helps make the latter into an "evaluative perception." However, it is upon pain's narrowly cognitive function that I now focus.

In feeling a toothache we are conscious of a pained condition of our consciousness as if it were a condition of a tooth (though it isn't). What functional facts can help explain more exactly how this can work? I need three paragraphs of preliminary thought to start my answer.

To be conscious in a sensory way of a tooth one must be conscious of a feature of the tooth. To be perceptually conscious of a feature of a tooth is (among other things) to know that the tooth has this feature. To know that something a, before the sense organs, has the feature B, one must do something like—or be in a state like—affirming that a is B. To do—or be in—this last, one must do something like attributing B to a. Hence to be in a sensory way conscious of a tooth one must do something like— or be in a state like—perceptually attributing a feature to this tooth. I conclude that in perceiving a toothache we "do" something like this: *attribute* to the tooth a pained condition.

I put "do" in scare quotes because of course this is not a deliberate action of ours. *Qua* activity it is not consciously *done*—we are not aware of engaging in the activity of attributing. This involuntarily *happens* to our perceptual consciousness. However, because the word "attribute" is an active and transitive verb, I align my sentences with its grammar and speak of us as *doing* the attributing. A similar point holds for perceptual affirmation.

If a second person, say a dentist, concludes from my behavior that I have a toothache, then either in speech or in silent thought she may attribute pain to my tooth (or if reflective, perhaps to me). Her attribution is not sensuous. It is a conceptual attribution: in making it she exercises only her concept of pain. In applying her mere thought of pain to my tooth (or to me) she thereby becomes only intellectually aware of pain as if in my tooth. The condition the dentist has ascribed to my tooth (or to me) is in fact a sensuous condition of my consciousness: in thought she ascribes to my tooth (or to me) the same condition—the pain—that I feel as if it is in my tooth. But only the thought of this pain figures in fixing the character (the content) of her conscious state, not the pain itself.

Now turn to me, suffering toothache. In perceiving my tooth's aching condition, I perceptually know it is aching, hence perceptually attribute aching to it. The aching appears to my sense of pain simply as pain in the tooth. Yet this pain is in fact a state of my consciousness. Therefore, in order for the pain to appear to me as if it belongs to the tooth, I must perceptually attribute to the tooth the very pain that is sensuously present in my consciousness. I call this a *sensuous attribution* of a *sensuous content* of my consciousness to my tooth. In this way I more than merely think of my tooth as if pained: I feel it as if pained. To be conscious of a pained condition of one's consciousness as if it were a pained condition of one's tooth is sensuously to attribute to the tooth the pain that qualifies one's conscious state.

We have now reached a functional account of pain's cognitive role in perceiving toothache: the pained quality of our consciousness of the tooth functions as the content of our sensuous attribution of this quality to the tooth. One represents to oneself a sensuous quality as if it belongs to one's tooth by the method of exemplifying this quality within one's sensory consciousness of the tooth.[2]

Understanding attention's role in this account is a key to grasping the full picture of how perceptual consciousness can be a consciousness both of its own quality and of something before our sense organs.

What *directly* holds our attention is pain, a state of our consciousness. This pain to which we directly attend functions as the content of a sensuous attribution of itself *to* the tooth. Our attention thereby reaches to the tooth. Our attention *to* is nothing but our consciousness *of*. Hence also our consciousness, of which our pain is an interior object and thereby a condition, "reaches" the tooth—it is consciousness of pain and also *of* the tooth *as* pained. Attributing to a tooth a sensuous content of consciousness (pain) carries to the tooth the attention that is immediately directed to the sensuous content of this perceptual activity of attributing. In being directly conscious of the sensuous content of our attribution of a pained condition to the tooth we thereby become *in*directly conscious of the tooth in our mouth and hence of its aching condition.[3]

4 Perceiving a Toothache by Sensuously Representing Trouble in a Tooth

I have now to start to make it evident that sense perception of toothache, while indirect, is also realistic. I make a beginning in this section, then carry this project forward another stage in the following section. This effort requires showing that we perceive by the sense of pain a genuine feature of a tooth. I have remarked that the obvious candidate for this feature is the tooth's aching. And I noted that this candidate suffers from at least the following two defects. First, since a tooth's aching consists in its causing us to feel (to be conscious of) pain, attentively to perceive its aching is to be conscious that it is causing us to be conscious of pain. Thus, feeling a toothache includes being conscious of being conscious. This seemed improbable. Nevertheless, this first difficulty I have resolved. The second I haven't. So I now turn to another apparent obstacle to realism in our presumed perception of toothache: that we perceive the aching condition of a tooth as if it is pain in the tooth, though it isn't.

D. M. Armstrong in *Bodily Sensations*[4] held that in feeling a toothache we perceive that "something-or-other untoward" is occurring in the tooth. I take Armstrong to have been proposing that in feeling a toothache we perceive that "something is wrong" in the tooth, or, for short, we perceive "trouble" in the tooth. However, according to Armstrong, in feeling a toothache we do not ascribe any specific nature to this trouble.

I am not perfectly sure that Armstrong is right in the positive part of his contention, but I think his view is plausible. So I propose that in feeling a toothache we percipients (at least we mature ones) perceive that there is trouble in a tooth. However, it is clear to me that Armstrong is wrong to deny that we perceptually ascribe a specific nature to the "trouble" in the tooth. Implicitly I have corrected this view already: felt *pain* is sensuously attributed to a tooth as if *it* is the nature of the trouble in the tooth.

What commonly causes a sufferer to feel pain as if in his tooth is the tooth's being diseased or damaged (or being under the threat thereof). A tooth's being diseased or damaged is an instance of trouble in the tooth. When we feel pain as if in a tooth we perceive trouble as if in the tooth and generally, in that case, there *is* trouble in it. Trouble in it is a genuine condition of a tooth that acts upon the tooth's pain receptors to cause us to perceive the trouble. Because we sensuously attribute pain as the trouble, but pain is not in the tooth, we misperceive the *nature* of the tooth's trouble. This is no more a barrier to our perceiving the trouble than it is a barrier to my attentively feeling with my hand (perceiving) the warmth of a radiator even though I fail thereby to discover that the nature of this heat is some function of internal vibration in the metal and, instead, misrepresent this nature as the felt sensuous quality, warmth.

I do not mean to suggest that it is easy fully to explain how the thermoperceptual achievement is possible. For the moment it is the mere fact of the achievement in the pained-perceptual case that I am exhibiting; therefore it is the mere fact of the thermoperceptual achievement I mean to cite as analogous. In the next section I go further in explaining how this pained-perceptual achievement of realism is possible.

With D. M. Armstrong's help, we have found a genuine condition of the tooth—trouble in it—that we can perceive by the sense of pain. Problems remain to be solved about how we can achieve this through sensuous misrepresentation. First, however, I need to secure the status of my account of perception of toothache as a representational theory. I need to make tolerably sure that within the framework I have built I am justified in speaking of pain as a nonconceptual, sensuous representation of trouble in the tooth.

A word about my use of the word "representational." One or more specialized definitions of this word and its affiliates have emerged in writings on perception during the last twenty years or so. I always mean nothing more recondite in regard to every form of attentive (hence conscious) sense perception than this: leaving aside our aversive reaction to pain and perhaps some comparable affective reactions to other bodily "sensations," the entire content of sense-perceptual consciousness is through and through representational—there is nothing in it that isn't characteristically functioning to represent (or "try" to represent) something before the sense organs. Furthermore, for all forms of attentive (hence conscious) sense perception every perception's representational content is of two integrated sorts: perceptually attributed nonconceptual, sensuous content is fused with perceptually attributed conceptual content (about which, more shortly). For the moment I postpone discussing perceptual attribution of conceptual content in order first to make it clear that the pain one feels in feeling toothache functions as a sensuous representation of a troubled condition of one's tooth.

Consider an analogy: I turn the handle of the kitchen water faucet to the "off" position, and I see a small stream of water continue to flow from the faucet. I push the handle hard, making it as tightly "off" as possible. A steady trickle of water continues to flow. I now see that there is something wrong with the (out of sight) valve that controls the flow of water. I do not see *what* is wrong. Given my background knowledge about water pipes, I see that something is wrong with a valve by seeing the water flow in circumstances in which I know that the handle has been turned tight in the "off" position. In seeing the water flow I perceive that there is trouble in the faucet valve.

However, it would be unnatural to describe this stream of water visibly flowing from the faucet as serving for me as a visual representation of the trouble in the faucet. I am perfectly conscious of the stream of water as one thing, the trouble in the valve as another. I do not perceptually take the valve itself to be either watery or flowing.

Now come back to my suffering from a toothache and thereby perceiving trouble within a tooth. Do I perceptually attribute a specific condition to the tooth as the nature of the trouble in it? I have given my affirmative answer: trouble in my tooth I perceive *as* the condition, pain. Since this pain is in fact not in the tooth but is a felt condition of my consciousness, the pain must therein represent for me the trouble in the tooth. Since this pain is obviously no concept and is rather a purely sensuous condition of my consciousness, it functions there as a purely sensuous representation of the trouble in my tooth.

5   Conceptual Attribution as Structuring Perceptual Content

It is now time to confront head-on this problem: since pain is not in our tooth, yet within perception we misrepresent the nature of the trouble in our tooth by sensuously representing the trouble's nature as pain, how, exactly, do we achieve perception of our tooth's condition *as* trouble?

Whether we call what we perceive in feeling toothache "something untoward" or "something wrong" or we call it merely "trouble" in the tooth, I think it plausible that Armstrong is right in proclaiming we do perceive something answering to a description along those lines. I think it further plausible that we can help explain the apparent fit of all these expressions with our perception of toothache by the hypothesis that perceptual *concepts* help to constitute this perception.

Perceiving that it is a tooth that is troubled requires exercise of a concept of a tooth, since the purely sensuous quality, pain, does not have for any part of its content, "tooth."

Moreover, perceiving that there is something untoward, something wrong, requires exercise of a perceptual concept having some such content as this: "needing to be changed, or treated, or fixed." The pain instantiated within our consciousness does function as the sensuous (mis)representation of the concrete nature of trouble in the tooth, but it can do even this only in virtue of a conceptual representation being tied to it. Our felt pain can represent a tooth as troubled only when a concept of a tooth and a concept of a condition needing to be changed or treated or fixed is united with the pain within a complex perceptual representation. Our perception of toothache integrates this sensuous quality of our consciousness—the pain—into an attribution to the tooth of a fusion of the trouble-concept's content and the sensuous content, the pain.

Perception of something wrong in a tooth also requires that the instantaneous aversion we feel in feeling the pain of toothache enter into the perceptual process itself. I offer this partial account: the simultaneous instantiation within consciousness of both a sensuous pain quality and a felt aversion to this quality so acts upon certain cognitive subsystems of our nervous system as to cause to be integrated, within the framework of a perceptual-attributional structure, a felt sensuous pain quality, a felt aversion, a perceptual concept of a tooth within one's mouth, and a perceptual concept of something needing changing or treating or fixing, so that a fused sensuous/conceptual/evaluative content—trouble-as-pain-in-a-tooth—is perceptually attributed to the tooth. This is our "evaluative perception" of a toothache.

If our aversive reaction were not integrated into the perceptual affirmation, we should not apply the *concept* of trouble to the tooth. If being-a-tooth, and being-in-this-mouth, and needing-changing-or-treating-or-fixing were not *conceptually* attributed to the tooth, we should not be perceiving a tooth as *a tooth* and as *troubled*. And if the pained quality of our consciousness were not itself *sensuously* attributed to the tooth, we should not be *feeling* trouble in the tooth.

Since pain is not in the tooth, our attribution of the sensuous quality, pain, as the nature of the trouble in the tooth is mistaken. How do we correct for this sensuous misrepresentation? A (partial) answer is now before us: there *is* a particular tooth in my mouth whose condition needs changing or treating or fixing, and within my perception of toothache the *conceptual* contribution to my ascription of trouble in a tooth is accurate.

This answer is completed in the next section, about causation within toothache.

The concepts whose exercise I have just assigned to the experience of feeling a toothache should not be identified with the capacity to use words or with something dependent upon this capacity. I do not mean to say that a person or other animal without language could not become perceptually conscious of toothache. How are we to think of these concepts? What, for example, is the nature of this perceptual concept of a tooth and of trouble in it which is independent of the use of words (or may be so)?

I take it that perceptual concepts of bodily parts and their conditions are practical in import: these concepts enable us to act noninstinctually toward our bodily parts as if they have the properties—or are in the states—that our perceptual concepts attribute to the bodily parts. Furthermore, because the practical import of objects and happenings before our sense organs, considered solely by reference to their natures, is invested in their causal powers, perceptual concepts attribute properties as causal powers of the objects and events to which they are attributed. Conceptual attribution of sensible properties as causal powers of things before our sense organs enables perception to position us for mapping out noninstinctual strategies of action directed toward parts of our bodies and toward things external to our bodies that are appropriate to those things possessing the causal powers attributed.

Indeed, the capacity of a perceptual concept to represent a thing as having a sensible property—hence as having a specific causal power—*consists* in the concept's enabling and disposing us (given goals that the property is relevant for reaching) to behave toward that thing as if it has that causal power. Thus, perceptually to attribute to a hole in the wall a square shape *is*, among other related powers and dispositions constituting the perceptual concept of squareness, to activate the power and disposition in the perceiver to fetch a square peg and push it into the hole in the wall in case the perceiver wants the hole plugged. The hole's capacity to receive a square peg in this way is a causal feature of its shape that partly constitutes the shape's physical squareness. If there were no conceivable behavior toward an object that would manifest the relevance for possible human purposes of its possessing a given property, then that property, simply *qua* itself, would have no causal content. For if it had causal content, then there would be relevance for some possible purpose of ours in our acting on it in such a way as to utilize its causal powers.

I need now to recognize a possible difference between experience of pain and other bodily "sensations" such as itches and tickles, on the one hand, and on the other, the perception of such sensible qualities of things and happenings as their colors, loudnesses, stenches, warmths, and the like. In order to do this, I need first to do something that is rather more important in itself: explain some sharp contrasts between the roles in perception of sensuous and conceptual contents. I am using the need to make a minor point about bodily feeling as motivation for making a major, contrasting one about perception in general.

For reasons I have space only to hint at, I have concluded that the purely sensuous contents of our perception of the class of sensible qualities made up of colors, loudnesses, stenches, warmths, and the like, do not contain, *qua* contents of sensuous attributions, any causal content. The hint at my reasons goes as follows. If we imagine ourselves making in isolation a *purely* sensuous attribution of the look of an object's color, of how a sound's loudness sounds to us, of the putrid smell of an odor, of the feel of a warm wind merely *qua* warm, and so on, we cannot imagine any behavior of ours that would be specifically adaptive exclusively to the purely sensuous content treated as if it is a feature of an object or event before our sense organs. That is to say, no matter what my goals may be, there is no behavior toward a thing experienced as if sensuously blue that will manifest the mere fact that I experience the thing as if sensuous blue belongs to it. Similarly, there is no behavior that might reflect how a sound's specific tone is heard, how a particular molecular emanation (an odor) smells, the mere feel of warm air, merely as such, and so on. But if there is no way to behave in relation to these sensuous qualities imagined as belonging to objects before our sense organs that, given our purposes, is specifically appropriate to these qualities being exactly those qualities of things before our sense organs—for example (ignoring simple color-matching behavior as not, in the relevant way, reflecting causal powers), if nothing in our possible behavior toward something perceptually taken to be sensuous blue can define a difference from its perceptually being taken to be sensuous yellow—then in sensuously attributing these qualities to things we thereby ascribe no causal power to the things we attribute them to. As contents of sensuous attributions to things before our sense organs, sensuous qualities of this large class have no causal content. Furthermore, for humans no instinctual behavioral responses are tied to the purely sensuous qualities attributed in our perceptions of colors, sounds, odors, warmth, and the like.[5]

Therefore, any information useful to action that is contained in perception of a given sens*ible* quality merely as such will be carried by *conceptual* attributions of causal powers or relations to it or to its alleged possessor. In view of the fact that its facilitating action constitutes the primordial purpose of perceptual cognition, and because purely sensuous attribution by itself is impossible to understand *as* attribution precisely because it serves no purpose of organizing our behavior so as to treat a thing as if it has the attributed property, I hypothesize (1) that the attributional structure within perception arises *originally* from *conceptual* attributions of causal agency to things before our sense organs, whereby perception's relevance for instrumental action is established, and (2) that the sensuous contents of perception get their capacity to participate in attributions of themselves to objects before the sense organs by being fused with conceptual attributions of causal powers. For example: our ability visually to attribute a color to a surface will have grown out of our ability within visual and tactual perception conceptually to attribute to surfaces their being surfaces, which is a causal feature of a visually perceived material object. In developing this perceptual–conceptual ability to ascribe to a thing the causal feature, "surface," attribution of a concept of surface will have made use of the perceptually available proto{sensuous *green*} that is waiting within our visual protoconsciousness to make sensuous *green* present to full visual consciousness as soon as the concept's attribution of "surface" to an object before the eyes can fully realize itself. It will fully realize itself by activating the potentially spread-out proto{sensuous *green*} within protoconsciousness and thereby transforming it into a spread-out sensuous *green* within consciousness which is capable of being perceptually attributed as the green matter of a surface before the eyes.[6]

In short, sensuous content cannot be attributed to anything without conceptual attribution of causal powers; but the latter cannot get a hold on the concrete reality before our eyes without a visually sensuous quality functioning as a sort of matter for the concept. It is in this sense that I believe perceptual concepts introduce a distinctively *cognitive structure* into the continuous stream of sensuous contents with which our everyday waking experience creates the presence to us of our bodies and of the world around us.

Now return to pain and the perception of toothache and to possible exceptions to the noncausal character of perception's sensuous contents. I think it possible that toothache's sensuous content, pain, along with the sensuous contents of itches, tickles, and the like, may be in one respect an exception. There may be indeed instinctual patterns of response associated with the attributions of these sensuous contents to parts of our bodies: for example, scratching an itch may be one. These sensuous contents of bodily "sensations," unlike the other perceptual sensuous contents just remarked, may therefore be found to have some measure of innate causal content. Nevertheless, I speculate that at least for mature human percipients, the account I have given of the role of perceptual concepts in attentive, hence in conscious perception of toothache is correct. Moreover (and now I need a very long sentence—but it *is* lucid!), since the richly cognitive structure of perception has been most fully developed because most needed in attentive, hence conscious perception of colors, sounds, odors, warmth, and the like, and in attentive, hence conscious perception of pressure by touch—though I've not spoken here of touch, see chapter 7 of my *Sensing the World*—in all of which, if I am right, the attributional structure of perception originates with attribution of perceptual *concepts* of causal agency, I speculate that, at least for mature humans, the same is true of attentive perception of aches and itches and tickles and the like, so that the cognitive structure of conscious perception even of aches and tickles and the like is also owing to the primordial attributional structure contributed by attribution of perceptual concepts of (here bodily) causal agency, into which structure sensuous contents such as pains and itchy or tickly feels are integrated by a kind of fusion, with the effect that these sensuous qualities too get attributed to objects and events before our sense organs.

6   Perceiving the Causation That Belongs to Toothache

Pointing out that we succeed in perceiving the trouble in our tooth *as* trouble by perceptually conceiving it as something needing changing, or fixing or treating does not do justice to our full perceptual representation of this trouble. I have in recent sections left out of focus the fact that in perceptually representing to ourselves the aching condition of our tooth as pain in the tooth, we represent this condition as an activity in the tooth. Recall that from the start we have been giving an account of perceiving a tooth *doing* something, namely, *aching*. Recall too that early on we noticed that a tooth's aching *is* the tooth's causing us to feel pain as if in the tooth. We also earlier on noticed that in attentively perceiving this aching we become conscious of the tooth's causing us to become conscious of pain (as if in the tooth). We can, therefore, now remind ourselves that we do perceive more of the nature of the trouble in the tooth than I have recently been remarking: we perceive a causal action of the tooth, its causing us to feel pain.

How, exactly, do we perceptually represent this causation? I answer that pain as the sensuous content of a perceptual attribution can serve to represent aching in a tooth only in virtue of its being fused with a perceptual *concept* with which we can attribute to the tooth the action of *causing* this very pain.

Think of this perceptual process as analogous to how we hear a bell as ringing when we listen with attention: we hear the bell *making* the sound we hear—we hear the bell as *sounding*. This requires that, within our auditory consciousness of the bell's sounding, there occurs a fusion of a perceptual concept of causal activity with a sensuous representation of tonal quality and loudness; this complex sensuous content we perceptually ascribe to the sounding as constituting both the character and the outcome of its conceptually indicated causal action. The "paining" of a tooth that we feel in feeling toothache is analogous to the sounding of a bell that we hear in hearing a bell ring. The sensuous representation, pain, is fused with a concept of the tooth as causally active. We perceive the tooth as the source of an action of itself, the "paining." Somewhat as, in hearing a bell sounding, we fuse the conceptually ascribed causal action of the sounding with its sensuously attributed quality and outcome, so too we experience the tooth's "making pain" as a "paining," fusing to the perceptual concept of a causal action a sensuous quality (pain) of our conscious state that is perceptually attributed to the tooth as if constituting both the quality and the interior outcome of the tooth's causal activity, the tooth's "paining."

There is, here, indeed, some sensuous misrepresentation of the concrete actuality of the tooth's action. Yet the tooth *is* causing the pain we feel—that is to say, cellular processes within the tooth are causing us to feel the pain; and with the aid of perceptual concepts we get right the abstract fact of causal agency in the tooth. Moreover, with the aid of the perceptually ascribed sensuous content, pain, we get concretely right a defining part of its effect—the pain we feel.

This completes the present account of how perception of toothache is, while indirect, realistic. Unhappily for me, it leaves objections still to be answered, only two of which I can tackle in this space. (See note 1 for reference to my treatment elsewhere of other objections.)

7   An Objection and a Complete Reply

Objection: It is a requirement of our very concept of sense perception that if we are to perceive a-tooth's-causing-us-to-feel-pain, that is, perceive a toothache, then the tooth's causing us to feel pain must itself play a causal role in generating perception of itself. This means that, since our feeling pain is a component of the complex condition perceived—that is, since it is a component of the condition, our-tooth-causing-us-to-feel-pain, our feeling pain must figure as part of the cause of our perception of toothache. However, this causal role is not possible on the present theory, because according to this theory, our feeling the pain of toothache exists only in the form of our sensuously attributing pain to a tooth; so attributing pain is itself a constituent of perceiving a toothache. But surely a constituent of a perception cannot cause the perception.

Reply: Our perceiving a tooth's aching comprises, in addition to our sensuously (but falsely) attributing pain to the tooth, our conceptually attributing tooth-hood to the tooth, trouble to the tooth, and causation of our pain to the tooth. I answer that it is plausible to suppose that in causing me to feel pain, the trouble in my tooth *thereby* also causes me conceptually to attribute to the tooth tooth-hood and to fuse the sensuous content, pain, with a felt aversion to it and with a conceptual attribution to the tooth of trouble in it and with a concept of causal action of the tooth, all of which yield the complete attentive perception of the "paining" that is toothache. In this way a tooth's causing me to feel pain, which is one constituent of the perception of toothache, brings about the remainder of the perception and thus brings about my full perceptual consciousness of the tooth's causing me to feel pain. This answers the objection.

8   Final Objection, and an Incomplete Reply

Objection: A condition of an object can be perceived by sense only if it exists wholly before our sense organs where it can contribute to the causal action of the object upon those organs wholly from before them and thereby help cause the perception to occur. A tooth's aching cannot exist independently of the sufferer's feeling the pain that figures in the definition of toothache; indeed, the sufferer's feeling this pain partly constitutes his toothache. But the sufferer's feeling pain does not exist before his sense organs. Therefore a tooth's aching cannot exist wholly before our sense of pain's organs. Therefore a tooth's aching cannot act upon our sense of pain's organs wholly from before them so as thereby to help cause us to perceive the tooth aching. Therefore, a tooth's aching cannot be something we perceive by sense.

Reply: This objection is so strong it requires a radical step: rejection of the opening premise—rejection of the proposition that if we perceive by sense a condition of an object, then the condition must exist wholly before our sense organs where it can contribute to the causal action of the object wholly from before those organs and thereby help to cause the perception to occur. Justification for the improbable rejection of the "wholly before our sense organs" portions of this premise requires an argument I have not here space to give in full because it needs as *its* premise the successful defense of a particular theory of the nature of a class of sensible qualities involving sounds, colors, odors, warmth, and the like. Hence the incompleteness of my reply.

I begin by giving a sharply abbreviated statement of a theory I have defended elsewhere for several senses, shortened here into an account of the perception and nature of a single sensible quality in the just-mentioned class, the loudness of the sound of a ringing carillon bell, which is here treated as stand-in for the class's other sensible qualities.

A sound is in fact a train of waves of air pressure stretching from the bell to the listener. Attentively to perceive this train of air-pressure waves by the sense of hearing is to become aurally conscious of its presence under some aspect. Of course we do not normally become aurally aware that the sound is a train of air-pressure waves. However, our auditory attention can be held by a sound before our ears only if at least one property of the sound holds it. In becoming conscious of the presence before our ears of a sound, we do become conscious of an instance of a property of the sound—for example, of a specific loudness. To become aurally conscious of an instance of loudness before the ears is to come to know by hearing that this loudness belongs to a sound before our ears. Standing in the open air, we can come thus perceptually to know by hearing it that this specific loudness belongs to a sound only by aurally attributing this loudness to the sound before our ears.

Now consider what the loudness is that we aurally attribute to the sound in hearing it. How the train of air waves sounds to a perceiver is the sensory impression it makes on the perceiver, the impression, *thus (loud)*—a sensuous quality belonging to the perceiver's conscious state. Though it is not so experienced by the perceiver, this sensory impression, *loud*, consists in fact in the perceiver's sensuously attributing to the sound before his ears the specific sensuous quality, *loud*, which is instantiated within his consciousness where it functions as the content of his attribution of it to the sound before his ears. By means of perceptual concepts he additionally attributes to the sound before his ears its causing itself to sound to him the way it does sound; finally, within this perception he conceptually attributes to the loudness its habitually sounding this same way when heard. Upon reflection, this habitually sounding so can be conceived as this (sort of) loudness's disposition so to sound when heard. It is therefore reasonable for us theorizers to hold that the sensible loudness we aurally perceive and become conscious of when we attentively listen to a sound is the causal action of the sound in bringing about in us a certain sensuous quality, *loud*, together with this (sort of) sound's disposition to do the same whenever heard.

Upon reflection we notice, of course, that the sens*uous* quality, *loud*, is not the sen*sible* quality we perceptually attribute to the sound itself, not the sound's heard loudness; for the sens*ible* quality is a present causation by the sound (the air-wave train) of the instantiation of the sens*uous* quality, *loud*, within our aural consciousness, together with this (sort of) sound's disposition to sound this way when heard. Lacking aural-perceptual knowledge of the real nature of the sound—its character as a train of air-pressure waves of a certain frequency and intensity—in auditory perception we conceptually mistake for loudness's nature the complex of sensuous qualities we attribute to the sound and to which, also, we conceptually assign the loud sound's causal power to sound to us the way it does sound.[7]

As premises in my reply to the present objection, then, I need the plausibility of a full argument for the conclusions just defended, plus the plausibility of similar arguments for all others in the relevant class of sensible qualities: for colors, smells, felt warmth, and the like; plus, if you will, I need an argument to the effect that other accounts of this class of sensible qualities are unacceptable. (I have developed such arguments for a close kin of this theory in *Sensing the World*—however, there are more accounts on the scene now than there were then.)

Assuming these premises achieved, we have the following situation. With our temperature sense, our hearing, our vision, our sense of smell, and the like, we perceive a certain class of sensible qualities which, as perceived, consist in the objects perceived causing in us then (and the sort causing habitually) certain sensory impressions; molecular velocities of a certain magnitude cause their containing macro-objects thermoperceptually to feel a certain way, trains of air-waves of a certain intensity and frequency cause themselves (and often their sources) to sound a certain way, surface reflectances and molecular emanations of specific sorts cause themselves or associated surfaces or bodies respectively to look and to smell certain specific ways, and so on. For brevity's sake speaking only of loudness as representative of the others, we perceive the loudness only by aurally perceiving the sound's causation of our own auditory impression and by aurally perceiving in the sound's loudness the habitual power of a such a sound to cause the same.

Now we observe that this attentive perception of causal action upon our conscious experience of objects of perception whose sensible properties we thereby perceive has the same sort of structure as our attentively perceiving a tooth's causing us to feel pain and thus perceiving the tooth's sensible property that consists in its aching. On the assumption that the above account of the nature of the specified class of sensible qualities is the best account, it is therefore reasonable to accept as fact that sense perception of a tooth's causing us to feel pain *also happens.*

This answers—incompletely—the objection that we cannot perceive toothache because it requires perceiving a condition that is not situated entirely before our sense organs where it can act upon those organs wholly from before them.[8]

9 Coda

Because perception's conceptual contents represent nothing but causal powers, but its purely sensuous contents do not, and because within perception every conceptually ascribed causal power's meaning for the percipient is purely practical and therefore points to the future, only our perceptual attribution of sensuous qualities to things before our sense organs makes both our bodies and the larger world a *presence* for us, and us present in a world.

Acknowledgments

Some parts of this essay strongly resemble parts of my "Chapter Two: Feeling One's Body" in my book, *Sensing the World* (Indianapolis, Ind.: Hackett, 1983). I thank the publishers for permission to make this use of that chapter.

I have had helpful comments on the present essay from Irving Singer and from Murat Aydede, for which I thank each of them.

Notes

1. Being conscious of something whose existence depends upon our being conscious of it, for example, a pain, poses problems many philosophers have remarked. I have not space here to attack these problems, but I take considerable trouble to solve them in sections 1–5 of chapter 8, *Sensing the World*, where I find it necessary to introduce ideas of transitions from linked protoattention (protoconsciousness) and protopain or protosensuous *green*, etc., into linked attention (consciousness) and pain or sensuous *green*, etc. Somewhat surprisingly, I find there that I can solve these problem only on the assumption—for me a welcome one because I first made it in a 1952 dissertation ("A Study of Cognition," Harvard University)—that all conscious and protoconscious states are brain states.

There is also a problem about the epistemic characterization—indeed a problem as to the very possibility—of direct consciousness of sensuous content where this consciousness of content cannot involve attribution of anything to anything, since attribution of sensuous content to external things makes use of *it* as an ingredient—that is, uses direct awareness of sensuous content as, so to say, the awareness of *what it is* that one sensuously attributes to something before the sense organs. I deal with this problem in section 5, chapter 8, *Sensing the World*.

2. Here I have been influenced by Nelson Goodman's account of reference by exemplification. See his *Languages of Art* (Indianapolis, Ind.: Hackett, 1976), pp. 52–57.

3. The importance for a theory of sense perception of the idea of sensuous attribution of a sensuous quality to an object before our sense organs cannot be exaggerated. I have here presented this notion as specifying a difference between pain perception of a bodily part and any merely ideational awareness of the same part. It is a principal thesis of my *Sensing the World* that, in this way of being internally different from merely ideational awareness, at least five other sensory modes of consciousness of physical objects and events resemble the pained mode.

4. D. M. Armstrong, *Bodily Sensations* (London: Routledge and Kegan Paul, 1962).

5. For my full argument that the sensuous contents of most perception, merely as such and without supplementation by perceptual attribution of conceptual contents, assign no causal powers to the sensuous qualities that they represent, see *Sensing the World*, chapter 8.

6. See note 1.

7. For detailed articulation and extended defense of a theory of sensible qualities and our perception of them that is similar to the theory outlined, see *Sensing the World*, chapters 2–8.

It is interesting that for sounds with frequencies below 18 cps we *hear* pressures and frequencies, as if we were feeling the sounds by touch, though it is demonstrable that we are not doing so. From this I infer that if all sounds had been of a frequency below 18 cps, we should never have had a use for the ideas of loudness and tonal quality of sounds—only the qualities needed for reporting the outcomes of feeling by touch—pressure and frequency of vibration—would have been needed to report auditory experience!

8. One might want to carry over from our conclusions about other sensible qualities the conjecture that, like them, a toothache as perceived includes also in its nature the afflicted tooth's habitually causing the same sensory impression while in this condition and holding our attention.

12   Categorizing Pain

Don Gustafson

This chapter contains four parts. In part I, I examine the widespread view that all pains have in common a "subjective quality" by dent of which they are categorized as pains. I find this view questionable. In part II, I romp through some of the history of pain theorizing (both "prescientific" and since circa 1650). Pain has been differently categorized against different cultural and intellectual backgrounds. By now, differently categorizing pain serves scientific inquiry and clinical, medical practice. In part III I also suggest that the category into which pains are put is warranted by the category's research and clinical utility. Since research and theorizing aims change and vary even at a given time, the categorization of pain is relative to changing goals of research and against the different conceptual backgrounds of changes in goals. In part III, I introduce some considerations showing that thinking of pains as sensations currently lacks utility. Some recent and current pain research points away from the simple sensation picture of pain. In part IV I suggest that pains are usefully categorized as emotionlike, that is, as conditions served by the limbic system with motivational or behavioral force on their own. I conclude part IV with a summary and a recommendation.

I

There is no one general way pains all feel. If this is so, we should wonder what defenders of it have in mind in the view that "qualia" define what kind of state or condition a conscious experience is, essentially, and that, there must be a single "pain quale" in virtue of which a state or condition is a pain, necessarily. What is the one feature of a feeling-sort that all and only pains have? Is the idea of such a thing credible? I think not. Let me suggest support for this claim.

The position to be tested here requires that each instance of pain has a specific quale in virtue of which each is typed as pain; no single pain quale, no pain. There must be a common feature (*the* pain quale) had by all cases of pain experience (cf. Chalmers

1996, p. 359, note 2: "I use the term ['qualia'] in what I think is the standard way, to refer to those properties of mental states that type those states by what it is like to have them . . ."). So we may ask what, *exactly*, is the felt quality all of the following have in common?

a dull ache in the gut

a sharp, catching pain in the side during a long run

a burning sensation in the feet felt by some diabetics

a crushing blow to the top of the thumb when it is caught in a slammed door

labor pain just before delivery

covade syndrome

a hangover headache

angina

phantom pain in an amputated foot

a gallstone pain

fibromyalgia syndrome

warm water on a sunburned back

a ruptured cyst or fibrous growth of the uterus

tic douloureux or trigeminal neuralgia

recurrent quasirheumatic limb pain peculiar to early youth

sympathetically maintained dystrophy

a radiating pain in the left side of the neck into the left arm

a sharp splinter under the thumbnail

a migraine (of one or another sort)

cancer pains (of one or another sort)

contact with a thorn or thistle or their poisons

a hand on a very hot surface

herpes zoster (hives)

tension-type headache

an ant or bee sting

endrometrosis pelvic pains

cluster headache

pancreatitus

carpal tunnel syndrome

a scorpion sting

phantom pains of numerous sorts

pericarditis, myocarditis, or endocarditis pains

corneal pains

testicular torsion pain

oral pains and dental pains, and periodontitis

acute appendicitis

opiate withdrawal pain

ulcerative gingivitis

And there are many more. This is a randomly selected list. The sheer cardinality of its members should suggest that it is unlikely that they should each have a single qualitative feature in common.

It will be said that these have in common that the host has a hurting, unwanted conscious experience. This is correct, for the most part. But suppose one hosts two or three of the pains in the previous list at the same time. Do they have a common, qualitative feel? Of course each is a hurtful state, but we were looking for a qualitative feature of the sensation hosted by the pained person or animal in virtue of which each conscious condition *is* unwanted. Our query was: what single sense quality do these conscious experiences, serially or at the same time, have in common in virtue of which they are all typed as painful, conscious, unwanted (for the most part) states? They are typed as a specific kind of conscious state, namely, pain, and it is in connection with this typing of them that we are seeking the common quality. For my part, I find no one thing that makes them all pains. Their qualities vary and vary in a multitude of ways.

This finding should not be surprising to anyone who has looked seriously at the language of pain expression and description. Consider merely the multiplicity of descriptors for pains (in English). This multiplicity makes it unlikely that we will find a single subjective quality for pain referenced by each descriptor.

Researchers often distinguish among the descriptors those that are (1) evaluative, (2) those that are sense quality oriented, and (3) those that are affective or emotive or motivational in character (Melzack 1975).

There is also diversity in each of these categories. Evaluative words range from mild to excruciating. A few such evaluative descriptors are:

amazing

distressing

miserable

horrible

intense

unbearable

—and so on. None of these seems to represent a sense quality. They are evaluative of the experience of pain, and they are often comparative with other states.

In category (3), affective terms, we find such words and their cognates as:

tiring

exhausting

suffocating

frightful

terrifying

punishing

vicious

killing

wretched

None is a subjective, sense quality term.

Only type (2) descriptors are more or less clearly sensory labels, thought one might question this in some cases. Terms that appear in the McGill Pain Questionnaire arise from a distillation of bits and pieces of pain language of many patients and subjects in a wide variety of clinical and experimental settings. These sensory words are further classed as temporal and include, among many others:

flickering

pulsing

throbbing

pounding

They are also classed as spatial. These include, to name just a few:

jumping

flashing

shooting

Punctate pressure words include:

pricking

boring

drilling

stabbing

Sensory terms also include incisive pressure terms, such as:

sharp

cutting

lacerating

The class of spatial words has a subclass of constrictive terms, including:

pinching

pressing

cramping

crushing

Traction pressure words include:

tugging

pulling

wrenching

Thermal words are prominent and a list contains anything from hot to searing. Brightness words include smarting and stinging and others. Dullness terms include:

dull

sore

aching

heavy

Such descriptors as tender, taut, rasping, and splitting are also used. The McGill Pain Questionnaire also includes terms such as:

radiating

penetrating

piercing

tight

numb

squeezing

tearing

cold and freezing

Many patients make use of likeness descriptors, for example, as if a hot poker is piercing my foot, or as if someone is standing on my chest, and on and on.

The list generated by the research that produced the McGill Pain Questionnaire shows an enormous variety of what might be called "sense quality terms." There just is no single term to be found here. There's an embarrassment of riches, too many different qualities from which to pick just one. And, there seems no intersecting, more general, term save for 'painful,' but this is the very word for which we are seeking a purely qualitative definition.

If it is claimed that the sole pain quality is the feature of being unwanted by the host, we should remember that not all injuries associated with pain are unwanted; battle injuries without pain are robust. More importantly, it seems completely plausible that pain is only often or typically an unwanted conscious state. One can take a variety of attitudes to pains. Some are said to be exquisite pleasures, by the masochists, for example, or by the religious ascetic who welcomes pain. And one can easily imagine one's delight in responding to a sharp pin in the leg, calf, heal, or toe with "that hurts!" as one is being checked by a neurologists for a possible spinal injury of a major sort. Wanted pain. Montanans caught out in thirty degree below zero weather with forty-five mile per hour winds might hope and even pray that their digits and limbs continue to hurt, to give pain. The worst happens when pain disappears—first red, then black, and then fingers or more are lost. Pain is wanted.

However hurtfulness is understood in a mature theory of the psychological and motivation character of pain, hurtfulness, too, does not appear to be a single feature when considered in the large variety of settings in which we find it in common experience. If we reduce it to a single parameter, as we might in theory, it will pretty clearly cover more than pain. It will encompass any condition the agent or animal takes to be untoward, counter to its good, a threat to its projects or ongoing activity, or a state to which its host needs to attend. It seems plain that humans categorize experiences as pains along several dimensions, with loads of differences as to sensory and affective dimensions. Pains seem too diverse to admit of a single defining or essential quale.

(There seems to me to be an error in the view that "there is something it is like to be in pain [or another 'mental' state]" if the hosting of this property defines what 'being in pain' means. This is because the understanding of "There is something it is like to be in pain [or whatever]" presupposes an understanding of "There is a condition of being in pain [or whatever]." If so, the first presupposes the second and thus "what it is like" cannot *be* what *being it* is. Anyone who does not see the error to which I am trying to point here probably has an antecedent view that "mental" conditions have subjective essences or something like this. In that case there needs to be an independent justification of this antecedent view. Of course, it will be an appeal to intuition of some sort, either introspection or conceptual intuition, or intuition concerning the meaning of predicates such as 'is in pain' and so on. A naturalistic program of the sort I am following in this chapter just does not accept such intuitions at the outset of inquiry. This program is to be judged by its utility and fruitfulness in supporting ongoing research and, in the case of pain, progress in clinical understanding and practice.)

II

The conception and understanding of pain has a history. Changes in this history suggest that the categories humans have found for their experiences are contingent on their beliefs and other views at their place in history and culture. The following discussion aims to illustrate these claims (cf. Keele 1957; Rey 1995; Dallenbach 1939; Brazier 1984; and Bonica 1990).

The history of pain and thinking about it is long and fascinating. A historian of pain can claim for early Greek, Egyptian, and Hebrew sources that "Primitive human thought . . . links pain closely with object intrusion" (Keele 1957, p. 2). Since pain was clearly associated with injury and with disease, itself an assault on one's body, it was natural to think of these "intrusions" from outside the body as pains. Such intrusions upset the continuity of the body/soul functions and the harmony on which they can be thought to depend.

Plato's *Timaeus* gives us one picture of sensation, affection, and pain. Evidently following the physical views of Democritus about atoms that constitute earth, air, fire, and water, Plato suggests that the triangular atoms of each of these substances impinge on the human body, ultimately on liver, heart, or brain. But firey atoms—sharp-edged and triangular—intrude on the body and are presented as burning pain, and not as sharp-edged firey atoms. These atoms and their vigorous motions produce "affection," of which each sensation type is an instance. Pain can then be conceived as an affection that is against nature, that is, is contrary to the natural state of the body/soul. In *Timaeus* Plato says, "When an affection which is against nature and violent occurs within us with intensity, it is painful, whereas the return back to the natural condition, when intense, is pleasant . . ." (cited in Keele 1957, p. 64). In short, pain is an affection or intrusion on a natural (homeostatic?) body/mind.

At this early stage of theorizing about and categorizing of pain, we find a distinction between what pain is, on the one side, and how it is presented in experience, on the other. This view can be seen to persist until about the eighteenth century when pain becomes conceived as just what it appears in experience as.

In contrast to the *Timaeus*, Aristotle picks out the heart as the vital organ of experience (as well as of nutrition, movement, and heat in the body)—see *The Parts of Animals* (3.4, 666a, cited in Keele 1957). And in Aristotle's *On the Soul* he importantly links the sense of touch with pain experience. Here the idea is not so much an intrusion on the body, as it is the notion of the active body coming into contact with external conditions. While pain is not yet fully internalized, as in eighteenth century and subsequent thinking, the introduction of the modality of touch as the key mode for pain experience is a step toward such an internal category for pain.

Corporeal touch can be rich in blood and directly connected to the heart, which is the seat of feeling. Aristotle theorized that excessive touch is what pain is, in its nature. If the blood of skin is hot, pain occurs. If the blood is cooled, pleasure results. Thus, "homeostatically" balanced, ". . . when their sensible extremes such as acid and sweet are brought into their proper ratio" but "in excess they are painful and destructive" (3.2, 426b, cited in Keele 1957), it is touch that is painful or pleasurable. (This connection between touch and pain remained a major notion in thinking about pain until the exact physiology of skin sensations is understood in the twentieth century.) On the other hand, Aristotle's conception of pain as linked to more-or-less homeostatic conditions and his consequent notion of the affective–emotional character of such conditions is a view that now has some currency. (This will be the theme of later parts of this essay.)

With Galen (A.D. 130–201) the nervous system becomes a key player in experience. Soft nerves receive impressions ("intrustions" again) and transfer these to the brain. Hard nerves, by contrast, serve for motor activity. The brain itself is said to be softer than any other nerves, so that it can receive sensations from any source. With Galen we get the first explicit account of the biological role of pain and the nervous system. In *De Usu Partium* he writes that "[n]ature . . . has had a triple end in view in the distribution of nerves: she wished to give sensibility to organs of perception, movement to organs of locomotion, and to all the others the faculty of recognizing the experience of injury . . . the distribution of verves is the perception of that which can cause harm" (Bk. 5.9; Galen quotations are from Keele 1957).

Galen's view, unlike Aristotle's, was grounded in observation of dissected animals, while Aristotle's was to him and many others a clear consequence of attention to common experience. The appeal to commonsense had no force against Galenian dissection and observation. Thus, Galen wrote that "one who knows by dissection the origin of the nerves that pass to each part will be better able to cure each part, deprived of sensation and movement . . ." (*De Locis Affectis*, 3.14, cited in Keele 1957). While Galen's picture of the operations of nerves is pneumatic, he clearly had in mind a mechanism in the nervous system and brain, serving pain. The conception of pain, I believe, takes another step toward "internalization" though now in the body itself.

Greek and Roman thought about pain and the nervous system was more or less lost in the West under Christian and church teaching. Among the Christian ascetics pain took on a religious role and it was not something to be avoided. The ascetic who engages in self-mutilation did not think of pain as unwanted because if it did not have its distressing character, it would not serve it religious function. It was the unwanted character of pain that was desired.

At the same time the tradition from Galen was alive and active in Arab thought (e.g., Avicenna). By the fifteenth century these sources were again available to, for instance, Leonardo da Vinci (1452–1519). Leonardo practiced dissection and produced wax casts of the areas and surface of the brain. Subsequent writers (Vesalius 1514–1564 and Eustachius 1520–1574) added details of the anatomy and parts of the brain, spinal cord, and other bodily systems. Galen's picture was reinstituted by, for instance, Vesalius with the view that "nerves originate in the brain. Nerves are to the brain as the aorta and its branches, to the heart . . ." (*Vesalius on the Brain*, cited in Keele 1957).

Against this background Descartes emphasized the role of nerves and brain in sense experience and pain. He presented a detailed account of what he took to be the hydrological and mechanical structure of the brain, giving pride of place to the infamous pineal gland. When reading Descartes's account of the medium of nervous action, namely, animal spirits, we should remember that this medium is of the same sort as blood since it arises from blood, though it has a less course form. This allows it to pass through fine pores of the brain and small ventricles of the brain. It is "animal" and thus is a common medium of all nervous systems. Somehow the medium has the properties of a taught string, acting as "one end of a rope one makes to strike at the same instant a bell which hangs at the other end" (*L'homme*, Foster translation, p. 262).

In this view of body and experience, we can see the presence of an important distinction between nociception and pain. Pain is experienced as a conscious experience in the soul or mind due to the soul–mind's access to a terminal "physical" activity in the brain. This activity roughly corresponds to a fragment of our current conception of the nociceptive system for pain. For Descartes's system falls totally on the bodily side of the chasm between mind and body. His view has factual consequences; for instance, it requires that pain *felt in* the hand, joint, head, and so on, is only mistakenly so felt. The "science" of pain, for Descartes, trumps the commonsense notion that pains are literally where they are felt to be. He points to phantom pain phenomena to support this requirement of his theory, since an amputated arm may have pain as if in the hand where there is no hand in which the pain is felt. (While felt pains may be instances of what Descartes calls "clear ideas," i.e., ideas incapable of being confused with other, different ideas, pains are not as well what he labels "distinct ideas," i.e., ideas the presence of which discloses the whole essence of the represented reality. The difference, between clear ideas and both clear and distinct ideas, allows Descartes room for recognizing that pains may not be, essentially, what they are clearly experienced as. Note, thus, that Descartes's view is somewhat different from the eighteenth-century notion that pains are conscious states that are transparently sensory impressions whose qualitative character defines them as pains.)

At the beginning of the eighteenth century the conception of pain still languished in the Aristotelian tradition of the pleasure/unpleasant emotional qualities, though inroads on the security of the tradition had been made by Descartes as against William Harvey's views about the heart and circulation of blood. In Erasmus Darwin (1731–1802) we still find pain thought of as a function of the strength of "sensoral motions" with those stronger than usual producing pain. So there was a mix of the Aristotelian notion of pain and pleasure (i.e., affective conditions) as continuous and of a similar kind, and the growing awareness of the role of the nervous system. Anatomy and the physiology of various parts of the nervous system became a focus of considerable research. This made the idea of pain as sensation, along with the other modalities of experience, seem apt.

By the nineteenth century physiology had become an experimental science. Psychophysical investigation was not long in arriving. Johannes Müller (1801–1858) developed the view that the central nervous system receives impulses from external things only through sensory nerves and such nerves are specific in the information they serve up. There is a specific form of energy for each kind of sensation. So changes in sight, hearing, smell, taste, and touch need to be correlated with specific physical differences in stimuli. But sensations are now conceived as inner states or events, with physical energies as their outer correlates. Since pain sense was still categorized as a special case of the sense of touch (along with heat, cold, itching, and the like), pain had not yet been thoroughly internalized as a state of consciousness alone. But the progressive internalization of pain had taken another step inward with the demise of the Aristotelian pain–pleasure conception and the idea that sensations of touch and pain were the *inner* correlates of external energies activating specific nerves.

Not all of the internalization of pain was due to changes in the early understanding of the physiology of the nervous system. Therapy for pain advanced when the use of opium and codeine (purified from uncertain mixtures) became available. These seemed to be acting on the consciousness of pain. Acetyl salicylic acid was produced and later came to the market as aspirin. In mid-century the anesthetic qualities of ether were known and ether came into use. Backed by the new physiological understanding of the sensory nerves, local and regional anesthetic pain modification came into use. But there seemed to be little understanding of the difference between anesthesia and sedation at that point. Substances that worked to relieve pain were thought to work on the sensory system and to block the pain pathways from leading to conscious sensations of pain. With the categorizing of pain as sensation, pure and simple, the internalization of pain was almost completed.

Finally, the nineteenth century produced an experimental and theoretical rivalry that was not settled until the last century. The "specificity" and the "intensive" theories of pain were to do battle well into the last century. These rivals produced specialized versions as test results of the main combatants generated data that seemed not to easily fit either. (In briefly telling this history of pain theory I shall follow in the tracks of Karl Dallenbach's 1939 account.)

After Johannes Müller's 1840 presentation of the theory of specific energies for nerves, systematic work on sensation and nerves got under way. On this view, qualitative difference in sensations depended on differences among specific nerves. Therefore, systematic categorization of sensation types and nerve types became the aim of research. Special areas of skin (more-or-less single cell size points) were correlated with sensations; separate spots of skin were discovered for warmth, cold, and touch. Von Frey (in 1894) mapped out spots sensitive for pain and touch; he also did histologic studies of skin to identify the specific end-organs that give rise to the different sensations. So, on this basis von Frey advanced the specificity theory of Müller so that the sense of touch was composed of the cutaneous modalities of touch, warmth, cold, and pain. Further, researchers with analgesics on animals and incisions in the spinal cord of animals discovered that touch could be spared but not pain or pain left in tact but touch not; this independence of sense experiences implies the specificity of their types. Discoveries of separate skin spots for warmth, cold, pressure, and pain that yielded their own specific quality was generally taken as sufficient evidence for the view that pain is a separate department of sense.

The classical classification of pain as part of an affective pain–pleasure continuum was now almost off the contending map in the science of pain. But the intensity or intensive theory was still in the running. This is, again, the view that pain is excessive or sufficiently intense stimulation of one or more of the other sense modalities. Touch is the model; more intense stimulation to touch transforms into pain, so there is thought to be a shift in quality of experience due to a change in quantity of stimulation. Accordingly, a summary account of the theoretical situation is that

[b]y the late 19th Century the debate about the sensory states of pain was divided into three main views: the classical one, pain as an emotional state; a more recent idea, pain as a concomitant of all sensations when sufficiently intense; and a concept with roots back to the great Moslem physician, Avicenna (980–1037 A.D.) of pain as specific sensation derived from particular afferent organs. Residues of all three of these interpretations persist to recent times in the somewhat less than orderly classifications of knowledge about this subject. (Perl and Kruger 1996, p. 182)

At the beginning of the last century, then, there were the contending theories of pain often called the "intensive" and the "specific sensory" theories. Each has vigorous defenders. As well, there was a vast minority who carried on with the tradition of pain as an affective quality on the pleasure–pain continuum (see Dallenbach 1939, pp. 338–339). Many of the defenders of the intensive theory were psychologists; the physiologists were supporters of the specific sensory theory. In time the last won the war. When the findings of Max von Frey were available to the larger world of science, it became clear that there are specific and distinct nerve endings serving different sensations. Warmth is correlated with Ruffini endings in the dermis; cold is correlated with Krause end-bulbs; pressure has Meissner corpuscles as subtending structure. Finally, pain is served by free nerve endings in the skin.

Skin physiology is complex, with layers and with differences between hairy and nonhairy (or glabrous) skin (see Greenspan and Bolanowski 1996). A number of types of endings are now recognized and each is a transducer for its own type of physical stimuli. What is important to note for present purposes is that anatomical and physiological findings settled a dispute at the level of psychology, between the intensive and specific sensory theories. As it turns out, much of the data thought to support the intensive theory is now explained by the specific nerve structure view and such processes as summation, sensitization, or adaptation.

Mapping of specific nerve endings to specific types of experience has been accomplished. Specific fibers for the transduction of noxious stimuli are identified. And the detailed structure and function of connecting neurons is known (Bromm and Desmedt 1995; Willis and Westlund 2004).

How does this upshot fit the internalization of pain story I have been sketching? The psychophysical work and the correlation of experience with underlying anatomy and physiology is predicated on the conception of the experiential side as types of sensations. Sensations are just the kinds of specific, individual items needed for secure correlation. Feelings, affects, complex emotions, and the like are not specific, sufficiently isolable, identifiable as "psychic atoms" so that psychophysical work could even begin. Touches, heat sensations, pressure sensations, and pains had to be thought of as individual, isolable units so that the scientific search for psychophysical correlations could progress. It is evident, from early work on psychophysics (Fechner 1860) that strength of stimulus (physical magnitude) and subjective intensity or strength are functionally related. Perhaps "just noticeable differences" are equal steps from threshold in sensation in every modality. Sensations are needed as the units in any such inquiry. And the category of sensation is already thought of as a type of inner, subjective condition, *in the mind!* The theoretical internalization of pain is almost complete.

A coda to the story is added with more or less purely philosophical theorizing. Work on the psychophysics of pain used introspective methods, either directly with subjects' responses to queries about subjective sensations or indirectly with subjects' (or patients') comparative judgments about pain severity or intensity. Introspection was thereby a legitimate data delivery method. And since the seventeenth and eighteenth centuries, philosophical idea of consciousness became "intuitive" as a subjective domain or "Cartesian Theatre." (This notion of consciousness is absent in Aristotle and his tradition until about the eighteenth century.) The conception of wholly subjective objects of introspection became a philosophical staple. Pain, in turn, became the paradigm subjective object of secure introspection. Incorrigibility of pain reports and expressions attained the status of a dogma.

This is my story of the internalization of pain. It is a history. Its episodes are contingent events. Our conception of pain is delivered to us by historical, scientific, and philosophical contingencies. If this is plausible, it is also plausible that we may wish to reconfigure our conception of pain or the categorization of pain if we find another more useful for current interests. After all, if the history just now sketched is possible, then there is no necessity in the idea of pain as an essentially hidden or inner subjective condition, accessible only by means of self-consciousness.

III

The task of this part of the discussion is to show that we are now in position to select the category for pain that best serves our scientific search for the nature or natures of our target phenomena. In fact, current research in the sciences of pain point us toward a conception of pain as an affectivelike, emotionlike, or needlike condition. (Perhaps Aristotle's model, transmogrified somewhat, will prove helpful at last.) I use the weasel terms "affective, emotion, and need*like*" rather than merely the unqualified categorical terms, since each of these categories is open to naturalistic reconfiguration in light of pragmatic demands of research and theorizing. Our findings are, in the nature of the case, *pro tem* only.

I cannot survey all of the findings that support a category type for pain other than sensation. Nor can I cite all the current theorists I think point us towards a richer, more complex category and away from simple sensation. What follows are samples of both the empirical findings against the simple sensation view and samples of theorists who have suggested a complex category for pains.

Why the simple sensation category for pain is implausible: sensations are the outputs of sensory modalities that may well be subtended by relatively modular systems. While higher levels of processing may affect full perception of ambient environments, the sensations that go to make for perceptions are themselves relatively immune from higher-level influence. The older philosophical notion of the passivity of impressions of sense, that they are just there in experience and not open to willful influences, is at work in the conception of their relative immunity to higher level influences. But this immunity from cognitive or other higher level effects is not present in the case of full-blown pain perception. And this suggests that the last is not a matter of simple sensation. There are a number of data that bear on this claim.

First, responsiveness to noxious stimuli is subject to anticipation effects and labeling effects. This should not be the case if the responsiveness in question is a matter of simple sensation. For instance, Pennebacker's characterization of anticipation and labeling effects is that "the anticipation of a change in internal state—even though there is in fact no change—is sufficient to cause the person to perceive and label the event as if the change had occurred" (Pennebacker 1982). A clinical example of this kind of effect is found in the fact that a perceived ambiguity of illness and the news of an indeterminate prognosis often increase the reported suffering or discomfort of patients. Many patients begin to experience increased pain when they learn the real and dire nature of their disease, even when the disease process has not progressed and may have become less dire. These facts suggest that pain is a more complex phenomenon than it would appear to be under the category of a simple sensation.

Second, the placebo effect for pain is robust. It can be dramatic, as in the case of a sham heart surgery's reduction of angina pectoris (Beecher 1961). Treatments of pain that are accompanied with the strongly held belief that they will be effective but that have no known analgesic action on their own are placebo responses if they reduce pain. One among the leading pain researchers of the last century writes with passion about the placebo response. "Some physicians think that anyone who responds to a placebo did not have a "real" pain: they are wrong. Some physicians think that a placebo is the same as no treatment: they too are wrong. Some think that only weak-minded suggestible people in minor pain respond: they are wrong. Even physicians respond to placebo! The placebo response is a powerful and widespread phenomenon" (Wall 2000, p. 127). Wall lists a few dramatic placebo reaction cases.

Wisdom tooth extraction is followed by inflammation swelling, resistance to movement by the patient, and pain and distress. For some time ultrasound has been used to treat inflammation. But trials have shown ultrasound to be no better than placebo; that is, it can be highly effective. In a double-blind test, health workers used ultrasound on the faces of patients with wisdom tooth extractions when some machines were actually functioning and some not. This was not known by the health workers or the patients; conditions for a double-blind study were met. There was no difference in upshot, that is, some swelling was reduced, inflammation subsided, and ability to move the jaw is enhanced. The treatment had good effect.

This placebo response has another interesting feature that goes to the question of the transient simple sensation picture of pain. In connection with the ultrasound treatment, when doctors taught patients how to apply the machine to themselves in just the way the doctors had, the pain was not reduced. The self-administered use of the machine had no beneficial result. The difference is that the white-coated, M.D.-identified health worker (whether or not an actual doctor) administering the ultrasound massage (even when the machine is not on) was crucial to the effect. It seems clear that the expectation of results contributes to the actual reduction of pain, up to and including its strong analgesic effect. Expectations are known to have effects in complex, emotionally loaded perceptual experiences. In transient simple sensations they seem to have nothing on which to act. They do act on pains. Accordingly, it is best to regard pains as complex emotional and perceptual events or processes.

Wall notes that the placebo response is also present in relief of headache and postoperative pain. The last is especially interesting because when patients are told by doctors that their pain-relieving drug will have some particular side effects, for example, depressed respiration in lung operations, when they receive the placebo pain treatment, they also experience the accompanying side effect. Further, if patients are divided into the group that requires a large dose of narcotic and a group that gets the same pain relief with a small dose of the same narcotic, then those who got pain relief from a small amount of the narcotic, also responded well to a small amount of the saline placebo injection.

There is, too, a "nocebo" response. Telling patients that a drug or treatment will have certain unpleasant side effects, when it actually does, and then using a placebo in place of the drug, will produce in patients the unpleasant effects. The placebo response is the satisfaction of a belief about the future, an expectation. There seems little hope of explaining this analgesic effect unless we begin with the conception of pain as a complex experience involving emotional and motivational features and cognitive sequels distant from sensations.

Third, phantom-limb pain is a dramatic example of centrally produced pain experienced *as if* in a limb or extremity. A phantom-limb is experienced after a limb has been amputated or sensory pathways have been removed. The spinal cord can be totally disrupted and a patient will still experience the presence of a limb or other body part below the level in the cord at which the disruption occurs. Pain often accompanies these experiences. Hence, phantom-limb pain. Such pain can be intractable in paraplegics. These clinical facts point to the etiology of phantom-limb pain in central brain processes or networks (Melzack 1995). Clearly the pains experienced are not simple sensations arising from free nerve endings in the skin. Whatever the full story about them, it is not a simple case of transduction, transmission, and qualitative sensation. Affective components are not separable from the pain experience. Cognitive sets or schema can affect such pain experiences. The (statistical) intercorrelations of these aspects of phantom-limb pain make it impossible to think of them as separable parts of a whole. Only under drug treatments of various sorts, or brain damages, and possibly hypnotic conditions can they dissociate, and then only relatively. They appear in experience as emotional distress.

Fourth, the cultural anthropology of pain shows great diversity of pain types, pain expression, the social roles of pain expression, and the relativity of the meaning of pain to its cultural background (Sargent 1984). This kind of relativity seems implausible if pain is categorized as simple qualitative feeling or sensation. As well, clear cases of emotions such as love, anger, fear, and grief also show the cultural relativity of emotions (Kleinman 1986, 1988). So there is a precedent for categorizing complex pain phenomena as emotional at base (Oatley and Johnson-Laird 1987).

These and data similar to these four types make it plausible that we categorize pain as something other than mere sense quality or sensation. These data also suggest that the category most likely to be useful is among the emotion, motivation, and need categories. In fact, researchers in the pain sciences have for some time urged this new classification. Philosophers' have lagged behind the sciences in this regard. Let me illustrate this trend.

It has been noted in part I of this essay that there was a long debate over the character of pain and that at the end of the nineteenth century the sensory view won the struggle. In 1894 Marshall asserted that "pleasures and pains can in no proper sense be classed with sensations," while Schiff in 1858 proposed that pain is a specific sensation (see Perl and Kruger 1996). But a brief cataloging of recent pain science shows that the sensory, simple sensation view has been wholly abandoned.

A list of pain researchers who have been most actively advocating against the sensation category while supporting the emotionlike, motivationlike, or needlike classification includes Melzack, Wall, Craig, Chapman, and others. (I will be selective in the citations for reasons of space.)

Let me begin with an important result from the review of relevant literature on the sensory and affective features of pain experience. Among the conclusions of Fernandez and Turk (1992) we find:

The differences in sensory versus affective measures [of pains] need to be related to measures of overall pain. This would permit a better appreciation of how each component contributes to pain as a whole.

Even if these conditions are met, one must be mindful that separation does not imply independence. Multivariate statistical studies reveal collinearity between the pain variables, SDT [Signal Detection Theory] indicies d' (a measure of neural sensitivity) and β (the reactive measure) can influence each other, and paired scaling may produce disparate but parallel functions for sensation and affect. This is not surprising because the physical sensation of pain . . . evokes unpleasant emotion, which in turn impacts on sensory perception. Emotional state related to the meaning of the situation can also affect pain perception . . . pain may be less like vision and audition but more like hunger and thirst, in which a certain unpleasantness/ emotional valence is confounded with the drive state (Wall 1979). (Fernandez and Turk 1992, p. 214)

These same authors talk of pain as the gestalt whose components are sensory and affective, no less one than the other. And, these "components" are not additive. A sensory rating of 6 and an affective rating of 7 will not give a pain rating of 13. "Instead, the additive relationship is qualified by parameters pertaining to the intercept and slope of regression lines for predicting overall pain from sensory and affective ratings" (ibid.). This means that in actual experience the so-called components cannot be independent in typical cases of pain phenomena. Accordingly, there is room for the emotion categorization of pain, since the simple sensation view of pain does not have these "components."

These authors are not the only writers to liken pain phenomena to need states. As early as 1979 and recently in his 2000 book Patrick Wall insisted that pain is not to be classed as sensation. In his most recent book Wall classes pain as a function of the full, up and running brain, with the sensory world making its contribution to virtually everything else that is present in the active brain and producing decisions "of tactics and strategies that could be appropriate to respond to the situation. We use the word 'pain' as shorthand for one of these groupings of relevant response tactics and strategies. Pain is not just a sensation but, like hunger and thirst, is an awareness of an action plan to be rid of it" (2000, p. 177).

Chapman has claimed that pain is "a state of discomfort and distress and not a transient event" (Chapman 1995—this view is also presented in Chapman and Nakamura 1999).

Furthermore, Craig has been among those pain researchers most insistent on the role of pain as emotion. In a summary statement in his 1995, Craig writes that

[p]ain has a powerful capacity to command conscious attention, even to the extent of disrupting other activities. . . . The biological primacy of painful experience and the threat to well-being it represents motivate defensive and recuperative action. This formulation of the role of emotion during pain is consistent with the current concepts that emphasize emotions as action dispositions which organize and direct appetitive/aversive experience and behavior. While pain is ordinarily construed as disruptive and disorganizing, it can be seen as focusing attention and mobilizing resources . . . in only recent decades has serious attention been devoted to emotional qualities of pain, relative to the sensory quality largely emphasized in the early 20th century. (Craig 1995, p. 303)

Accordingly, there are on offer various permutations of the emotion categorizing of pain, in contrast to the classification of pain as one among the sensations. Which we select should depend on its utility for purposes of guiding research and therapy.

Finally, the psychologists Bolles and Fanselow (1980) proposed a model of fear and pain, stressing that both these conditions are motivational systems and that each, when activated, reduces the influence of the other. The idea is that noxious stimuli activating the nociceptive systems thereby activate the fear system that motivates "fight or flight" behavior. At the same time, Bolles and Fanselow propose, the fear system may activate the pain suppression systems. The fear response is useful to the animal in removing it, efficiently and effectively, from harm's way. Having done so, the pain system is a second motivational system that puts the animal into its recuperative phase. The two motivational systems, on the model, inhibit each other. Many experimental findings and some clinical observations are explained by the model. (But see the commentary by Chapman and others in Bolles and Fanselow 1980, pp. 303–315.)

While D. D. Price (1999) has emphasized the emotional component of pain experience, he has a sequential processing model of pain in which nociceptive input is first and engenders pain sensations. Second in the processing sequence is cognitive appraisal and emotional response. While these stages or dimensions of a pain experience might be teased apart in experimental conditions, they are both required for pain experience, on Price's model. As well, there is feedback from the complex, first stage, of the brain processing to arousal and somatomotor activation; and arousal and autonomic changes may impact both nociceptive inputs and sensations themselves. Price's view is emphatic on the role of many processes beyond nociceptive and resulting sensation. He claims that

[p]ain sensation may be a salient but not the sole determinant of the affective state during pain. Thus, consistent with modern views of the mechanisms of emotional feelings formulated by Damasio (1994), Arnold (1970) and Price, *et al.* (1985), the emotional state that accompanies the immediate unpleasantness of pain represents a synthesis from several psychophysiological sources. These include pain sensation, arousal, and authnomic, neuroendocrine, and somatomotor responses, all in relationship to meanings of the pain and the context in which it occurs. (Price 1999, pp. 49–50).

This model takes us far beyond the older idea of pain as transient pure sensation. Not even acute pain examples (a burn of the hand or blow to the thumb, etc.) cited by philosophers and others are conceived as simple sensations in current pain sciences.

I could continue to cite sources from recent pain research, both experimental and clinical, showing the move away from pain as sensation toward categorizing pain as a complex emotional and motivational condition. This condition is activated by nociceptive stimuli, chronic neuropathies, inflammatory processes, spontaneous (or ectopic) discharge of pain neurons, wind-up activity in the dorsal root neurons, spontaneous thalamic activity, and others. These sources can be, in turn, modified by activation of the motivational and hormonal systems. It is clear that we have come to appreciate the complexity of pain phenomena and the multiple neurochemical and gene expression systems in which these phenomena inhere.

In the major medical problem pain presents, we cannot rest with accounts of the specific mechanisms underlying nociceptive, inflammatory, and neuropathic pain. If we adopt the category of emotion condition for pain, then we may look more closely at the mechanisms involved in the emotion response. This work is under way. In a statement of the projects required, it has been noted that "the notion that there is a class of drug, a universal analgesic, that can intrinsically reduce all pain, is obsolete and has to be abandoned. Pain is heterogeneous in terms of etiological factors, mechanisms, and temporal characteristics. Consequently, treatment must be targeted not at general symptoms, the pain, or its temporal properties, acute or chronic, but rather at the underlying neurobiological mechanisms responsible" (Scholz and Wolff 2002).

IV

In summary, in part I, I tried to show that there is no one thing pain is *in experience*. From the variety of pain types and pain descriptors, it is implausible that there is a single defining subjective sensory quality or a single affective quale that anything that is a pain must exhibit. This is not to deny the existence of various subjective qualitative features of pains; it is to reject the view, evidently held by not a few philosophers, that there is a defining quality of a sensory sort that makes anything having it a pain.

The strategy of part I was broadly empirical. I think of it as presenting a range of data, the best explanation of which is that there is no one thing and hence no single transient sensory quality that defines pain. That being the case, our intuitions about pain categorized as a sensation should lose some steam.

The aim of part II was to show that, in fact, in our conceptual history we have had a variety of ways of thinking about pain, using different categories and descriptions. My "seven league boot travel" through history was meant to show this. It also suggested that there was a thread running through the history of pain, from a condition conceived as arising from external, that is, nonmental or nonpsychological sources to pain as an inner, conscious state, a state of consciousness. Early on pain is a kind of external condition; to attribute pain, under this conception, is to make reference to external sources, such as assaults, intrusions. Because some painful conditions did not appear to arise from the person's or animal's external environment, since they were in fact the result of disease conditions that could not then be seen or located, the intrusion conception might have been extended to refer to assaults from the unseen realm, of gods, or spells cast, or some such. This may be thought of as the first phase in the internalization of pain. Finally, by the eighteenth century the notion of the transparent, inner, consciousness realm had been incorporated in our thinking, so the internalization of pain as conscious, inner, or subjective sensation became dogma.

Since this is a story about the contingencies of history, its recognition ought to make it easier for us to accept the idea that we may wish to make decisions about how we wish to categorize our target. I urged that such decisions had been made in the eighteenth through the early twentieth centuries by considerations of utility for research programs (psychophysics, for instance) and clinical treatments (empirically discovered analgesics). So it is open to us to categorize pain phenomena in ways useful to current and future research and clinical practice. In particular, we need to think of pain as an emotionlike condition. Thus we may begin to unravel the nature of the mechanism of pain, so categorized, and learn the neurochemical features of such conditions. Then we may design ways of changing their production of our pains.

My methodology throughout has been to set out some facts and urge that the best explanation of them is, as a first step, accepting pain as a kind of emotionlike or need-like condition. The considerations are broadly empirical, as is fitting for a naturalistic program such as this. If they seem to conflict with common ways of thinking and speaking, with common, unscientifically informed, thought, I can only suggest that I don't share the sense of conflict. Since the daily lingo of pain expression, complaint, sympathy, aid, and comfort is not a theoretical and scientific way of speaking and thinking, such lingo may go its merry way without concern for the science of pain. But it's not concerned with such theoretical and empirical matters anyway so it's not in a position to conflict with the story of the neurophysical and neurochemical mechanisms for pain (see Gustafson 2000 for more on the supposed conflict between the science of pain and current common ways of thinking of one's pains). But some philosophers continue to give intellectual pride of place of common pain conceptions and the intuitions they produce. Thereby these thinkers generate their own worry about taking the science of pain seriously. Perhaps it will be useful if they begin by rejecting the sensation conception of pain and remove from circulation the so-called intuitions accompanying this conception.

References

Arnold, M. B. 1970. *Feelings and Emotions*. New York: Academic Press.

Beecher, H. K. 1959. *Measurement of Subjective Responses*. New York: Oxford University Press.

———. 1961. "Surgery as Placebo." *Journal of the American Medical Association* 176: 88–93.

Bolles, R. C., and M. S. Fanselow. 1980. "A Perceptual–Defensive–Recuperative Model of Fear and Pain." *Behavioral and Brain Sciences* 3: 291–328.

Bonica, J. J., ed. 1990. *The Management of Pain*. 2d ed. Vols. 1–2. Philadelphia–London: Lea and Febiger.

Brazier, A. B. 1984. *A History of Neurophysiology in the 17th and 18th Centuries*. New York: Raven Press.

Bromm, B., and J. E. Desmedt. eds. 1995. *Advances in Pain Research and Therapy*, vol. 22. New York: Raven Press.

Chalmers, D. 1996. *The Conscious Mind*. New York: Oxford University Press.

Chapman, C. R. 1995. "The Affective Dimension of Pain." In Bromm and Desmedt 1995.

Chapman, C. R., and Y. Nakamura. 1999. "A Passion of the Soul: An Introduction to Pain for Consciousness Researchers." *Consciousness and Cognition* 8: 391–422.

Craig, K. D. 1986. "Social Modeling Influences: Pain in Contest." In *The Psychology of Pain*, 2d ed., ed. R. A. Sternbach. New York: Raven Press.

Craig, K. D. 1995. "From Nociception to Pain: The Role of Emotion." In Broom and Desmedt 1995.

Dallenbach, K. M. 1939. "Pain: History and Present Status." *American Journal of Psychology* 3: 331–347.

Damasio, A. 1994. *Descartes Error*. New York: Avon Books.

Descartes, R. 1664 (1901). *L'homme*. Translated by M. Foster. *Lectures on the History of Physiology during the 16th, 17th and 18th Centuries*. Cambridge: Cambridge University Press.

Fechner, G. T. 1860. *Elemente Der Psychophysik*. Leipzig: Breitkopf and Hätel, 2.

Fernandez, E., and D. C. Turk. 1992. "Sensory and Affective Components of Pain: Separation and Synthesis." *Psychological Bulletin* 112(2): 205–217.

Greenspan, J. D., and S. J. Bolanowski. 1996. "The Psychophysics of Tactile Perception and Its Peripheral Physiological Basis." In Kruger 1996.

Gustafson, D. 1998. "Pain, Qualia, and the Explanatory Gap." *Philosophical Psychology* 11: 371–387.

———. 2000. "On the Supposed Utility of a Folk Theory of Pain." *Brain and Mind* 1: 223–228.

Keele, K. D. 1957. *Anatomies of Pain*. Springfield, Ill.: Charles C. Thomas.

Kleinman, A. 1986. *Social Origins of Disease and Distress*. New Haven: Yale University Press.

———. 1988. *The Illness Narratives: Suffering, Healing, and the Human Condition*. New York: Basic Books.

Kruger, L., ed. 1996. *Handbook of Perception and Cognition*. 2d ed. San Diego: Academic Press.

Melzack, R. 1975. "The McGill Pain Questionnaire." *Pain* 1: 277.

———. 1995. "Phantom-Limb and the Brain." In Bromm and Desmedt 1995.

Pennebacker, J. W. 1982. *The Psychology of Physical Systems*. New York: Springer-Verlag.

Perl, E. R., and L. Kruger. 1996. "Nociception and Pain: Evolution of Concepts and Observations." In Kruger 1996.

Oatley, K., and P. N. Johnson-Laird. 1987. "Towards a Cognitive Theory of Emotions." *Cognition and Emotion* 1(1): 29–50.

Price, D. D. 1999. *Psychological Mechanisms of Pain and Analgesia*. Seattle: IASP Press.

Price, D. D., et al. 1985. "A Quantitative–Experiential Analysis of Human Emotion." *Motivation and Emotion* 9: 19–38.

Rey, R. 1995. *The History of Pain*. Cambridge, Mass.: Harvard University Press.

Sargent, C. 1984. "Between Death and Shame: Dimensions of Pain in Bariba Culture." *Social Science and Medicine* 19: 1299–1304.

Scholz, J., and C. J. Wolff. 2002. "Can We Conquer Pain?" *Nature Neuroscience* 5 (suppl.): 1062–1067.

Wall, P. D. 1979. "On the Relation of Injury to Pain." *Pain* 6: 253–264.

———. 2000. *Pain: The Science of Suffering*. New York: Columbia University Press.

Willis, W. D., Jr., and K. N. Westlund. 2004. "Pain Systems." In *The Human Nervous System*, 2d ed. G. Paxinos and J. Mai. New York: Elsevier.

13 The Experimental Use of Introspection in the Scientific Study of Pain and Its Integration with Third-Person Methodologies: The Experiential-Phenomenological Approach

Donald D. Price and Murat Aydede

1 Introduction

Understanding the nature of pain depends, at least partly, on recognizing its subjectivity (thus, its first-person epistemology). This in turn requires using a first-person experiential method in addition to third-person experimental approaches to study it. This paper is an attempt to spell out what the former approach is and how it can be integrated with the latter. We start our discussion by examining some foundational issues raised by the use of introspection. We argue that such a first-person method in the scientific study of pain (as in the study of any experience) is in fact indispensable by demonstrating that it has in fact been consistently used in conjunction with conventional third-person methodologies, and this for good reasons. We show that, contrary to what appears to be a widespread opinion, there is absolutely no reason to think that the use of such a first-person approach is scientifically and methodologically suspect. We distinguish between two uses of introspective methods in scientific experiments: one draws on the subjects' introspective reports where any investigator has equal and objective access. The other is where the investigator becomes a subject of his own study and draws on the introspection of his own experiences. We give examples using or approximating both strategies that include studies of second pain summation and its relationship to neural activities, and brain imaging–psychophysical studies wherein sensory and affective qualities of pain are correlated with cerebral cortical activity. We explain what we call the experiential or phenomenological approach that has its origins in the work of Price and Barrell (1980). This approach capitalizes on the scientific prospects and benefits of using the introspection of the investigator. We distinguish between its vertical and horizontal applications. Finally, we conclude that integrating such an approach to standard third-person methodologies can only help us in having a fuller understanding of pain and of conscious experience in general.

2   Foundational Framework and Some Preliminaries

Understandably, there has been resistance, among psychologists and neuroscientists, against self-consciously using introspective (first-person) methods in their scientific studies. For it has been thought self-evident that the deliverances of introspection are not intersubjectively accessible, hence verifiable, and what is not intersubjectively verifiable cannot be the subject matter of science because science is in the business of studying objective reality. The objectivity of science consists, at a minimum, in the intersubjective availability of its subject matter, in that no one is epistemically privileged with regard to gathering evidence about the object of the study. Thus, no one has any special epistemic authority over evidence that others cannot in principle enjoy. But it is claimed that the objects of introspection (in the most general sense of what is being introspected) are essentially subjective in just this sense: necessarily it is accessible only to the person who does the introspecting. What I introspect is not, indeed cannot, even in principle, be epistemically available to anyone else, at least not in the way available to me. I occupy a uniquely privileged position with respect to the contents of my mind that no one else can occupy. And, this, it is claimed, renders introspection scientifically dubious.

This distrust, of course, is not solely directed upon the method of access in question, that is, introspection, but arises also with respect to what is thus being accessed, that is, the objects of introspection. The intuition is that anything that can be accessed *only* through introspection cannot be a proper object of science. For if the introspected contents were physical, it would seem to be objectively available to anyone at least in principle. But if they are not physical, then they can only be the kinds of things that find their place in a Cartesian dualist framework. Hence we had the behaviorist reaction in the first part of the twentieth century to the previously dominant psychology based on classical introspection (Titchener 1908). Behaviorism can be seen as a response to this kind of worry: if the contents of mind can only be accessed via introspection, then these contents cannot be part of the natural world that is equally accessible (and, in principle, in the same way) to anyone who wishes to study it. Thus, if there is to be a *science* of psychology, the subject matter of study had better be objectively available. This is why psychology was defined by behaviorists as the science of *behavior* (conceived as response to environmental stimuli).

Although behaviorism is now dead—at least, as an explicitly promoted psychological doctrine—the basic dilemma that had led to it is still with us, and continues to trouble scientists who wish to study the human phenomenal mind. The revival of cognitivism (or mentalism in general) in psychology after behaviorism was primarily due to the fact that we were able to figure out a way to study the cognitive mind as an intermediary between stimuli and behavior by using third-person methods without exclusively relying on subjects' verbal behavior. The ontological status of the cognitive states implicated or reported were kosher because they could be modeled as states of information processing systems or as computational states defined over mental representations realized in the brain. What made them kosher, of course, was our ability to see how they could turn out to be objectively accessible brain structures, which in turn made verbal reports—at least in controlled experimental situations—relatively reliable first-person expressions of objective facts. It was, in other words, our ability to see how the objects of introspection as implicated in verbal reports (and as indicated in nonverbal behavior—button pressing, etc.) could turn out to be facts that are in principle objectively, that is, intersubjectively, accessible. Thus introspection was in principle dispensable, and its proper place was the context of discovery—not the context of justification. Conversely, once we were able to see how the objects of introspection could be part of nature, there was no reason to shy away from introspection. Indeed, the subjects' verbal reports about their own cognitive states have routinely been taken as evidence for the cognitive models postulated.

But, of course, this mentalistic turn was a revolution only in our understanding of the *cognitive* mind. Hence its scope was restricted to *cognitive* psychology, which is, largely (and very roughly), the study of belief fixation for the guidance of action on the basis of perception which employs computationally structured mental representations. Functionalism replaced behaviorism and restored the reality and causal efficacy of the cognitive mind. Nothing comparable, however, has happened in our scientific understanding of conscious experience: the *experiential* or *phenomenal* mind seems as mysterious as ever.[1]

Understanding conscious experience poses serious problems for scientists working in the relevant fields. Although these problems are general for any theorist who studies perceptual phenomena, the situation that the pain scientist finds himself in is unique for a number of reasons. First, pain, unlike most conscious experiential states such as visual, auditory, tactile experiences, have an immediate affective and emotional aspect to it, which underlies its intimate personal as well as clinical urgency. Second, most other experiences are predominantly (some say, exhaustively) representational. Having such perceptual experiences is important to us insofar as they carry information about our immediate environment. Thus, in such cases our immediate interest and attention are not focused on the experiences themselves, but rather on their objects, that is, what they represent, and how they represent what they do. Similarly, the perception scientist can, for the most part, focus his attention on the scientific study of how information about the environment is extracted from the impinging stimuli and processed in a way that results in the recognition of the distal stimuli responsible for the proximal ones. Theorizing about this process, as we know from our best extant theories, does not heavily involve making significant assumptions about the conscious nature of experiences and their phenomenal characteristics that the subjects undergoing such perceptual processes have conscious access to.

Not so with pain and with its scientific study. Even though pain is an experience that may plausibly be taken to represent tissue damage, or at least, potential damage, our immediate concern is, in the first instance, the experience itself, that is, the pain, and not what the experience might be representing—if it does represent anything. Notice that pain here is not the object of our perceptual experience, but rather, it is the experience itself (Aydede 2001). And, of course, this experience may in turn have a perceptual object itself, that is, be the perception of some potential bodily damage. This fact makes immediately clear why the pain scientist is in a peculiar and difficult situation with respect to the subject matter of his study (Aydede and Güzeldere 2002). A vision scientist doesn't have to discuss and often avoids discussing conscious visual experiences and their qualities as they are consciously available to the subject having the experience. This is because she can do her job quite well by focusing attention on how the computational processes result in reconstructing the distal scene and make it available to the conceptual system. No visual scientists would comfortably claim that the subject matter of their study is the visual experience as *consciously* revealed to the agent having the experience.

The pain scientist, on the other hand, has no choice but to identify the subject matter of his study with the conscious experience, namely, pain—and its qualities and dimensions.[2] It is this fact that generates the peculiarity of the situation that the pain scientist finds himself in. It is also a difficult situation due to the reasons given above in the opening paragraphs. For conscious experiences seem accessible only through introspection, from an essentially first-person perspective. And to say that pain scientists study *pain* is to say that they study a phenomenon accessible only through introspection, which is to say that they study a phenomenon that is not intersubjectively available. This has the odd prima facie consequence that the science of pain is not really scientific.

This is not the case, of course, because the pain scientist takes himself to study a perfectly objective phenomenon, the *brain activity* associated or correlated with pain (or, as some would like to put it, the brain activity that causally generates the pain). This approach is objective, and all the standard empirical methods and conventional scientific procedures apply. This does, however, seem to commit the pain scientist to some sort of metaphysical dualism, according to which, even though pain phenomena may lawfully depend on brain activity, they are nonetheless ontologically distinct (nonphysical, psychic) phenomena, accessible only through the special epistemic faculty of introspection. Not only that, but it also makes the pain scientist vulnerable to the charge that he is making a false advertisement when he claims to scientifically study pain; for it appears that this is not what he does. What he does, if we take the observation just made seriously, is to study the physical correlates of pain but not the pain itself: pain experience, being an essentially subjective phenomenon, cannot be studied scientifically—or so it seems. This is, more or less, the *intuitive framework* that reflects the uneasy relation that exists between pain and its scientific study.[3] The modern pain scientist has always felt the tension that this peculiarity of his subject matter has imposed on him.

But is this tension unavoidable? We think not. In fact, we think that the modern science of pain is in the unique position of providing a paradigm for scientifically studying the conscious experience per se (i.e., taking the experience as the explanandum). First, let us start by clarifying a common mistake that might have been created by the above description of what pain scientists actually do study.

Let us take, for a moment (to be rejected later), the intuitive framework at its face value and assume that some form of metaphysical dualism is true, say, epiphenomenalism—a version of property dualism that takes all and only the physical events to be causally efficacious. This would include the nomological causation of conscious experiences and their phenomenal qualities, that is, qualia, which are themselves causally impotent (on this view, qualia are "nomological danglers," Feigl 1967). We suspect that many pain scientists might, at least implicitly or half-consciously, be assuming some such metaphysical doctrine—as we said, this is, after all, the intuitive framework naturally suggested by the very nature of the subject matter of study and by the demands of being scientific. Taken at face value, this framework, however, does not in fact imply that the subject matter of pain science is solely the brain activity. On the contrary, it *implies* that it is the brain activity *as related to pain phenomena* (you have to put the emphasis in the right place!). The latter is a form of subjective experience. It follows immediately that insofar as pain science is in the business of discovering the brain mechanisms *underlying pain experiences* accessible only through introspection, pain science is in fact irremediably committed to using first-person methods. This necessarily includes introspection along with whatever third-person methodologies are required to do brain science. In other words, if your scientific interest in the brain reflects an interest in knowing about the seat of *mental* activity, then you cannot avoid introspection. You want to know which experience types or qualia are produced by which brain activity: there is no telling if you do not know, at a minimum, when certain experiences occur and what their qualities are. And knowing *that* is to engage in introspection.

Indeed, apart from vexing issues about dualism, if we look at the history of scientific pain research, we find precisely this, namely, that introspection has been used extensively. Psychophysical studies of pain, for instance (as with psychophysical studies of any experience type conducted on humans), routinely take the subjects' reports (verbal or otherwise) to indicate the occurrence of sensations. In other words, psychophysics routinely relies on the introspection of the subjects. Here, relying on introspection has literally no alternatives: psychophysics *is* the scientific study of the lawful relations between stimulus properties and sensations. But we find the same thing if we look at the electrical and chemical stimulation experiments conducted on humans. Recording from single cells via microelectrodes similarly relies sometimes on introspective reports of patients in uncovering the relations among stimuli, brain activity, and sensory phenomena.[4] It is true, most of these more intrusive types of experiments were done on animals. But insofar as we take the findings as indicating something about the *psychology* of animals, we do that indirectly by relying on what the similar structures underlying human experiences reveal. Conversely, we take the findings of such experiments on animals as indirect evidence for the human case, which waits to be confirmed directly on the basis of introspection. This confirmation would occur whenever experimenting on humans becomes possible or whenever an occasion involving a patient with central nervous system damage arises that allows the scientist to study the case that would otherwise be impossible to study experimentally. In all these studies, introspection plays an indispensable role—directly or indirectly. What is more striking, however, is that with the recent advance of brain imaging techniques, we can now study, directly and with an unprecedented and increasing accuracy, the complex relationships that exist among stimuli, brain activity, and sensations in conscious subjects. This is largely the result of the fact that they can introspect and report their introspections in real time. We will give examples of each of these below.

So, it is simply not true that a successful science of pain can or should dispense with first-person methods, with introspection. For the sake of convenience, we have so far assumed—with what seems to be suggested by the intuitive framework anyway—that epiphenomenalism is true and that pain scientists seek to discover, among other things, lawful correlations between brain activity and experiences, essentially implicating introspection. (Of course, this conclusion can, *mutatis mutandis*, be generalized to brain science at large and all types of conscious experience.) We have shown that this assumption is consistent with the actual historical practice in pain research.

But being consistent is one thing and being demanded by it is another. Although introspection and first-person methodologies in general may be necessary for pain science, this does not in fact logically commit the pain scientist to any form of metaphysical dualism—or so we would like to argue. As we have seen, when it comes to conscious experience and its qualities, it is difficult to make sense of how they can be physical in nature. This is bolstered by the intuition that introspection seems to be the *only* available method of access to qualia, and that if qualia were physical in nature, this should not be so: they should in principle be epistemically accessible to anyone in the same way—at least in principle, if not in practice. The idea that something can be physical but not epistemically accessible to any one in the same way, even in principle, strikes all of us as odd. So we are naturally *tempted* by this to adopt a dualist position. On the other hand, epiphenomenalism about qualia, the only tenable form of dualism in our opinion, is no less bizarre, and being scientists/naturalists, dualism strikes most of us as profligate, and given the traditional scientific commitment to physicalist objectivism and its methods, intellectually offensive. This is the tension or dilemma confronting the pain scientist. So, what to do?

The solution is to replace metaphysical dualism with epistemological dualism while preserving metaphysical physicalist monism. However, it is not clear how or whether this can be done successfully, although present efforts in this direction seem quite promising.[5] Although we will not pursue this line in any detail here, we would like to briefly point out the outlines of the fundamental idea. Introspection is a way—apparently the only way—of coming to know about our experiences and their qualities directly. As such, it is an *epistemological* activity, albeit an essentially first-person activity. However, strictly speaking, what we get at through introspection may, for all we know and appearances to the contrary, be completely physical in nature. It may be that what the brain scientist gets at through third-person methodologies turns out to be *identical* with what we get at through introspection, namely brain activity of the relevant sort. There are not two fundamentally distinct but correlated sorts of activity here, one physical and the other irreducibly psychic or phenomenal.[6] Rather, there is only one kind: the brain activity with all and only physical properties. So a scientifically respectable monism is preserved. But given the kind of cognitive organisms we are, it is essential that we have a way of getting at some of our own experiential activity directly and immediately. Indeed, such an access should be expected whenever one is dealing with epistemologically sophisticated intentional organisms such as ourselves that have to informationally interact with their environment in real time. The epistemological advantages of knowing the perceptual sources of information flowing into the central conceptual repertoire of the organism to fix its behavioral strategies are quite clear: such organisms will be able to deal with their environment (including, social environment) far more successfully than those without introspective capabilities (Armstrong 1968). So perhaps introspection gives us a way of accessing certain of our own brain activity in a direct and immediate way without telling us what the complex physical properties of such brain activity are—those properties that would be revealed by a third-person scientific inquiry. As an analogy, think of our visual color detection, we are capable of recognizing instances of red, of discriminating them from other colors, all the while not having the slightest clue about what the nature of the physical surface properties are that we are thus picking out or visually responding to. Color *science* tells us what those surface properties are. Suppose they are, say, certain triplets of surface spectral reflectances. Our visual system, then, responds to such physically complex properties as spectral reflectances without giving us any information about their complex internal structure in any usable form: we perceive colors as simples. Analogously, it may be that introspection manages to convey information to us about our own brain activity underlying/realizing experiences without revealing to us what complex physical features of this activity are that we are directly and immediately responding to.

This is all speculative. However, at least it gives the flavor of a way of dealing with the above dilemma.[7] It is attractive precisely because, if it works, it saves the scientists from both the epistemological Scylla of being essentially committed to a first-person methodology (which is to be found nowhere in science but in psychology) and the metaphysical Charybdis of dualism. Scientists can have their cake and eat it too, in short. Most importantly this approach would legitimatize the indispensable role of introspection in the scientific study of pain by showing how it is in fact complementary to third-person methodologies within a fully naturalistic framework. There would remain no mystery about why the brute correlations discovered (or, to be discovered—see below) by the scientist hold between what the first-person and third-person methods of access get at: for they are access to one and the same phenomenon, namely brain activity.

To wrap up, the indispensability of introspection, and first-person methods in general, in the scientific study of pain naturally suggest that some form of dualism is correct about the phenomenal mind. We take this dualism to be an epistemological one, which consists in the different forms of access to one and the same phenomena, namely, to the brain activity realizing pain experiences.[8] Metaphysical dualism certainly entails epistemological dualism. But endorsing epistemological dualism leaves the option of accepting metaphysical dualism wide open. Thus, assuming that we can make philosophical sense of epistemological dualism within a physicalistic monism in the way suggested above and the intellectual costliness of metaphysical dualism in a scientific approach, we tentatively endorse physicalism and thus reject metaphysical dualism.[9] This meshes well with efforts to legitimize first-person methods within a scientific framework: it makes subjectivity metaphysically, and thus scientifically, kosher. Moreover, this solution, if it works, makes it possible to objectively study the subjectivity of conscious experience itself.

3 Two Ways in Which Introspection Is Used in Scientific Experiments

So far we have indiscriminately talked about introspection and a first-person methodology in general as if it had a unified use in scientific practice. Neither have we talked about what introspection involves, or what kind of activity it is. However, attempting to answer the latter question would go beyond the scope of the present paper. Suffice it to say that there are basically two kinds of proposals conceiving introspection either as a form of inner sense (the higher-order perceptual model—HOP—or simply, the perceptual model),[10] or as a form of higher-order thought (HOT, or the conceptual model).[11] For the purposes of this paper, we would like to operate with an intuitive notion of introspection, which simply takes introspection to be some kind of inner perception eventually yielding conceptually articulated knowledge about one's own experiences and their qualities. In other words, introspection can incorporate elements from both perceptual and conceptual models.[12] But nothing crucial will depend on this in what follows. We would like to expand on the former question in the remainder of this paper, detailing extensively the methodological and practical ramifications of the distinctions proposed.

In a typical psychophysical experiment done on adult humans, part of the data the investigator collects consists of verbal reports of his subjects following a previously determined set of rules and questions. It is assumed that what the subjects report are the dependent variables to be measured by manipulating the independent variables, usually stimulus parameters. In this kind of experimental paradigm, the data collected are considered objective from the perspective of the investigator in that any other investigator can easily have access to the same set of data in two distinct ways. He can check out the results of the other investigator himself or he can replicate the experiment either on the same set of subjects or on others. This is typical of a third-person approach. There is no question about the investigator's engaging in introspection himself. However, that part of the data set that consists of verbal reports of subjects is typically taken to indicate the occurrences of sensory states and their qualities.

Here it is important to be clear about how the investigator takes this "indication" relation. The most plausible thing to say is that the subjects' reports *express* or *describe* what goes on in their mind or sensorium. In other words, they come to directly observe/believe that such and such sensations are occurring to them and they *express* these observations/beliefs by *verbally reporting* them as instructed.[13] But how do they form these observations/beliefs? The obvious answer is that they form them on the basis of introspection, that is, directly noticing what is in their experience and expressing those contents through words or magnitude judgments. In other words, implicitly (or, explicitly depending on the experimental design) the subjects are instructed to notice what is in their experience and report whether a sensory state of a certain type is occurring to them and what its intensities or qualities are.

Of course, if the investigator has behaviorist inclinations, the interpretation of the "indication" relation will likely be different. The verbal reports will be taken to be pieces of mere behavior that are not semantically relevant, but whose features are interpreted as the dependent variables controlled directly by stimuli (instead of *expressing* the occurrence of internal dependent variables, i.e., sensations). There is another option that such a behavioristically inclined investigator may take. She may interpret the verbal reports as indicating the occurrence of certain types of brain states correlated with stimulus properties while being silent as to whether the occurrence of these brain states are identical or correlated with sensory/mental states. Here the notion of indication is that of *natural indication* (as opposed to a semantic one) in the sense in which smoke indicates fire.

Behaviorism has failed partly because it has prohibited the postulation of mental states mediating stimuli and behavioral responses. Indeed, it did not allow for postulating *any* intermediary as a theoretically significant parameter, be it mental or physical (brain states). We take this failure seriously, and feel justified, along with most psychologists we believe, in engaging in realist mentalistic talk and in making mentalistic hypotheses within a naturalistic framework. We, therefore, take the most plausible interpretation of what the subjects are doing in such experiments, namely that they are reporting their introspective observations about their own sensations and their qualities—just as we said above.

If this is right, then this is one sense in which introspection is indispensable in pain science, and neuropsychology in general. Interestingly, while most researchers in the field take this sort of use of introspection as perfectly objective and acceptable, they nevertheless have doubts about the legitimacy of using introspection in a more direct way, namely, by becoming subjects themselves. We have already touched upon some of the reasons of why that is the case, and tried to deflect the worries. Our point here, however, is that there is no *principled* methodological difference between investigators' drawing on subjects' introspective reports (which is indispensable anyway) and drawing on their own—if the context of the latter is properly understood and some potential methodological pitfalls are preempted. As we said in the beginning, we will propose an experimental paradigm where both kinds of introspection are used in a complementary fashion, and argue that this paradigm is superior to the one that licenses *only* the subjects' (≠investigators) introspective reports as legitimate. We call this approach the *experiential* or *phenomenological* approach.[14]

4 Studies of Pain

4.1 Studies of "First" and "Second" Pain From a First-Person Perspective

We can start by asking whether there exists any precedent for scientific investigators observing and analyzing their own experiences of pain. There have been both phenomenological and psychophysical studies of pain in which at least some of the investigators were subjects in their own experiments. As an example for discussion, several studies throughout the last century sought to characterize the subjective experience of "first" and "second" pain by having the investigators conduct experiments on themselves. First and second pain results from a sudden noxious stimulus to a distal part of the body, such as the hand or foot. The 0.5- to 1.5-second delay between the two pains occurs as a result of the fact that nerve impulses in C-axons travel much slower (0.5 to 1.5 meters/sec) than those in thinly myelinated A-axons (6–30 meters/sec). The first study utilized a paradigm wherein two investigators independently mapped the body regions wherein they experienced first and second pain in response to brief intense electrical shock (Lewis and Pochin 1938). They did so prior to knowing the results of the other investigator. The body maps of both Lewis and Pochin were nearly identical, thereby providing convergent confirmation of their results. Both maps showed that first and second pain could be perceived near the elbow but not the lower trunk even though both sites were about the same distance from the brain. The reason for this difference only became known later, when impulse conduction velocities of peripheral and central pain-related nerve cells became characterized. C-fibers that supply the trunk have a short conduction distance to the spinal cord, whereas C-fibers that supply the skin near the elbow have a long conduction distance. Once both "trunk" and "elbow" C-fibers reach the spinal cord, they synapse on nerve cells that have fast-conducting axons. As a result of differences in peripheral conduction distance and time, first and second pain can be discriminated at the elbow but not the trunk.

A study by Landau and Bishop (1953) similarly used themselves as investigator-participants to determine the qualities of pain related to selective stimulation of peripheral nociceptors supplied by A- and C-axons. Their study was much more experiential-phenomenological than that of Lewis and Pochin. Using standard stimuli, they attempted as much as possible to simply notice the qualities and intensities of first and second pain prior to reflecting on the causes or interpretations of these two types of pain. Both investigators observed that first pain was sharp or stinging, well localized, and brief, whereas second pain was diffuse, less well localized, and had qualities of dullness, aching, throbbing, or burning. The latter quality was prevalent when skin C-nociceptors were selectively stimulated. Second pain was longer lasting than first pain and was accompanied by a feeling of unpleasantness that differed from that of first pain. The distinct unpleasantness of second pain was associated with a sense of vagueness and poor localizability and with its dull, diffuse, and long-lasting sensory qualities. Both Landau and Bishop independently observed these sensory and affective qualities through passively noticing their own direct experience.

There are three important points to be made about these first two studies. First, they arrived at very straightforward observations about the experiential nature of specific types of pain and even made cogent inferences concerning their mechanisms. Second, the observations have since been incorporated into our body of knowledge of pain and have been replicated in several studies using more conventional experimental designs and methods drawing on other subjects' first-person access to their own experiences. Finally and most critically, the results were obtained through investigators' introspecting their own experiences of pain and other sensations.

Some studies have combined experiential and psychophysical methods.[15] Subjects of these studies included one of the present authors (D.D.P.) as well as those who were unfamiliar with first and second pain or the hypotheses of the study. The result of these studies replicated and extended the observations of Landau and Bishop. First, the investigators experienced the same kind of sensory and affective qualities reported by them and thereby replicated their observations. Furthermore, untrained subjects also reported them without our provocation or suggestion that such qualities existed. Thus, knowledge of first and second pain is both first- and third-person. This knowledge was then extended in psychophysical experiments wherein subjects reported on and separately rated the intensities of first and second pain. Using series of four computer-driven heat pulses (2.5 sec duration, peak temperature 52°C) or four 5–9 mA electrical shocks, subjects' mean ratings of first pain were not statistically different throughout each series (Price et al. 1977, 1994). Unlike first pain, second pain progressively increased in mean intensity and duration throughout a series of shocks or heat pulses when the interstimulus interval was less than three seconds but not when it was five seconds. All of these results also were confirmed in direct experience. In the case of one of the present authors that addressed the question (D.D.P.), second pain did indeed become stronger, more diffuse, and more unpleasant with repeated heat pulses or repeated electrical shocks.

4.2 Psychophysical-Neural Parallels of First and Second Pain

We now turn to studies where direct comparisons can be made between the types of psychophysical data just described and neural activity at different levels of the nervous system, including the human nervous system.[16] When the same types of heat pulses or electrical shocks described above are applied to the skin of monkeys, individual neurons within the spinal cord dorsal horn respond with a double response (i.e., two sets of impulse discharges). The earlier of the two is related to synaptic input from A-nociceptors and the delayed response is related to synaptic input from C-nociceptors. Similar to first and second pain, the first response does not significantly change throughout a series of heat pulses, whereas the second delayed response increases progressively both in magnitude and duration. Similar to second pain, temporal summation of the delayed neural response was observed when the interstimulus interval was three seconds or less but not five seconds. Moreover, this summation must occur within the spinal cord dorsal horn, because similar experiments conducted on peripheral A- and C-nociceptors show that their responses do not increase with stimulus repetition (Price et al. 1977). Thus, temporal summation of second pain depends on mechanisms of the central nervous system (i.e., dorsal horn neurons) not changes in peripheral receptors.

These psychophysical–neural parallels have been confirmed not only in the case of single neurons of the spinal cord dorsal horn but also in the case of neural imaging at the level of the somatosensory region of the cerebral cortex (Tommerdahl et al. 1996). Using a brain-imaging method of intrinsic optical density measurements (OIS), Tommerdahl and colleagues imaged neural activity within the primary somatosensory cortex of anesthetized squirrel monkeys as their hands were repetitively tapped with a heated thermode. These taps reliably evoke first and second pain in human subjects. Their method of neural imaging has a high degree of both spatial and temporal resolution, measuring local cortical neural activity within 50–100 microns and sampling neural activity that has occurred within a third of a second. Heat taps produced localized activity in two regions of the primary somatosensory cortex, termed 3a and 1. When heat taps were presented at rate of once every three seconds, delayed neural activity within these regions occurred in response to each tap and grew progressively more intense with each successive tap. This temporal summation of this neural response paralleled human psychophysical experiences of second pain in several distinct ways. Both types of responses summate at the same rate of stimulus repetition and have a similar growth in intensity during a series of heat taps. The perceived skin area in which second pain is perceived and the area of cortical neural activity both increase with repeated heat taps.

Since the psychophysical experiments utilized human subjects and the neurophysiological experiments utilized anesthetized monkeys, more definitive conclusions about how these patterns of neural activity relate to the conscious experience of second pain summation await future experiments in humans. But given the results so far, there is reason to think that robust correlations would be established. Furthermore, interpretations of the results of such experiments can be improved by several conditions of the experimental design. First, the subjective data and neural data should be obtained in the same subjects. Second, the neuroimaging methods should have a high level of spatial and temporal resolution. Third and finally, the subjects of such experiments should include both the investigators and subjects unfamiliar with the hypotheses. It is very clear that including investigators among the subjects in such experiments cannot fail to help interpret the data in a more accurate and detailed way. Given the results so far, and certain constraints we will describe later, there is no reason to shy away from taking the experiential-phenomenal method here (in addition to standard methods).

4.3   Relating Brain Activity to Sensory and Affective Dimensions of Pain

A good example of an experiment that supports the feasibility of this type of experimental strategy is that conducted by Rainville and his colleagues (Rainville et al. 1997). The experiment was not about first and second pain, but about the subjective qualities of pain sensation and pain unpleasantness, and how they relate to each other. Participants of this study rated pain sensation intensity and pain unpleasantness of moderately painful immersion of the left hand in a 47° C water bath. Two experimental conditions included one in which hypnotic suggestions were given to *enhance* pain unpleasantness and another in which suggestions were given to *decrease* pain unpleasantness. Suggestions also were given in both conditions to the effect that, unlike pain unpleasantness, pain sensation would not change. They found that suggestions for enhancement of unpleasantness increased magnitudes of both pain–unpleasantness ratings and neural activity in the anterior cingulate cortex (area 24) in comparison to the condition wherein suggestions for decreased unpleasantness were given. Neural activity in S-1 somatosensory cortex, like subjects' mean ratings of pain sensation intensity, were not statistically different across the two experimental conditions.

A second similarly designed study used hypnotic suggestions to modify the intensity of pain sensation. In this experiment, the suggestions were effective in producing parallel changes in ratings of pain sensation intensity and neural activity in S-1 somatosensory cortex (Hofbauer et al. 2001). It is important to recognize that the stimulus intensities were exactly the same across experimental conditions that produced different subjective magnitudes of unpleasantness or pain sensation. These experiments reflect a strategy designed to identify neural structures differentially involved in two separate dimensions of pain experience. It is necessarily simplistic because sensory and affective dimensions of pain cover broad and complex experiential territories. However, this type of experiment not only clearly and essentially draws on the introspective reports of subjects used in the experiments, but also strongly suggests that adopting the experiential-phenomenal method could only help here. In particular it could help clarify, better describe, and expand the results of the specific experiments from which the results are obtained. It could only help suggest new ways of extending the experimental design in question to new experiments by formulating new hypotheses. The latter possibility is a reliable sign of the fertility and productivity of any methodology used in scientific experiments and hypothesis forming.

4.4 The Significance of Parallels between Pain Phenomenology and Central Neural Activity

A single brief heat tap can lead to multiple pains with diverse sensory qualities, partly as a result of the frequency with which it is presented. The sensory qualities of pain depend not only on features of the stimulus and on the transducing properties of sensory receptors but also on integrative mechanisms of the central nervous system. For example, there is nothing about the physical properties of heat taps or even the physiological characteristics of peripheral C-nociceptors that would allow someone to predict temporal summation of second pain. Yet the subjective qualities of this phenomenon can be characterized and systematically related to central neural mechanisms, including those at the highest levels of the central nervous system.

Similarly, there is nothing about the stimulus in Rainville et al.'s study that allows one to predict changes in the affective or sensory qualities of pain produced by the suggestions given to the subjects. Sensory and affective qualities of pain covary with patterns of neural activity in the central nervous system and not just the physical characteristics of stimuli. Given this high variability between the stimulus characteristics (or, features of the peripheral nervous system) and qualities of the resulting subjective experience, it is very clear that the relationships between these qualities of pain experience and neural activity can be explored *only* through careful analysis of both subjective experience and neural activity. It should by now be obvious that drawing on introspection is an indispensable condition of such an exploration.[17]

It is very important to recognize that parallels between experience and neural activity do not prove that the neural activity sufficient for a given subjective quality of pain exists within one specific brain region, such as the somatosensory cortical area (second pain) or anterior cingulate cortical area 24 (pain unpleasantness). However, activity in these regions may represent a beginning or necessary stage of processing that is required for such qualities. It is not difficult to envision how knowledge of patterns of activity in these brain regions will eventually form at least part of a coherent explanation of how patterns of brain activity entail the existence of experiences with a subjective epistemology for their possessors.

4.5 Possibilities for a Refined Analysis of the Relationships between Pain and Brain Activity

Our evolving knowledge of the relationships between brain activity and experiential states such as pain depends equally on improvements in methods of analyzing neural activity and on methods of investigating human experience from a first-person perspective. Complete explanations of these relationships require an integration of experiential methods, such as phenomenology and psychophysics, and neuroscience.

As we touched upon previously, first-person experiential methods are extensively and essentially used in psychophysics. Throughout the history of psychophysical research, we see, moreover, that experiential methods and introspective observation have been constantly improved upon. Paradigms of detection, direct scaling, and differential scaling of different experiential dimensions have continued to improve throughout the last century.[18] Psychophysical observers can be trained to detect very small differences in sensory qualities and intensities and to differentially judge magnitudes of different dimensions or qualities of their own experience (e.g., as in the Rainville studies described above). These paradigms and response abilities can be applied not only to sensory phenomena but also to several dimensions of human experience in general, including the relationships among higher-order mental states (such as beliefs, desires, hopes, expectations, etc.). Thus, the well-known power law in psychophysics applies not only to sensory phenomena, such as sound and pain, but to such dimensions as expectation, desire, and emotional feeling intensity (Price and Barrell 1984). In fact, studies have directly scaled these three dimensions and found lawful interrelationships between them.[19]

Less well known, however, are extant methods for observing the contents of experience and improvements in methods of analyzing experiences. These include Eastern meditative practices, phenomenology, and explicit experiential paradigms. Buddhist meditative practices offer a method for observing what one's mind is doing as it does it, to be present with one's mind. As Varela, Thompson, and Rosch (1991) point out, "The purpose of the mind in Buddhism is not to become absorbed but to render the mind able to be present with itself long enough to gain insight into its own nature and functioning." This method is compatible with phenomenology, the study of how phenomena are presented in our experience prior to analysis or explanation. Phenomenologists offer reflections on direct experiences that are nearly totally missing in mainstream psychology and cognitive science.[20] For example, there is relatively little emphasis within mainstream psychology on understanding the experiential structure of phenomena such as pain or emotional feelings. Consequently, there is often a lack of any search for detailed structural analyses within these kinds of experience. In fact, within the framework of a large part of psychology, experiences such as pain or anxiety are usually discussed only in relation to the possible external conditions under which they are present or to the physical processes that seem to generate them. Yet there is no search for experiential dimensions internal to these phenomena. Obviously, both the study of pain, and the conscious experience in general, might substantially benefit from an integration of improved experiential-phenomenological approaches and conventional methodologies of neuroscience. There is no reason at all that justifiably prohibits a systematic first-person approach to studying these phenomena involving the investigators themselves as subjects.

5   The Experiential Approach and Method: The *Horizontal* Phase

Recognizing the limitations of traditional phenomenology and classical introspectionism, Price and Barrell (1980) and Barrell and Barrell (1975) developed an experiential approach and method for the study of human experience, one that utilizes both the basic principles of phenomenology and psychometric methods (e.g., psychophysics). This approach and method allows for the discovery of common factors or dimensions within specific types of experience such as anger, anxiety, and pain, as well as for the characterization of the interrelationships among these common factors. The focus of this kind of approach is "horizontal" in that it is the interrelationships among the elements of one's total experience (broadly construed) as revealed to one's introspective consciousness that are under investigations—without an attempt to relate them to brain structures. Later, we will sketch the second, "vertical," phase of this approach where experiments are designed to discover the correlations between the findings of the first phase (experiential elements and regularities) with the elements of brain activity. However, the horizontal phase can stand on its own, and we expect to find quite robust and interesting results simply by studying the elements of conscious experience.

The horizontal experiential paradigm itself consists of several stages that include (1) questioning and observing; (2) describing experiences from a first-person perspective; (3) understanding experiences through discovering common factors and their inter-relationships (i.e., anxiety, pain, etc.); and (4) application of quantitative methods to test generality and functional relationships between common factors. Of these four stages of research, the first three are unique in that the investigators are the subjects of their own research questions (i.e., coinvestigators) and the last stage utilizes accepted psychometric methods, derived mainly from psychophysics, to test hypotheses in other human observers. The last stage is no different in principle from conventional psychological/psychophysical research. However, it is the combination of these stages that produces direct knowledge about experiential phenomena. It is a paradigm that could be directly applied to the study of consciousness in general and to pain in particular. We now turn to the explanation of each of these stages.

(1) Questioning and observing experiences An experiential approach begins with the investigator or a group of investigators posing a general question about their experiences such as "What is it like to experience performance anxiety?" or "What is it like to experience the unpleasantness of a specific form of laboratory pain, such as immersion of the hand in a heated water bath?"[21] The question should be general, clearly understandable, and specifically directed to *how* a given phenomenon is experienced. The focus is to be on *how* we experience something rather than the specific objects of experience or the stimulus conditions in which the experience occurs. Thus, if the question concerns the experience of pain unpleasantness, we would attempt to passively notice the qualities of sensations, thoughts, and feelings that would occur during the experience as opposed to our explanation of *why* the experience occurred in the first place. This approach has historical roots in phenomenological analysis which asserts that the process of experiencing rather than the specific objects of experience leads to an understanding of the essential structure and factors within different types of experience.[22] For example, noticing how we experience performance anxiety would include our thoughts and feelings that occurred during or just before performance and not just the specific targets or "stimulus objects" of experience (e.g., the particular audience or the particular acts of performance). However, noticing *how* we experience the audience and our thoughts and feelings about our prospective performance is more likely to reveal the content of experience that is *common* across different experiences of performance anxiety (Barrell et al. 1985). Most important, the investigators in the study are to pose the question to themselves, that is, to their own experience of the phenomenon.

Once a question is formed, the investigators can assume a particular stance toward the subject of interest in order to gain information about it. The stance consists of passive attention, simply noticing whatever is happening in one's experience.[23] It is simply "being *with*" the experience and includes an intermittent attention to whatever is occurring or has just occurred without a strong, hard focus and without explanation or judgment of the experience itself. This process is similar to locating an object in one's peripheral visual field, rather than in its focal region. In this respect it is a form of introspection different from the one promoted by classical introspectionists, who insisted on active attention to one's experiential elements, which often resulted in the disappearance of or changing the elements thus attended.[24] In our paradigm, one simply just allows oneself to notice occasionally whatever is going on in experience without interpretation or judgment. This requires acceptance of whatever is happening. Once acceptance occurs, we are more willing to look at the processes of experience impartially. Otherwise, we tend to experience through the filters of our own biases. Passive attending is not interpretative or judgmental nor, as we said, a form of classical introspectionism. Rather, it occurs in a present-oriented context, that is, either in the present or a "reliving" of past situations. However, since it is often difficult to attend passively exactly at the same moment one is experiencing a phenomenon, this form of attention often consists of immediate retrospective attention to what has just occurred in experience. This form of observation has been used in certain forms of psychotherapy, particularly gestalt therapy (Perls, Hefferline, and Goodman 1951), and yet is also consistent with underlying principles of phenomenology.[25]

(2) Describing experiences Once we develop the ability to passively attend and notice how we experience a given phenomenon, such as anxiety or a specific type of laboratory pain, we can also develop a method and style of describing these experiences. The experiential method contains a way of reporting experience that is consistent with the principles of phenomenology in that it requires reports of observers to be first-person, present-tense accounts. Thus, the report refers to immediate experience and is written in the first-person present tense, since even reports of previous experiences can result from "reliving" those situations. A (necessarily simplistic) experiential report of a few moments of a specific type of experimental pain serves to exemplify this type of account:

My (D.D.P.) hand was immersed in a 47° C water bath (Rainville et al.'s 1997 experiment) when the following experience occurred: Intense burning and throbbing in my hand. Feel bothered by this and slightly annoyed. Is it going to get stronger? Feeling of concern. Hope my hand isn't going to be scalded.

This type of report can be generated from an immediate retrospective reliving of this experience, but is more likely to be distorted if it were obtained long after it occurred. With continued "snapshots" of experiences such as this, one would begin to observe commonly experienced types of thoughts, sensations, and emotional feelings that characterize the experience of a specific type of pain in a particular individual.

(3) First-person understanding of experiences With continued reliving or direct experiences of pain in a given experimental or clinical situation, one would find commonly experienced factors.[26] Thus, thoughts and feelings such as "Is it going to get stronger? Feeling of concern. I hope my hand isn't going to be scalded" might be characterized by the statement "I think and feel a concern for future consequences related to this pain." This expression more concisely describes the sense of what is experienced. Part of the process of so-called phenomenological reduction consists in an attempt to capture experiences in this more concise way.[27] Similarly, the statement "Feel bothered by this and slightly annoyed" could in this way be reduced to the expression that "I have a feeling of intrusion related to this pain." This process represents a distillation of the sensations and their general significance that are present in an individual's experience during a specific type of experimental pain. This type of account can be a starting point for understanding experiences as described below.

Thus, following several experiential descriptions of what is present in several situations in which the phenomenon (e.g., pain) takes place, we then become interested in understanding the factors that make up this type of experience. We begin this process by analyzing our data in such a way as to determine what is present in *all* situations in which this phenomenon occurs. The aim of this analysis is to identify the *common factors* (or what phenomenologists call structural invariants) of experience, an aim of structural phenomenology.[28] At this time, we must be careful to reflect in such a way that no preconceptions are brought in for the purpose of interpretation. The assumption is that the answers to our question can be directly identified within the data. We are to reflect on the data, continuing to ask, "What is present in my experience during the situations wherein I am feeling pain during immersion of the hand in this experiment?" "Which of these common elements or factors are necessary or sufficient for this experience?" Both *definitional* hypotheses and *functional* hypotheses can be generated from this type of reflection. Definitional hypotheses posit the experiential factors that are commonly present during a type of phenomenon, such as pain from a heated water bath, and functional hypotheses are statements about their interrelationships.

To use our analysis of pain as an example, we can formulate the following definitional hypothesis on the basis of this analysis. Experiential factors that are present in my experience of this form of pain include (1) an intense burning throbbing sensation in the hand, (2) an experienced intrusion or threat *associated with this sensation*, and (3) a feeling of unpleasantness *associated with this felt intrusion or threat*. The sense of intrusion can apply simply to the present without concern for future threat or it can also include the latter. The experiential example given above contains both factors. When the factors of intrusion or threat are present, they are accompanied by a felt sense, one that is normally experienced as a state of the body.[29] This felt sense constitutes the pain–unpleasantness. As Gendlin (1962) has pointed out, the felt sense of an emotional feeling is its key feature: it seems to reflect the physical and biological state of the body in the same way that the experience of thirst and hunger reflect biological states.[30] For example, the felt intrusion and threat during pain seems to be *about* the experienced integrity of the self, body, and consciousness that is implicitly there in experience prior to explanation (i.e., the experience carries the information without this information being necessarily conceptualized and thus made conscious *to* the subject-investigator as such prior to its becoming the target of introspection). We could also formulate hypotheses that are about the functional relationships between these three factors. For example, felt unpleasantness should increase as a function of experienced intrusion or threat. Experienced intrusion, in turn, should increase as a function of the intensity of burning, throbbing sensation.

Once factors for a given type of experience are agreed upon by a group of subject-investigators, the investigators can formulate *functional* hypotheses about the interactions between them. To do so, the investigators can utilize several specific experiences in which a factor, such as the perceived degree of intrusion or threat varies in magnitude from one specific experience to another. The relative magnitude of pain–unpleasantness may then be observed to increase in relationship to the degree of experienced intrusion or threat. The end result of this analysis are quantitative and qualitative expressions of how the experiential factors relate to each other, that is, the generation of functional hypotheses.

An aim of our experiential method is to generate qualitative descriptions of given kinds of experience, descriptions that express phenomena as they reveal themselves to the experiencing person. They are to be agreed upon and well understood by all those capable of having the experiences in question. Each final phenomenological-experiential account of a kind of experience (e.g., pain) is to contain precise explicit statements about what is essential for a kind of experience, omitting particulars. Important differences exist between our method and similar approaches used by others (Van Kaam 1959). For one, the experiential method requires the investigators themselves rather than other participants to produce the descriptions. Another difference is that our method more exactly specifies the mode of observing and questioning. A third difference is that our method includes a strategy of determining the necessary and sufficient factors for the type of experience in question. Finally, it allows the investigators to discover the functional relationships of common factors of given kinds of experience. Thus, the experiential method can generate both *definitional* and *functional* hypotheses that can be subjected to quantitative methods of testing.

(4) Application of quantitative methods Both definitional and functional hypotheses can be tested with accepted psychometric methods in conventional psychological experiments. These methods and experiments would be similar in principle to those already used, particularly in psychophysical studies. They would involve controlled observations of ratings of experiential dimensions by participants who are commonly used in experimental psychology (e.g., undergraduate college students). The participants of the experiments would *not* be the investigators and would not have knowledge of the hypotheses of the study. Thus, the last stage of the horizontal experiential paradigm would involve a third-person epistemology (from the investigators' perspective) but one that is complementary to the first-person exploration of stages 1–3 described above. Thus, in the example of experimental heat pain cited above, psychophysical observers would rate not only pain sensation intensity and pain unpleasantness, but also their experienced intrusion and concern for future consequences. They could also rate other experiential factors, such as anxiety and annoyance. Further experiments involve experimental manipulations of specific dimensions of experience to directly test functional hypotheses. What is unique to this horizontal phenomenological approach, in strong contrast with traditional psychological methods, is the conceptual shift in which the independent variable becomes the experiential dimension to be manipulated as opposed to the external conditions used to produce changes in the experiential dimension. A manipulation of a given dimension, such as perceived intrusion or perceived threat, can be achieved by a variety of manipulations, and the consistent functional interaction between the manipulated dimension and the hypothesized dependent dimension provides a test of the functional relation between the two experiential dimensions.

6 Interfacing Subjective Experience and Neural Activity: The *Vertical* Phase of the Experiential Approach

Although the results obtained from the studies following the horizontal phase of the experiential paradigm would have intrinsic scientific value of their own, the real significance of these results would emerge when they are integrated with neuroscience. This is the vertical phase where an attempt to correlate the results of the first phase with brain activity is made. There is an important sense in which the horizontal phase is designed with an eye to present the phenomenal structure of specific types of experience and its elements in such a clear and precise way as to make them ready to be "hooked up" to the structural elements of the neural activity underlying/realizing those experiences. In other words, the combination of the horizontal and vertical phases would give us (if anything would) the long sought-after mind–brain correlations, and thus prepare the way for explaining mind–brain relationship within a metaphysically monistic framework while preserving the subjectivity of the mental.

To illustrate more vividly the potential of such a two-phase approach, let us reconsider a hypothetical experiment that would be an extension and refinement of the one conducted by Rainville et al. (1997) described above. This kind of imaging study could be interfaced with the horizontal experiential paradigm described above to provide a much more elaborate characterization of both subjective experience and patterns of neural activity that covary with the different and subtle subdimensions of pain unpleasantness. Investigator-participants could identify these subdimensions using the first-person experiential approach described above (the horizontal phase). These subdimensions might include, for example, factors such as perceived intrusiveness and concern with future harm. Subjects of neural imaging experiments could then rate these subdimensions under different experimental conditions designed to generate variation in their magnitudes or to selectively modulate their magnitudes, as in the case of Rainville et al. (1997). Patterns of cerebral cortical activity that covary with different subjective dimensions of pain could then be identified. Each investigator would make two types of observations. The first would be an observation of the different sensory, cognitive, and affective qualities of pain from his or her embodied perspective. The second would be an observation of the spatiotemporal map of brain activity associated with this subjective experience. Furthermore, it may be possible to nearly simultaneously observe changes in both subjective experience and patterns of brain activity. Once a reliable correspondence is established between a dimension of experience and specific neural activity by using first- and third-person approaches, one needs to test between-subject variations in this correspondence.

7   Conclusion

The demonstration that similar experiential descriptions between individuals are associated with similar patterns and levels of critical brain processes would support the
view that we share qualia. Indeed, under naturalistic assumptions, such results would
make a strong—albeit nondemonstrative—case for the claim that there is a looser but
perfectly legitimate sense in which qualia are intersubjectively accessible to all those
who share, to close enough approximation, the neural hardware (perhaps even including some higher animals).[31]

In practice, of course, we do assume that we share qualia. We assume that our experiences instantiate identical or similar phenomenal qualities, and in this sense and in
this sense only, qualia are intersubjectively accessible and thus objective, insofar as we
have independent third-person ways of confirming their occurrences in others—which
we do. Both the horizontal and vertical phases of the experiential method proceed on
these assumptions (as do, as a matter of fact, all the other conventional psychophysical and brain imaging experiments conducted on subjects and their introspective
reports). But once the confirming results come in (if they do) we can use a boot-
strapping strategy to argue that our initial assumptions were justified. Indeed, this is
how most scientific enterprises usually go. You start with certain assumptions, and on
the basis of the success of your results, which usually depends on the explanatory
power of the model that would confirm the empirical data if it were correct, you then
turn back and claim that your assumptions were and are justified.

The kind of questions that are usually asked by philosophers might then be empirically addressed. For example, are there lawlike correlations between the elements of
subjective experiences and patterns of brain activity? Philosophers and scientists alike
have usually assumed that the answer to this question is affirmative. But of course, if
so, it seems an empirical matter to find out by doing science. If the results are as
expected or assumed, then this can be taken as giving strong support for the metaphysical supervenience of the mental on the physical—if, as we mentioned before,
philosophical sense can be made for the naturalistic claim that subjective conscious
experience can be identical to objective brain activity. So far we have tentatively
assumed that the naturalistic claim *does* make philosophical sense. However, if it turns
out that naturalism about consciousness cannot be sustained philosophically, we
would certainly be surprised and quite puzzled for reasons given in the beginning
sections of the paper; but very little would be lost in terms of the value of our methodological proposals.[32] For the two-phase experiential paradigm we have proposed can
still be carried out on dualistic assumptions. We would then be asking questions not

about metaphysical supervenience but only about natural or nomological supervenience where experiential elements are only (inexplicably) correlated with specific types of neural activity—if they are.

There are also questions of more neuroscientific nature. For example, does the encoding of a specific pain quality occur in multiple brain areas simultaneously or in just one place? And if it occurs only in one place, how do other brain areas access that encoded representation in a way that would explain the unified nature of the subjective experience in question (the binding problem)? Presently, these questions would be viewed to be difficult (perhaps, even impossible) to answer by many neuroscientists. Answering these would of course rely on more technologies than just brain imaging techniques. However, increasing improvements both in neuroscience technology and in methods for investigating human experience could lead to the conditions that could initiate a satisfactory inquiry into a *complete* understanding of consciousness—perhaps using pain as a model due to its peculiar status as mentioned in the beginning.

Acknowledgments

The authors are grateful to Pierre Rainville, Robert D'Amico, and Eddy Nahmias for their helpful comments on an earlier version of this paper.

Notes

1. Block (1995) marks the distinction by claiming that although we are now in a position to scientifically penetrate the *access* consciousness, which is causally, functionally, computationally, or otherwise physically, explainable, no one has any idea about how to scientifically study the *phenomenal* consciousness, which is not explainable at all in similar terms. Similarly, Chalmers (1996) thinks that the *psychological* (functional/intentional) aspects of the mind are in principle amenable to scientific treatment, but the "hard problem," on his view, is the explanation of conscious experience, the phenomenal mind, which resists scientific understanding. Indeed, for Chalmers, the phenomenal aspects of the mind are not physical at all, and that is the reason why phenomenal consciousness cannot be naturalized or physically explained. See Güzeldere and Aydede 1997 for a criticism of the access versus phenomenal consciousness distinction.

2. The "official" definition of 'pain,' recognized by the International Association for the Study of Pain (IASP), goes like this:

Pain: An unpleasant sensory and emotional experience associated with actual or potential tissue damage, or described in terms of such damage. *Note*: Pain is always subjective. Each individual learns the application of the word through experiences related to injury in early life. . . . Unpleasant abnormal experiences (dysaesthesia) may also be pain but are not necessarily so because, subjectively, they may not have the usual sensory qualities of pain. Many people report pain in the absence of tissue damage or any likely pathological cause; usually this happens for psychological reasons. There is no way to distinguish their experience from that due to tissue damage if we take the subjective report. If they regard their experience as pain and if they report it in the same ways as pain caused by tissue damage, it should be accepted as pain. This definition avoids tying pain to the stimulus. . . . (IASP, 1986, p. 250)

For critical discussions of this definition see Price 1988, 1999; Aydede and Güzeldere 2002.

3. This uneasy relation does not exist only for pain, but arises in all other intransitive bodily sensations (itches, tickles, etc.) and their scientific study.

4. Experiments involving single-cell recordings or recordings from a population of cells were usually designed to find out the *connectivity patterns* among nerves or nerve bundles. For this purpose, correlations with sensations were not essential and not sought after.

5. Treating subjectivity as merely an epistemological phenomenon lacking any antiphysicalist metaphysical implications has been on the rise since the early 1990s. Among such physicalists are Loar (1990/1997); Papineau (1993); Sturgeon (1994); Pereboom (1994); Lycan (1996); Hill (1997); Levine (1993, 2001); Tye (1995, 1999); Perry (2001); and Aydede and Güzeldere (2005, forthcoming). Those who argue against physicalism on the basis of subjectivity and phenomenal character of conscious experience include Kripke (1970/1980); Nagel (1974); Jackson (1982, 1986); and Chalmers (1996).

6. The well-known Frege puzzles in the philosophy of language concern a similar phenomenon. One can know that Mark Twain is clever without knowing that Samuel Clemens is clever, even though what makes both true is precisely the same state of affairs. There are no separate states of affairs or facts here, but only one fact that can be expressed either by saying "I know that Mark Twain is clever" or by saying "I know that Samuel Clemens is clever." I may even know one and the same fact without realizing that I do: I may think that there are two facts here if I do not know that Mark Twain is Samuel Clemens. Similarly, the idea in the main text is that there are two very different ways of representing one and the same phenomenon, say, pain: one under its scientific description, the other introspectively.

7. For an extensive and detailed elaboration of this kind of approach to experience and introspection, see Aydede and Güzeldere 2005, forthcoming.

8. We haven't distinguished between claims of identity of the physical with the mental, on the one hand, and the claims of metaphysical supervenience of the mental on the physical, on the other. The terminology of 'realization', 'implementation', etc. usually indicates that a supervenience claim is being made. Identity is stronger than supervenience in that the latter is a claim about one-way entailment: if the mental is only metaphysically supervenient on the physical then more than one physical kind or different kinds of brain activity can realize or implement a given mental kind. For sensation types like pain it is more likely that a form of type–identity claim is true, so we will operate under this assumption, but nothing crucially depends on this for what follows. The reader may make the necessary changes in the text if supervenience strikes one as more plausible.

9. One of us, DDP, hasn't decided yet whether he agrees or disagrees with physicalistic monism but he unquestionably accepts epistemological dualism. Any potential disagreement between DDP and MA about this doesn't affect the main position of this paper.

10. Locke 1693/1975; Armstrong 1968; Lycan 1996; Lormand 1996.

11. Rosenthal 1997, 2000. Among the proponents of HOT, some, like Dretske (1995) and Shoemaker (1994), take introspection to be inferential, i.e., not direct and immediate.

12. See Aydede and Güzeldere 2005, forthcoming for more details about a general account of introspection of experiences along these lines. See Aydede, in prep., for an account of introspection of bodily sensations including pain.

13. Here, we use 'belief' in the standard philosophical sense as that cognitive state (like knowledge) that underlies the assent one makes to a descriptive statement. Our use here is not the ordinary layman's use in the sense of having an opinion that reflects one's biases prejudices, etc.

14. This approach has some close affinities with the approach of Varela and his colleagues (Varela, Thompson, and Rosch 1991; Varela 1996), which is sometimes called "neurophenomenology" for reasons that will become clearer as we proceed. See Flanagan 1992 and Aydede and Güzeldere 2002 for a defense of similar approaches.

15. Barrell and Price 1975; Price 1972; Price et al. 1977, 1994.

16. See Price 1988 and 1999 for reviews.

17. It is noteworthy that these findings make a strong *prima facie* case against pure representationalist theories of pain, or as philosophers sometimes put it, against externalist perceptual theories of pain, according to which the phenomenal content of a pain experience is exhausted by its external informational/representational content (e.g., Dretske 1995, 1999; Tye 1995, 1997, this volume). Such a position requires the theorist to find a feature of the stimulus or its immediate effect on the body (tissue damage or impending tissue damage) for every discernable phenomenological quality of pain as its representational content. As we have seen, however, the second pain summation phenomenon clearly does not correlate with any stimulus property or even with anything in the peripheral nervous system. This makes it hard to see what the phenomenology of second pain summation can have as its representational content. Aydede (2001) uses the affective dimension of pain in an attempt to make a case against purely representational theories of pain, but the second pain summation demonstrates that the same kind of argument can also be mounted by using sensory–discriminative qualities of pain.

18. See Gesheider 1998 and Stevens 1975 for reviews.

19. Price and Fields 1997; Price, Barrell, and Barrell; Price, Riley, and Barrell 1985, 2001.

20. Bakan 1967; Buytendyck 1961; Merleau-Ponty 1962; Husserl 1952.

21. See Barrell and Barrell 1975; Van Kaam 1959; Price and Barrell 1980.

22. See Husserl 1952; Merleau-Ponty 1962; Van Kaam 1959.

23. See Perls, Hefferline, and Goodman 1951; Barrell and Barrell 1975; Price and Barrell 1980.

24. See Boring 1961; Titchener 1908, 1924.

25. See Husserl 1952; Merleau-Ponty 1962; Van Kaam 1959; Varela, Thompson, and Rosch 1991.

26. See Price and Barrell 1980; Van Kaam 1959.

27. See Van Kaam 1959; Price and Barrell 1980.

28. See Merleau-Ponty 1962; Van Kaam 1959.

29. Note that it is not necessary for one's experience simply to *have* this felt sense of unpleasantness that it be experienced *as* threatening or intrusive, in that one need not have the conceptual resources to characterize one's experience as such. The claim here is that the experience is threatening and intrusive when and only when it is unpleasant—although it will also, as a matter of fact, be normally conceived as such by the passive introspection of subject-investigators who have the relevant conceptual resources.

30. See Damasio 1994, 1999; Price 2000.

31. For intriguing remarks to similar effect, see Feigl 1967; and Nagel 1974.

32. Even though systematic and pervasive failure of finding out mind–brain correlations could contribute to giving up naturalism, what we have in mind here is rather the failure of philosophical arguments for naturalism.

References

Armstrong, David. 1968. *A Materialist Theory of the Mind*. London: Routledge and Kegan Paul.

Aydede, Murat. 2001. "Naturalism, Introspection, and Direct Realism about Pain." *Consciousness and Emotion* 2(1): 29–73.

———. In prep. "Introspecting Pain and Other Intransitive Bodily Sensations." University of Florida, Department of Philosophy.

Aydede, Murat, and G. Güzeldere. 2002. "Some Foundational Problems in the Scientific Study of Pain." *Philosophy of Science* 69 (suppl.): 265–283.

———. 2005. "Cognitive Architecture, Concepts, and Introspection: An Information-Theoretic Solution to the Problem of Phenomenal Consciousness." *Noûs* 39(2): 197–255.

———. Forthcoming. *Information, Experience, and Concepts: A Naturalistic Theory of Phenomenal Consciousness and Its Introspection*. Oxford: Oxford University Press.

Bakan, D. 1967. *On Method*. San Francisco: Jossey-Bass.

Barrell, J. J., and J. E. Barrell. 1975. "A Self-Directed Approach for a Science of Human Experience." *Journal of Phenomenological Psychology* (fall): 63–73.

Barrell, J. J., D. Medieros, J. E. Barrell, and D. D. Price. 1985. "Anxiety: An Obstacle to Performance." *Journal of Humanistic Psychology* 25: 106–122.

Barrell, J. J., and D. D. Price. 1975. "The Perception of First and Second Pain as a Function of Psychological Set." *Perception and Psychophysics* 17(2): 163–166.

Block, N. 1995. "On a Confusion about a Concept of Consciousness." *Brain and Behavioral Sciences* 18(2): 227–247.

Boring, E. G. 1961. *A History of Experimental Psychology.* New York: Appleton–Century–Crofts.

Buytendyck, F. J. J. 1961. *Pain.* London: Hutchinson.

Chalmers, D. J. 1996. *The Conscious Mind: In Search of a Fundamental Theory.* Oxford: Oxford University Press.

Damasio, A. 1994. *Descartes' Error.* New York: Avon Books.

———. 1999. *The Feeling of What Happens.* New York: Avon Books.

Dretske, F. 1995. *Naturalizing the Mind.* Cambridge, Mass.: MIT Press.

———. 1999. "The Mind's Awareness of Itself." *Philosophical Studies* 95:103–124.

Feigl, H. 1967. *The "Mental" and the "Physical": The Essay and a Postscript.* Minneapolis: University of Minnesota Press.

Flanagan, Owen. 1992. *Consciousness Reconsidered.* Cambridge, Mass.: MIT Press.

Gendlin, E. T. 1962. *Experiencing and the Creation of Meaning: A Philosophical and Psychological Approach to the Subjective.* New York: Free Press.

Gescheider, G. A. 1998. *Psychophysics: The Fundamentals.* New York: Lawrence Erlbaum.

Güzeldere, G., and M. Aydede. 1997. "On the Relation between Phenomenal and Representational Properties." *Brain and Behavioral Sciences* 20(1): 151–153.

Hill, Christopher. 1997. "Imaginability, Conceivability, Possibility, and the Mind–Body Problem." *Philosophical Studies* 87: 61–85.

Hofbauer, R. K., P. Rainville, G. H. Duncan, and M. C. Bushnell. 2001. "Cortical Representation of the Sensory Dimension of Pain." *Journal of Neurophysiology* 86(1): 402–411.

Husserl, E. 1952. *Ideas: General Introduction to Pure Phenomenology.* New York: Macmillan.

IASP. 1986. "Pain Terms: A List with Definitions and Notes on Usage." Recommended by the International Association for the Study of Pain (IASP). *Pain* (suppl.) 3: 216–221.

Jackson, Frank. 1982. "Epiphenomenal Qualia." *Philosophical Quarterly* 32: 127–136.

———. 1986. "What Mary Didn't Know." *Journal of Philosophy* 83(5): 291–295.

Kripke, Saul. 1970/1980. *Naming and Necessity.* Cambridge, Mass.: Harvard University Press.

Landau, W., and G. H. Bishop. 1953. "Pain from Dermal, Periosteal, and Fascial Endings and from Inflammation." *Archives of Neurology and Psychiatry* 69: 490–504.

Levine, Joseph. 1993. "On Leaving Out What It's Like." In *Consciousness*, ed. Martin Davis and Glyn W. Humphreys. Oxford: Basil Blackwell.

———. 2001. *Purple Haze: The Puzzle of Consciousness*. Oxford: Oxford University Press.

Lewis, T., and E. E. Pochin. 1938. "The Double Response of the Human Skin to a Single Stimulus." *Clinical Sciences* 3: 67–76.

Loar, Brian. 1990/1997. "Phenomenal States." In *The Nature of Consciousness: Philosophical Debates*, ed. Ned Block, Owen Flanagan, and Güven Güzeldere. Cambridge, Mass.: MIT Press.

Locke, John. 1693/1975. *An Essay Concerning Human Understanding*. Amherst, NY: Prometheus Books.

Lormand, E. 1996. "Inner Sense Until Proven Guilty." Draft. University of Michigan.

Lycan, William G. 1996. *Consciousness and Experience*. Cambridge, Mass.: MIT Press.

Merleau-Ponty, M. 1962. *The Phenomenology of Perception*. New York: Humanities Press.

Nagel, Thomas. 1974. "What Is It Like to Be a Bat?" *Philosophical Review* 83: 435–450.

Papineau, David. 1993. "Physicalism, Consciousness, and the Antipathetic Fallacy." *Australasian Journal of Philosophy* 71: 169–183.

Pereboom, Derk. 1994. "Bats, Brain Scientists, and the Limitations of Introspection." *Philosophy and Phenomenological Research* 54(2): 315–329.

Perls, F., R. F. Hefferline, and P. Goodman. 1951. *Gestalt Therapy*. New York: Dell.

Perry, John. 2001. *Possibility, Consciousness, and Conceivability*. Cambridge, Mass.: MIT Press.

Price, D. D. 1972. "Characteristics of Second Pain and Flexion Reflexes Indicative of Prolonged Central Summation." *Experimental Neurology* 37: 371–387.

———. 1988. *Psychological and Neural Mechanisms of Pain*. New York: Raven Press.

———. 1999. *Psychological Mechanisms of Pain and Analgesia*. Seattle: IASP Press.

———. 2000. "Psychological and Neural Mechanisms of the Affective Dimension of Pain." *Science* 288(5472): 1769–1772.

Price, D. D., and J. J. Barrell. 1980. "An Experiential Approach with Quantitative Methods: A Research Paradigm." *Journal of Humanistic Psychology* 20(3): 75–95.

———. 1984. "Some General Laws of Human Emotion: Interrelationships between Intensities of Desire, Expectations, and Emotional Feeling." *Journal of Personality* 52: 389–409.

Price, D. D., J. E. Barrell, and J. J. Barrell. 1985. "A Quantitative-Experiential Analysis of Human Emotions." *Motivation and Emotion* 9: 19–38.

Price, D. D., and H. L. Fields. 1997. "The Contribution of Desire and Expectation to Placebo Analgesia: Implications for New Research Strategies." In *The Placebo Effect*, ed. Ann Harrington. Cambridge, Mass.: Harvard University Press.

Price, D. D., H. Frenk, J. Mao, and D. J. Mayer. 1994. "The NMDA Receptor Antagonist Dextromethorphan Selectively Reduces Temporal Summation of Second Pain." *Pain* 59:165–174.

Price, D. D., J. W. Hu, R. Dubner, and R. Gracely. 1977. "Peripheral Suppression of First Pain and Central Summation of Second Pain Evoked by Noxious Heat Pulses." *Pain* 3: 57–68.

Price, D. D., J. Riley, and J. J. Barrell. 2001. "Are Lived Choices Based on Emotional Processes?" *Cognition and Emotions* 15(3): 365–379.

Rainville, P., G. H. Duncan, D. D. Price, B. Carrier, and M. C. Bushnell. 1997. "Pain Affect Encoded in Human Anterior Cingulate but not Somatosensory Cortex." *Science* 277: 968–971.

Rosenthal, David. 1997. "A Theory of Consciousness." In *The Nature of Consciousness*, ed. Ned Block, Owen Flanagan, and Güven Güzeldere. Cambridge, Mass.: MIT Press.

———. 2000. "Introspection and Self-Interpretation." *Philosophical Topics* 28(2): 201–233.

Shoemaker, Sidney, 1994. "Phenomenal Character." *Noûs*, 28: 21–38.

Stevens, S. S. 1975. *Psychophysics—Introduction to Its Perceptual, Neural, and Social Prospects*. New York: John Wiley.

Sturgeon, Scott. 1994. "The Epistemic View of Subjectivity." *Journal of Philosophy* 91(5): 221–235.

Titchener, E. B. 1908. *Experimental Psychology of the Thought Processes*. New York: Macmillan.

———. 1924. *The Psychology of Feeling and Attention*. New York: Macmillan.

Tommerdahl, M., K. A Delemos, C. J. Vierck, O. V. Favorov, and B. L. Whitsel. 1996. "Anterior Parietal Cortical Response to Tactile and Skin Heating Stimulation." *Journal of Neurophysiology* 75(6): 2662–2670.

Tye, M. 1995. *Ten Problems of Consciousness*. Cambridge, Mass.: MIT Press.

———. 1997. "A Representational Theory of Pains and Their Phenomenal Character." In *The Nature of Consciousness: Philosophical Debates*, ed. N. Block, O. Flanagan, and G. Güzeldere. Cambridge, Mass.: MIT Press.

———. 1999. "Phenomenal Consciousness: The Explanatory Gap as a Cognitive Illusion." *Mind* 108: 705–725.

Van Kaam, A. L. 1959. "Phenomenal Analysis: Exemplified by a Study of the Experience of Really Feeling Understood." *Journal of Individual Psychology* 5: 66–72.

Varela, F. 1996. "Neurophenomenology: A Methodological Remedy for the Hard Problem." *Journal of Consciousness Studies*. 3(4): 330–349.

Varela, F., E. Thompson, and E. Rosch. 1991. *The Embodied Mind: Cognitive Science and Human Experience*. Cambridge, Mass.: MIT Press.

Peer Commentary on Donald D. Price and Murat Aydede

14 Introspections without Introspeculations

Shaun Gallagher and Morten Overgaard

We will divide up our remarks to address three topics discussed in Price and Aydede: pain, introspection as it is currently practiced, and the proposed methodological introspection. We will leave aside any discussion of the interesting metaphysical questions raised in their paper.

1 Pain

We have little to say about pain itself. It is clear that although pain is distinguishable from other experiential states, Price and Aydede do not isolate it from those other experiential dimensions (especially the affective). Thus they write: "First, pain, unlike most conscious experiential states such as visual, auditory, tactile experiences, have an immediate affective and emotional aspect to it, which underlies its intimate personal as well as clinical urgency" (this volume, p. 245). Our ordinary experience of pain, then, is not of some pure painfulness—and we should even say that pain infects our visual, auditory, tactile experiences. When I am in pain, my experience of the world via these various modalities is affected—and indeed, these perceptual experiences do have affective and emotional aspects to them that may be the result of the pain itself. So we agree that pain does not function as an object of our perceptual experience, but infiltrates that experience itself. "Notice that pain here is not the object of our perceptual experience, but rather, it is the experience itself" (ibid., p. 246). But this also motivates the following question: to what extent does the pain that one tries to introspect actually affect the introspection? This is just the opposite of what might be the more expected methodological question: to what extent does introspection affect (change) the experience of pain (or any experience)?

2 Introspection as Already Practiced

The authors argue that introspection is indispensable and that it is already a part of third-person studies. But what is the nature of the introspection that is already practiced in science? There does seem to be a very basic agreement among scientists interested in introspection that it involves a direct attending to the subject's own consciousness, no matter how one would go about defining introspection from there. To quote just a few of the "first-generation introspectionists," Knight Dunlap writes " 'Introspection' is usually defined in terms which are equivalent to the expression *consciousness scrutinizing itself*" (1912, p. 404). Angell writes: "It [introspection] consists simply in the direct examination of one's own mental processes" (1908, p. 5). Stratton considers it to be the "direct acquaintance with the state of our minds which all of us to some extent possess" (1914, p. 2). Finally, William James characterizes it as: "the looking into our own minds" (James 1890, p. 185). Price and Aydede, however, suggest that it is beyond the scope of their paper to examine the nature of introspection, and they are satisfied with some combination of HOP and HOT models (this volume, p. 251). They suggest that nothing crucial about their analysis depends on this issue. We disagree, especially in regard to the first part of their paper. For purposes of discussion we will stay with the model they propose, but we note that there is certainly much more to say about whether HOP or HOT models genuinely capture the concept of introspection. What's important here is that whether one appeals to HOP, HOT, or some alternative phenomenological model of introspection, introspection is understood as a reflective second-order cognitive act that thematizes first-order phenomenal experience, and makes that experience the object of reflection.

Price and Aydede claim that, concerning this kind of introspection, "there is absolutely no reason to think that the use of such a first-person approach is scientifically and methodologically suspect." Especially with respect to what they characterize as the introspection already practiced in experimental settings, however, there may be some justified suspicion. Indeed, and on a more basic level, before we get to the issue of scientific reliability, we think that Price and Aydede's analysis may reflect a not uncommon confusion about the very nature of introspection and how it is practiced. Indeed, in controlled scientific experiments that require verbal reports, it is not clear at all that introspection is used in any strict methodological fashion. There are really two points here. First, whether the practices that Price and Aydede call introspective are really introspective; second, whether those practices are above suspicion.

According to Price and Aydede, introspection is already practiced in experimental science because "the subjects' verbal reports [or nonverbal behaviors like button-pushes] about their own cognitive states have routinely been taken as evidence for the cognitive models postulated" (this volume, p. 245). First, one might argue that *all* reports given by subjects are, at least indirectly, about their own cognitive (mental, emotional, experiential) states. If one instructs a subject to push a button, or say "now" when she sees the light come on, then the subject is reporting about the light, but also about her visual experience. Even if one instructs the subject in a way that carefully avoids mention of an experiential state: "Push the button when the light comes on," the only access that the subject has to the fact of the light coming on is by way of her experience of the light coming on. In this sense the first-person perspective is inherent in all experiments that depend on subjects' reports. One might follow this to its logical conclusion, that even scientific observations made by the experimenter, usually considered as third-person data, presuppose and are tied to first-person reports of the scientist. The scientist might say "The subject's premotor cortex was activated 300 msecs before the subject raised her arm." But that could easily be a report on the scientist's experienced perception of the instrument that measured the timing of the activation. It would be odd, however, to say that this third-person fact was based on introspection, although it is a first-person report of the scientist. More generally, however, and less extremely, it is odd to say that the first-person reports of a subject are necessarily introspective, although this is precisely what Price and Aydede claim.

For example, I may ask the subject to say "now" when she sees the light come on. How precisely does the subject know when she sees the light come on? Does she reflectively introspect her experience looking for the particular visual state of seeing the light come on? Or does she simply see the light come on and report that? One might ask, "How could she possibly report that she sees the light come on if she doesn't introspectively see that she sees the light come on? Is it possible that we can report on what we experience without employing introspection?" There is a long tradition in philosophical phenomenology (specifically the tradition that follows Husserl) that answers in the affirmative. We can report on what we experience without using introspection because we have an implicit, nonintrospective, prereflective self-awareness of our own experience. At the same time that I see the light, I know that I see the light. This knowledge of seeing the light is not based on reflectively or introspectively turning our attention to our own experience. It is rather built into our experience as an essential part of it, and it is precisely that which defines our experience as conscious experience. On this view, I consciously experience the light coming on just as I see the light coming on. I don't have to verify through introspection that I have just seen the light come on, since my first-order phenomenal experience is already something of which I am aware in the very experience of it.

First-person reports of this kind, then, are not introspective reports. They are prereflective experiential reports. So it is not correct to say that from a first-person perspective "conscious experiences seem accessible only through introspection" (this volume, p. 246) or "introspection seems to be the *only* available method of access to qualia" (ibid., p. 249). This applies to pain as well. It is not the case that our access to pain phenomena is "only through the special epistemic faculty of introspection" (ibid., p. 247). Indeed, introspection on pain is usually motivated only because we already know that we are in pain—and we have that knowledge prereflectively, and can report it on that basis, without introspecting it. A stimulus is applied. The experimenter asks, "Is that painful?" I do not have to introspect to say "yes." I do not have to "observe/believe that such and such sensations" are happening to me (ibid., p. 252). I can report directly and immediately on my experience of pain because my access is directly and immediately in the pain experience.

In addition, not all reports are about consciously experienced states. In philosophy and especially in cognitive science, there seems to be agreement, at least to a large extent, that not all mental states are conscious (Marcel 1983). Examples such as blindsight and subliminal perception serve to illustrate that subjects may perform a number of tasks that we normally do not hesitate to call "mental" even though subjects report no conscious awareness of them. Therefore, it seems logical to conclude that not all reports about mental states are introspective, that is, not all of them are about consciousness.

The second point concerns how reliable or how methodological such experiential reports are. In general and for many cases, these kinds of reports do seem very reliable. If an experimenter applies a stimulus that causes a relatively high degree of pain, or a sensory stimulus that is well above threshold, the subjects' reports that they experience the stimulus as painful or as clearly present seem above suspicion. Reliability may decrease, however, when the stimulus is closer to threshold, and may depend on the mode of report, or other subjective factors that qualify the report. Marcel (1993), for example, has shown that requests for quick reports of close-to-threshold stimuli using different modes of report (verbal, eye blink, button push) elicit contradictory responses. At the appearance of a just noticeable light stimulus, subjects will report with a button push that they did see the light and then contradict that with a verbal report. This kind of data, and more generally, uneven or inconsistent data can motivate two different strategies. Most often, following established scientific procedure, data are averaged out across trials or subjects, and the inconsistencies are washed out. Less often, scientists are motivated to take this first-person data seriously and to employ introspective methods to investigate it. The second part of Price and Aydede's paper turns to what is genuinely the use of introspection in such contexts.

3 Methodologies of Introspection

So far we have argued that not all verbal reports on experience are introspective reports, and introspection is not the only access we have to experience. Thus we take issue with Price and Aydede's claim that "Introspection is a way—apparently the only way—of coming to know about our experiences and their qualities directly" (this volume, p. 249). Even if, however, many first-person reports found in scientific experiments are first-person, nonintrospective experiential reports, we do not mean to rule out the usefulness of introspection. The use of introspective reports, that is, self-reports that thematize experience, can certainly provide more information about the subject's experience. In regard to pain, for instance, introspective reports can specify the qualities and subjective measurements of pain. In addition, if done in a methodically controlled way, introspection can address issues pertaining to the reliability of some nonintrospective experiential reports.

It will be fruitful to compare and contrast three different models of how to employ a methodical introspection in experimental situations: (1) Price and Aydede's experiential model; (2) what has been called a "new introspectionism," developed by Overgaard, his colleagues, and a number of other researchers (e.g., Marcel 1983, 2003; Jack and Roepstorff 2002; see comments by Frith 2002; Gallagher 2002); and (3) the method of neurophenomenology developed by Varela (1996), Lutz, and their colleagues (Lutz et al. 2002). All three share the same ambition to improve techniques for subjective reporting in order to gain more insights into the "subjective pole" in the comparison between objective neural states and subjective conscious states.

3.1 The Experiential Model

Price and Aydede suggest an approach consisting of two stages. First, one is to use a "horizontal approach" in which an investigator or subject introspectively examines what some or other subjective state feels like. Price and Aydede advise us to avoid speculations about why something was experienced and to focus specifically on *how* it feels instead, thus avoiding interpreting or judging one's own experiences. To do this, they argue that one should notice experiences passively without controlling attention so that one observes ongoing thoughts, emotions, or perceptions as if they were "seen in the periphery of one's visual field." Price and Aydede suggest not only "simple" kinds of experiential states as objects for the introspective examination, but, apart from their main example of pain, they suggest performance anxiety as a kind of mental state that can be studied with their approach.

The introspective examination consists not only of "inner observation" but also of a description or verbalization of the observations. Of course, aside from the possibilities presented by those who are poetically blessed, one cannot describe, say, the sensation of coldness with many words. Our ordinary, prosaic linguistic practices have not sufficiently evolved to describe a subjective state in such a way that someone who never experienced coldness himself would get an idea of what that sensation is like simply through the description. To address this scarcity of words, Price and Aydede want their subjects to describe associations and thoughts that may arise in connection with the relevant experience: "Intense burning and throbbing in my hand. Feel bothered by this and slightly annoyed. Is it going to get stronger? Feeling of concern. Hope my hand isn't going to be scalded."

Price and Aydede suggest that scientists should use themselves as subjects. The example of a description of associations that arise when lowering one's hand into cold water was in fact given by one of the authors of the paper. The argument for using oneself, as investigator, as a subject, seems to be that the reports of the investigators are as "subjective" as are the reports of naive subjects, and, in this sense, just as valid as experimental data. However, one should not forget that investigators must be assumed to always have certain hypotheses and results they hope to find, and thus they are likely more biased as subjects. Using the same argument as Price and Aydede, one could say that given that investigators and naive subjects have the same status in terms of validity, one could reduce the possibility of the confounding effects of interpretations and judgments by using naive (though probably trained) subjects only.

Price and Aydede believe that their approach is compatible with experimental methods found in psychophysics. To integrate their rather open method of describing experiences, however, they find it necessary to ask subjects to scale the presence of, say, "a throbbing sensation of pain," "fear of bodily harm," or some other state described by the subjects themselves during the "open description." This would in essence make possible a quantification of the descriptions, and, as such, it would make the reports commensurable with cognitive neuroscience. This is the second, "vertical," aspect of their approach. In classical cognitive neuroscience, one uses stimulus input as an experimental variable. Subjects would be presented with two or more different kinds of stimuli and in order to find the neural activations for perceiving one kind of stimulus, the neural activations caused by the other kinds of stimuli would be subtracted from the first. Thus, the reasoning goes, one will find the essential features involved with perceiving the first kind of stimulus. When using neuroscientific techniques there seems no way around using such a subtractive method even though it has been severely criticized (Friston et al. 1996; Overgaard 2004). However, one does not need to define one's variables based upon different stimuli. One could keep stimulus features constant and only vary the instructions as now seen in an increasing number of studies, or one could define the experimental conditions based on the subjective reports themselves. This latter strategy is suggested by Price and Aydede, and it points to a way of integrating the open, subjective reporting strategy with the methodology of cognitive neuroscience.

Price and Aydede do not stand alone defending this kind of view on subjective reports and their integration with neuroscience. On several occasions, the authors appeal to introspectionism and phenomenology as research directions with similar goals. Within both traditions, in the last decade, new developments have occurred that deserve a comparison with the suggestions of Price and Aydede.

3.2 New Introspectionism

A "new" introspectionist approach to subjective reporting has been put forward in Marcel 1983, 2003; Jack and Roepstorff 2002; Overgaard 2003a; Overgaard, Nielsen, and Fuglsang-Frederiksen 2004; and is further developed in Overgaard and Sørensen 2004; and Ramsøy and Overgaard 2004.

In Overgaard 2003b, an outline for introspective reporting is described. It is suggested that one should perform consciousness studies on a metaphysically neutral ground, and that an important reason for the shortcomings of classical introspectionism was the commitment to certain metaphysical claims, for example, the belief in "elementarism." It is suggested that one should use introspective reporting in experiments on consciousness, given that nonintrospective reporting may reflect nonconscious processing. Furthermore, it is suggested that one should reconsider the use of stimulus conditions as the only variable in experiments in cognitive neuroscience, and instead use differences in instructions or in the subjects' own reporting as the categories of analysis. This line of thinking corresponds very well with Price and Aydede's reflections on the use of subjective reporting in experiments as well as in analysis of data.

Ramsøy and Overgaard (2004) presented subjects with a visual identification task using varied durations, and asked subjects (1) to guess what was shown on the computer screen, and (2) to scale how clearly they experienced the image. The steps of the scale, including their definitions, were made by the subjects themselves with the instruction that there should be a 1:1 correspondence between experienced differences and reported differences. After a pilot experiment, in which the subjects developed the scale and became accustomed to using it, the actual experiment was run. The subjects ended up using more or less the same scale (named the Phenomenal Awareness Scale or PAS), and for reasons of analysis, the investigators decided to merge the scales to include only the points of the scale that were shared by the subjects. This strategy corresponds almost completely to the proposals of Price and Aydede in the "horizontal stage" of analysis, and in the transformation from the horizontal to the vertical strategy.

In Overgaard, Nielsen, and Fuglsang-Frederiksen 2004, PAS was used with different subjects when using a similar visual display, and coupled with transcranial magnetic stimulation and EEG. This study aimed to identify the involvement of the ventral projection streams from primary visual cortex in visual consciousness.

In effect, the approach of "new introspectionism" (described in more detail in Overgaard 2003b) shares all important features with the approach of Price and Aydede. The subjects started out using their own terminology, which then was used for the purpose of scaling the subjective reports, and finally, it was integrated with neuroscience to search for neural correlates of consciousness.

There are some minor differences between the two approaches as well. Price and Aydede seem more optimistic about which mental phenomena one can study with their suggested approach. Yet, during the horizontal stage, a phenomenon like performance anxiety would give rise to many different associations and thoughts, with the result that it would be almost impossible to tell whether the very different subjective reports basically reflect identical conscious states. Even in the example of the experience of coldness, as mentioned above, the "spontaneous utterings" of the subjects are so relatively different that the investigators must perform some amount of interpretation of the reports in order to create categories suitable for quantitative analysis. Such a post hoc analysis, of course, shares all the problems of the creation of reporting categories in advance of an experiment.

A further line of research using introspective reports aims to identify differences between reports that are specifically about "how" something is experienced and reports about "what" is experienced. In Overgaard and Sørensen 2004, subjects were presented with a simple design for visual stimuli. On one of three possible locations, a simple figure followed by a mask would appear. The figure was either a triangle, a square, or a circle, or some variation of one these figures (e.g., a half-circle, an upside-down triangle, or a combination of two figures). The colour was either blue, green, or red, although the hue varied. The subjects were to identify the displayed figure by pointing at the corresponding figure drawn on three scales: one for stimulus shape, one for color, and one for the location. The scale of shapes consisted of a display of thirty-four different figures, some of which were included in the data material. The scale of colors consisted of eight different levels of hue for the colors. As with the shapes, only some of the colors were actually included as stimuli. The scale of positions displayed the fixation cross in the middle and the three different locations where the stimulus could occur. The responses of the subjects were treated as being either "correct," "incorrect," or "near correct." "Near correct" responses partially matched stimulus in a manner that was only partially correct (e.g., when they pointed at the same color as the one presented, but in a brighter or darker tone). The results of the data analysis showed that subjects in the nonintrospective condition had significantly more correct *and* incorrect responses, whereas the introspective subjects most often were "near correct." In addition, subjects in the introspective condition tended to be more liberal about their reports of, say, color, while the subjects in the nonintrospective condition tended to show a more conservative style conforming to specific color categories.

These results open up questions such as to what degree and how precisely introspection might change (visual) experiences. It seems necessary to address the issue if or how this knowledge should change our way of using introspective reports.

3.3 Neurophenomenology

A third approach, neurophenomenology, as espoused by Francisco Varela (1996), follows the phenomenological tradition initiated by Husserl. This view involves training both the scientists and the experimental subjects in phenomenological method, including use of the phenomenological reduction, that is, the setting aside or "bracketing" of opinions or theories that a subject may have about his experience. This method involves shifting our attention from *what* we experience to *how* we experience. This correlates well with Price and Aydede's advice to avoid speculations about why something was experienced and to focus specifically on *how* it feels.

Lutz et al. (2002) employ the neurophenomenological method to study the highly variable successive brain responses to repeated and identical stimulations in many empirical testing situations that target specified cognitive tasks. Their hypothesis is that this variability is generated in mental fluctuations due to the subject's attentive state, spontaneous thought processes, strategy decisions for carrying out the task, and so on. These *subjective parameters* include distractions, cognitive interference, and so on. To control for such subjective processes is difficult and they are usually averaged out across a series of trials and across subjects. Lutz and his colleagues decided to take these subjective parameters more seriously. They combined a process of trained phenomenological reflection with the dynamical analysis of neural processes measured by EEG in a paradigm involving a 3-D perceptual illusion. Importantly, Lutz and his colleagues used the introspective first-person data not simply as more data for analysis, but as contributing to their analytic framework.

Phenomenological training in this experiment consisted in training subjects to deliver consistent and clear reports of their experience through a reflective introspection. The goal of phenomenological reflection is to gain intuitions of the structural invariants of an experience, not to average them out. Phenomenological reflection can be either self-induced by subjects familiar with it (not unlike Price and Aydede's proposal that the scientist use herself as the subject), or guided by the experimenter through *open questions*—questions directed not at opinions or theories, but at experience. Again, this resembles the "open description" discussed by Price and Aydede. Rather than employing predefined categories, and asking "Do you think this experience is like *X* or *Y* or *Z*?" the open question asks simply "How would you describe your experience?" Open questions posed immediately after the task help the subject to redirect his or her attention toward the implicit strategy or degree of attention he or she implemented during the task. Subjects can be reexposed to the stimuli until they find "their own stable experiential invariants" to describe the specific elements of their experiences.

In a series of preliminary or practice trials, the subjects developed descriptions (refined verbal reports) of the subjective parameters while engaged in a depth perception task. Subjects thus became knowledgeable about their own experience and developed descriptions of experiential invariants on the basis of open questions, reporting on the presence or absence or degree of distractions, inattentive moments, cognitive strategies, and so on. On the basis of these first-person introspective descriptions, descriptive categories were formulated a posteriori to create phenomenologically based clusters that are then used as analytic tools in the main trials. For example, with regard to the subject's experienced readiness for the stimulus, the results specified three readiness states: *steady readiness* (SR), in which subjects reported that they were alert and well prepared as the task began; *fragmented readiness* (FR) in which subjects reported that they were prepared less "sharply" (due to a momentary tiredness) or less "focally" (due to small distractions, etc.); and *unreadiness* (SU) in which subjects reported that they were unprepared as the task began. Subjects then used these categories to report their readiness state during the main trials as the experimenters recorded the electrical brain activity. The first-person reports were correlated with both behavioral measures (reaction times) and dynamic descriptions of the transient patterns of local and long-distance synchrony occurring between oscillating neural populations. Using these correlations, Lutz et al. were able to show that distinct subjective parameters correlate to specific dynamic brain patterns just prior to presentation of the stimulus. The results were significantly different relative to results based on averaging across trials.

The experimental protocol used in Lutz et al. 2002 thus employs a practical phenomenological method. The subjects are asked to provide a description of their own experience using an open-question format, and thus without the imposition of predetermined theoretical categories. They are trained to gain introspective intimacy with their own experience. Their first-person introspective reports are then intersubjectively and scientifically validated both in setting up the phenomenological clusters and in using those clusters to interpret results that correlated with objective measurements of behavior and brain activity.

4   Conclusion

These three approaches share a number of common features.

1. Use of preliminary trials or pilot experiments to train subjects and to develop introspective or phenomenological descriptions of experience or subject-developed scales (Lutz; Overgaard).

2. A pushing aside of theories or speculations in favor of attending to how experience is happening (Price and Aydede's avoidance of speculation; Varela's phenomenological reduction).[1]

3. The use of open questions to develop a description of the experience (Price and Aydede's "open description"; Lutz's open questions).

4. The formulation of common categories that transform first-person introspective descriptions into intersubjectively verified and commonly understood reports (Lutz's phenomenological clusters; Overgaard).

5. The use of these phenomenologically generated categories not just as data, but also as part of the analytical instrument (Price and Aydede; Overgaard; Lutz).

6. The integration of first-person data with third-person behavioral, psychophysical, and neurological measurement (EEG, TMS, PET, fMRI) in search of correlations among experiences, brain activity, and behavioral responses (Price and Aydede; Overgaard; Lutz).

Putting all of these elements together may provide a fuller and more detailed conception of how a methodical introspection could work than found in any one of the models. Perhaps a more integrated model that recognizes the precise difference between introspection and first-person, prereflexive, experiential reports is now called for.

Notes

1. Price and Aydede's use of subjects unfamiliar with the distinction between first and second pain or the hypotheses of the study would not be sufficient to rule out all theories or speculations they might have about the experience. The use of naive subjects contrasts very clearly with their suggestion to include investigators among the subjects, which "cannot fail to help interpret the data in a more accurate and detailed way" (p. 256). In both cases, a proficient practice of phenomenological reduction would be appropriate. We note that Price and Aydede's characterization of the phenomenological reduction as related to making descriptions more concise ("Part of the process of so-called 'phenomenological reduction' consists in an attempt to capture experiences in this more concise way" [p. 262]) is not part of the traditional definition of the reduction. Conciseness is not the issue in this context. Rather, one understanding of the phenomenological reduction is what motivates us to put aside any discussion of the metaphysical considerations that Price and Aydede seem at pains to address. All such considerations are simply bracketed as irrelevant to the experimental process. The "indispensability of introspection, and first-person methods in general, in the scientific study of pain" can be totally divorced from having to make any metaphysical decisions about the nature of the mind, dualism, identity theory, and so on, and is totally neutral in this regard. We do not have to adopt a metaphysical position in order to justify the use of introspection. Nor do we have to try to make subjectivity "metaphysically kosher" in order to put it to work in science. Price and Aydede miss the main point of the phenomenological reduction if they think that these metaphysical issues do have to be addressed in this context.

Acknowledgments

Morten Overgaard's work is supported by a grant from Carlsberg Foundations.

References

Angell, J. R. 1908. *Psychology*. 4th ed. New York: Henry Holt.

Dunlap, K. 1912. "The Case against Introspection." *Psychological Review* 19: 404–413.

Friston K. J., C. J. Price, P. Fletcher, C. Moore, R. S. Frackowiak, and R. J. Dolan. 1996. "The Trouble with Cognitive Subtraction." *Neuroimage* 4: 97–104.

Frith, C. 2002. "How Can We Share Experiences." *Trends in Cognitive Sciences* 6(9): 374.

Gallagher, S. 2002. "Experimenting with Introspection." *Trends in Cognitive Sciences* 6(9): 374–375.

Jack, A. I., and A. Roepstorff. 2002. "Introspection and Cognitive Brain Mapping: From Stimulus–Response to Script–Report." *Trends in Cognitive Sciences* 6(9): 333–339.

James, W. 1890. *Principles of Psychology*. New York: Dover.

Lutz, A., J. P. Lachaux, J. Martinerie, and F. Varela. 2002. "Guiding the Study of Brain Dynamics by Using First-Person Data: Synchrony Patterns Correlate with Ongoing Conscious States during a Simple Visual Task." *Proceedings of the National Academy of Sciences USA* 99(3): 1586–1591.

Marcel, A. J. 1983. "Conscious and Unconscious Perception: Experiments on Visual Masking and Word Recognition." *Cognitive Psychology* 15: 197–237.

———. 1993. "Slippage in the Unity of Consciousness." Pp. 168–180, in *Experimental and Theoretical Studies of Consciousness*, ed. G. R. Bock and J. Marsh. Ciba Foundation Symposium 174. New York: John Wiley.

———. 2003. "Introspective Report: Trust, Self Knowledge, and Science." *Journal of Consciousness Studies* 10(9/10): 167–186.

Overgaard, M. 2003a. "On the Theoretical and Methodological Foundations for a Science of Consciousness." *Bulletin fra Forum for Antropologisk Psykologi* 13: 6–31.

———. 2003b. "Theoretical and Empirical Studies of Consciousness." Ph.D. diss., University of Aarhus.

———. 2004. "Confounding Factors in Contrastive Analysis." *Synthese* 141(2): 217–231.

Overgaard, M., J. F. Nielsen, and A. Fuglsang-Frederiksen. 2004. "A TMS Study of the Ventral Projection Streams from V1 with Implications for the Finding of Neural Correlates of Consciousness." *Brain and Cognition* 54(1): 58–64.

Overgaard, M., and T. A. Sørensen. 2004. "Introspection Distinct from First Order Experiences." *Journal of Consciousness Studies* 11(7–8): 77–95.

Ramsøy, T. Z., and M. Overgaard. 2004. "Introspection and Subliminal Perception." *Phenomenology and the Cognitive Sciences* 3(1): 1–23.

Stratton, G. M. 1914. *Experimental Psychology and Its Bearing on Culture*. New York: Macmillan.

Varela, F. J. 1996. "Neurophenomenology: A Methodological Remedy for the Hard Problem." *Journal of Consciousness Studies* 3(4): 330–349.

15 Sensations and Methodology

Robert D'Amico

Price and Aydede defend an experimental design in the field of neurophysiological psychology wherein correlations are sought between the data of brain activity as measured, observed, or detected in third-person reports and the conscious experiences as such of the experimenter presented in first-person reports. The data in the form of first-person experiences of the experimenter are then generalized and replicated, according to the authors, by the verbal reports and reactions of other experimental subjects, viewed as having experiences analogous to the experimenter's own.

The position and expectation of the authors is that such experiments would constitute a way to study psychophysical correlations, they could thereby be used to refine and revise the explanatory hypotheses for such correlations (in the tried and true way of scientific research), and thereby replicated and confirmed. Are there any reasons to think such an experimental design is improper? Are there any methodological grounds for thinking so (grounds of what constitutes proper and replicable scientific research) or any broader, quasiphilosophical grounds for thinking such an experimental design is not "kosher," as the authors put it?

The central intent of Price and Aydede is to remove both the methodological resistance to the experimental design imagined above within what they call the "intuitive framework" that brain scientists adopt when considering such proposals. They also dismiss some of the broader philosophical objections by defending the experimental design as, in effect, neutral between physicalism and dualism, seen as the competing metaphysical frameworks for current philosophy of mind.

Their specific focus in this article is on the study of pain phenomena. They take such phenomena as paradigmatic of the study of conscious sensations, experiences, or qualia. They trace some of the intuitive framework that provides the main resistance to the study of sensations by such first-person reports and the use of introspection to the rise of behaviorism in psychology, especially in the post–World War II period, which saw both a massive expansion of psychological research and an effort to dismiss the introspective tradition in psychology.

The comments to follow are mostly about this "intuitive framework" and its resistance to introspection. I concede the authors' challenge to dogmas about experimental design reflecting latent or unquestioned behaviorist assumptions. I do think, however, there may be more than that lying within the worries about the experimental design they propose. My suggestions may explain why such resistance outlives behaviorism. My comments are then about the more difficult, broader issues.

I will at the end turn to a briefer section on the authors' appeal to the phenomenological tradition in their defense of how to carry out the first-person experimental component. My aim in that second part is of necessity limited. I only wish to set out some further details about the phenomenological tradition for the purpose of perhaps shedding more light on the very issue that is behind my first set of comments, namely how to conceive of the relation between philosophical and scientific questions in this area.

The main idea of Price and Aydede is that introspection of conscious mental experience can, by way of verbal reports of such experiences, be scientific evidence for objective facts of the matter, namely the objective facts concerning the systematic correlations in human behavior that the authors call "psychophysics." It is the study of lawful relations between external stimuli (both bodily and environmental) and conscious sensations. They hold that given this project there is no good reason to believe that the first-person reports are not compatible with the evidence of the traditional third-person experimental techniques. Some areas of neurological and psychological research simply have conscious experience as an object of study and if they wish to be scientific, then they may proceed as outlined above.

The resistance to this proposal rests, in my view, upon an asymmetry between the two sets of data. The data concerning the brain states (the data gained through brain imaging techniques, for example) can be revised and refined. I mean that given any hypothesis concerning such states or given any results of applying these techniques to detect those states, we can picture how those results would lead to a thoroughgoing revision and refinement of the initial hypotheses used to collect the data. I do not mean that the brain-imaging technique itself would be challenged or that the data would be found to be in some way corrupt (though both of course could occur), but the experimenter would have the option to reinterpret the data in a certain way as a response to the results. For example, the evidence for a correlation could be treated as needing to be revised because a brain state was wrongly identified and, as it turns out, should now be taken to correlate with a quite different sensation. Or perhaps the initial identification simply needs to be refined since the data can be interpreted as showing that distinct brain states had been mistakenly grouped together and further testing may specify the right match. I don't believe there is anything very contentious or creative here (at least on my part), since what I am describing is very much the art and skill of experimental testing practiced throughout the sciences.

But the first-person reports about the sensations in the above design, or more precisely the experimenter's direct sensations as such, simply cannot be pictured as either revisable or refineable in this way. We cannot imagine, I suggest, that the experimenter could be led by some persistent or even robust brain imaging results to the conclusion that such and such an experience that he or she is now having is not in fact a pain experience after all; or that the diffuse pain felt in that testing was in fact a sharp pain after all, given the experiment's third-person results.

I assume that the experimental design could already rid us of ambiguous pain sensations. Thus I do not mean that it is just hard to make the qualitative discrimination in question. Also, we could imagine or picture an experimenter revising or refining the interpretations of the verbal reports of other experimental subjects, within limits I suppose, to accommodate persistent third-person results about the subject's brain activity. For example, there may be a way to recommend that experimenters change their understanding of their subject's verbal reports in such and such an experimental situation given some persistent or robust imaging data. What I am claiming we cannot imagine is that the experimenter engaged in making these reports says to himself (in the spirit of revision and refinement), "I am not feeling a diffuse pain after all, what I had been feeling all long is a sharp pain. I apparently misunderstood my own self-report."

What is behind the asymmetry is the simple but deep point that the having of an experience (putting aside the verbal reports of others for the moment) is conceptually, as we might put it, not available for further revision or refinement. In fact the authors *almost* assert this point when they state: "Insofar as pain science is in the business of discovering the brain mechanisms underlying pain experiences accessible only through introspection, pain science is in fact irremediably committed to using first-person methods" (this vol., p. 247).

What would justify treating that data of introspection as fixed in this way and specifically not subject to revision or refinement? The authors downplay this worry since they present the experiment as committed to a complex method only part of which includes first-person reports. But if my comments above are defensible, the experimental design is committed not just to including first-person reports, but to the stronger position that introspective data are autonomous from revision by third-person reports. Here is where I think we find the core of resistance to this proposal and perhaps the core intuition behind the "intuitive framework."

There is perhaps a philosophical doctrine behind this asymmetry. The so-called "transparency of consciousness" has, for instance, been appealed to in the history of philosophy as both an intuitive certainty to some and a surreptitious dogma to others. But such a limit could also be thought of as simply stipulated by the experimental design. The authors may be suggesting as much since the stipulation would state that when doing psychophysics of a certain sort one ought to treat first-person reports as closed to further revision and refinement.

But the asymmetry behind this stipulation cries out for some defense. For instance, in research into decision making and rational choice, which often parallels the design proposed by the authors above, results can still involve the refinement and revision of the very first-person reports that I claim cannot occur in the sensation cases. For example, the experimenter may attribute various beliefs not currently noted by the subject or introduce some auxiliary mechanism, such as belief perseverance, which allows for reinterpretation of the subject's reported reasoning in a fashion that also allows for the refinement and revision cited above. In such cases of experimentation there is no stipulation to simply take the subject's response, even noted introspectively, as fixed and protected from further methodological revision and refinement. I have suggested that the design parallels break down when first-person reports are of sensory qualia. But it is precisely that feature that the authors have selected as the key to this new field of psychophysics.

There are examples in science of results being stipulated as invariant (most famously perhaps, the speed of light in the theory of relativity), but these cases emerge as conclusions from broad theoretical reflection, not as a methodological constraint on how the data is collected to begin with.

I stressed that the authors echo my point when they proclaim that they are "irremediably committed to using first-person methods." But what makes for such an "irremediable" commitment in my view is that the proposed discipline of psychophysics, as presented in the article, bridges two fundamental conceptual schemes for understanding the world: the scheme for physical objects and that for persons. Though such schemes share features in common, they diverge fundamentally with respect to what is involved in conceiving of a being as "minded" versus merely thing-like. Specifically with regard to my point above, denying "mindedness" to a being (or what amounts to the same thing, namely questioning the report of how a sensation seems to that being) amounts to denying that the conceptual scheme for persons is appropriate to that being. But this shift in concepts is a quite different matter from discovering contingent matters of fact holding for an area of reality, and it does not result from further refinement and revision. Rather, conceptual schemes are the preconditions, as we might put it, of thinking about or inquiry into any object or person whatsoever.

In a footnote, Price and Aydede present their approach as involving "two very different ways of representing one and the same phenomenon, say, pain: one under its scientific description, the other introspectively" (this vol., p. 268, n. 6). I do not dispute that way of presenting their approach, but if I am right the "two very different ways or representing" are not alternative descriptions of some objective matter of fact, rather they are two different kinds of representation that make possible the determination of what are the objective matters of fact. It cannot be assumed, uncontentiously, that the extension of these concepts are meshing or coreferring. Therefore, they also cannot be treated as simply matters to be harnessed together by way of correlations, no matter how strongly supported. The search for correlations is a way to study the ubiquitous regularities in the world, but to represent them as correlations already requires their proper conceptualization.

Am I just restating the position of dualism (which the authors claim can peacefully coexist with their design)? I do not believe that is my point. I am not making any proposal about what ontologically is or is not the stuff of the world, I am making a point about how we understand and make sense of the world.[1] I am confident that the world is knowable to a reasonable degree of confidence, but that optimism does not support a ban on conceptual constraints over inquiry. These constraints are, for example, like those of logical form, or like those of transitivity, identity, and number. Thus I am not building into what I have said any quirky views about epistemology or ontology when I speak of conceptual schemes.

Of course I am quite sure that brain scientists would be very unhappy with my way of defending their intuitions against first-person reports. While I am convinced that something like this idea lies within their "framework" and lies behind their resistance to the experimental design of Price and Aydede, I did not seek to capture the rough and ready reliance upon empiricism throughout the sciences.

Price and Aydede also appeal to the phenomenological tradition in their account of how the so-called horizontal approach to reporting of sensations from the first-person perspective can be the result of what the authors call "passive attending." This point opens a rather large matter, which I can only comment upon briefly. First, there is not a consistent view of the so-called phenomenological tradition. Second, there is not a consistent view among those identified with this tradition as to what the phenomenological method is and specifically what it has to do with ongoing psychological research. Third, there is a separate question that the authors are not directly concerned with, though I believe it to have been the key motivation for the so-called phenomenological movement; is philosophy autonomous from the social and natural sciences (and specifically autonomous from psychology)?

The article centrally appeals to Maurice Merleau-Ponty who does treat phenomenology as a way to reform psychological research. Edmund Husserl, on the other hand, is often clearer and more persuasive in holding that psychology is neither a foundation for, nor a replacement of, philosophy. Thus Husserl presents the phenomenological method, which I will not try to characterize now, as a way to do philosophy, not as a competitive method for the sciences. In fact, Husserl holds that in the scientific study of the "psychophysical nexus of nature" (Husserl 1965, p. 86) all psychological facts are simply objective, determinable facts of nature. But Husserl also holds that there are "decisive arguments to prove that physical natural science cannot be philosophy . . . can never in any way serve as a foundation for philosophy" (ibid., p. 87). He then proposes a way to do the philosophical work of clarification he thinks precedes, and is thereby presupposed by, the sciences, including psychology. He specifically distinguishes philosophical from scientific inquiry by showing that philosophical results hold independently of contingent matters of fact about the world.

While I grant that this issue is not one Price and Aydede discuss, I believe it bears upon my criticism of their experimental design. First, Husserl is defending the notion that there are conceptual constraints upon having any experience whatsoever. He thinks these "discoveries," with respect to what makes any experience whatsoever possible, are not empirical matters of fact. Second, Husserl thinks, more controversially, that the proper way to investigate these conceptual conditions for any possible experience is by careful attention to mental experience as such, while bracketing judgments about the reality of either the object of experience or any correlated matters of fact with respect to experiences. It is this controversial proposal that leads Husserl to call his method "phenomenology." Finally, Husserl's approach leads him to think that what is most important to criticize in the sciences is the uncritical adoption of philosophical positions within scientific research. The authors' notion that the method they propose is neutral between physicalism and dualism is rightly in the spirit of Husserl's notion of bracketing such assumptions, but Husserl's concern is with what conditions scientific research a priori, not what is methodologically required. Thus Husserl does not think the phenomenological method, whatever else it is, is a method for the psychological sciences or any science at all.[2]

I am not recommending Husserl's approach, which I consider flawed, but only pointing to some degree of convergence with my criticism above. Also I am stressing that both the phenomenological tradition and the so-called phenomenological method remain matters of considerable controversy.[3]

In conclusion, by appealing to the phenomenological method or tradition Price and Aydede do not end up supporting their argument about the role of first-person reports. Their overall viewpoint remains within what Husserl would have considered a "naturalistic assumption," namely that psychophysical correlations explain and account for the self-certainty of mental states. But they agree with Husserl that such correlations, if they were borne out by empirical inquiry, would be, at least to some extent, philosophically neutral. But Price and Aydede make their case for that result on the basis of the kind of data involved and what is methodologically required in scientific research, not by appealing to the autonomy of philosophy from the sciences.

Notes

1. While the view I sketch requires a great deal more defense, I am at least in good company, since both Donald Davidson (1980) and John Searle (1984) hold that physicalism (of a certain sort) is compatible with the claim that, as Davidson puts it, "mental and physical predicates are not made for one another" (Davidson 1980, p. 218). Searle develops a view, the details of which need not delay us here, distinguishing self-referential concepts from physical concepts. Thus, this idea of distinct conceptual schemes is not committed to dualism (though it is compatible with it), and yet it does not permit empirical evidence of correlations of the sort the authors defend. See also D'Amico 1997.

2. Of course Husserl (1965) uses the word "science" when discussing philosophy in a way that was traditional in German philosophy of his time. Though confusing in English, when Husserl calls philosophy a "rigorous science" he is not proposing that his method for the study of philosophy is a method for either the natural or social sciences, quite the contrary.

3. For a taste of this dispute see Steven Crowell 2002, which is in part a criticism of D'Amico 1999. I have an as yet unpublished response to Crowell under the title "Misunderstanding the Problem of Philosophical Methods."

References

Crowell, Steven. 2002. "Is There a Phenomenological Research Program?" *Synthese* 131: 419–444.

D'Amico, Robert. 1997. "Impossible Laws." *Philosophy of the Social Sciences* 27: 309–327.

———. 1999. *Contemporary Continental Philosophy*. Boulder: Westview Press.

Davidson, Donald. 1980. *Essays on Actions and Events*. Oxford: Oxford University Press.

Husserl, Edmund. 1965. *Phenomenology and the Crisis of Philosophy*. New York: Harper Torchbooks.

Searle, John. 1984. *Brains, Minds, and Science*. Cambridge, Mass.: Harvard University Press.

16 Pain: Making the Private Experience Public

Robert C. Coghill

The experience of pain provides a compelling model system for the study of how neural mechanisms support conscious events. It possesses a set of unique attributes including limited representational content, neural correlates that are objectively assessable, and marked subjective variability. Furthermore, the development of a better understanding of both the pain experience and methods for communicating that experience has substantial practical importance. Subjective reports of pain are currently the best indicator for treatment in clinical situations, and alteration of that subjective experience is a central goal of pain therapy.

1 Limited Representational Aspects of Pain

As Price and Aydede point out, many perceptual states (i.e., vision) are predominantly representational, while pain is not the object of a perceptual experience, but is the experience itself. In other words, while being burnt with a 51°C stimulus, one is not aware of the thermal qualities of the stimulator, but is simply experiencing pain. In contrast, while one is being stimulated with a 40°C warm stimulus, one attributes the sensation of warmth to the external object (i.e., the stimulator).

The minimal representational content of pain versus the highly representational content of vision is clearly reflected in their respective nervous-system mechanisms. Visual systems are associated with enormous amounts of neural hardware (the entire occipital lobe and significant portions of parietal, temporal, and frontal lobes). Much of this hardware is almost certainly devoted to highly complex signal processing involved in the construction of a neural representation of a highly complex outer world. Distinct features are processed as visual information proceeds serially through various regions of the occipital cortex and rostrally into the parietal and temporal lobes. In the case of pain, the core afferent information—how rapidly or badly is the tissue being damaged and where on or in the body is that damage occurring—is relatively simple and very likely requires minimal signal processing to generate appropriate responses and experiences. For example, minimal transformations in neural information related to the intensity of a noxious stimulus occur as it is transmitted from a peripheral nerve, to the spinal cord, and to the brain. Importantly, neural responses at all levels in this afferent transmission closely mirror subjective reports of pain magnitude (Price 1999). Furthermore, the simplicity and absence of major serial transformations of core information about noxious stimuli may contribute to the preservation of subjective availability of pain following significant damage to numerous central nervous system sites that are important to somatosensory processing (see Coghill et al. 1999 for a more complete discussion).

2 Operational Dualism in the Study of Pain

Since the days of Sherrington, neurophysiologists who study pain have engaged in a form of operational dualism in which they draw clear distinctions between the conscious experience (*pain*) and the reduced physical (i.e., neural) mechanisms responding to and encoding information about actual or impending tissue damage (*nociception*) (Sherrington 1908). For example, after an injection of capsaicin (the active ingredient in chili peppers) into my skin, I experience substantial, intense, unpleasant burning *pain* in skin regions in and around the injection site. In this instance, this particular experience is initially caused by the activation of a population of a certain type of peripheral nerve cells (C-polymodal *nociceptors*) that project into the spinal cord. This *nociceptive information* is processed by *nociceptive neurons* and eventually distributed to numerous subcortical and cortical sites within the brain where (presumably) the conscious experience of *pain* arises.

As physical events (neural discharges, neurotransmitter release, biochemical changes in the intracellular milieu of neurons), nociceptive processes can be readily examined with objective methods. However, accessing the first-person experience of pain from an objective third-person perspective has remained difficult. Objective, external referents that are readily available to third-person observers often provide misleading or inaccurate information about the first-person experience. Painless yet massive tissue damage can occur in individuals with congenital analgesia, who for a variety of neurological reasons are unable to experience pain (Baxter and Olsezewski 1960). Similarly, normal individuals injured during situations of great duress (i.e., combat) may not report pain despite substantial tissue damage (Beecher 1959). Conversely, patients with certain forms of central nervous system damage may develop chronic pain (termed central pain) and have the perception of pain and tissue damage arising from a perfectly healthy body region (Dejerine and Roussy 1906).

Using a purely behaviorist approach to infer a first-person experience of pain from a third-person perspective has long been rejected by scientists who study pain or nociceptive mechanisms because the presence of nociceptive responses does not necessarily indicate that the subject is experiencing pain. For example, nociceptive responses in the form of reflex withdrawals can be elicited when a noxious stimulus is applied to the hindpaw of a rat or dog (or lower limb of a human). When the lower (lumbo-sacral) spinal cord is disconnected from the brain by surgical transection at a higher (thoracic) level, these reflex withdrawals are still present. Moreover, these spinally mediated reflexes are surprisingly well coordinated (Sherrington 1910). The withdrawals still become more vigorous as the stimulus becomes more intense. The leg opposite to the stimulated site extends as if to better support the body as the stimulated leg is withdrawn. These withdrawal responses can also be exquisitely modulated by stimulation of other body regions to easily produce the outward appearance of a logical, conscious decision about the best motor plan for escaping from the noxious stimulus (Morgan, Heinricher, and Fields 1994). Obviously, in the case of the spinal cord transected human subject, such reflex withdrawals occur with no verbal report of a pain experience, indicating that the nociceptive information has not been sufficiently processed to elicit a subjectively available conscious experience.

Given that both the physical condition of the body and the behavior of the subject provide a third-person observer with a vastly incomplete and quite possibly erroneous view of that subject's experience of pain, the pain scientist is compelled to follow two possible courses of action. The first course of action consists of disregarding the experiential aspect of pain in its entirety and focusing solely on nociceptive processing. This course is taken by the vast majority of scientists who use animal models to study the pharmacology, neurochemistry, anatomy, and neurophysiology of nervous system nociceptive mechanisms thought to contribute in some way to the experience of pain. As Price and Aydede emphasize, the second course of action involves studying some aspect of the experience of pain using methods that rely *necessarily* on introspection.

3 Introspection and Psychophysics

The history of pain research is rich with examples of investigations that were based on first-person introspective methods in which the subjects were the experimenters. However, such methods have fallen out of favor as the field has become progressively more aware of how interindividual differences can skew the results of studies based on a highly limited number of individual subject-experimenters. Advances in the field of psychophysics contributed substantially to the decline of the purely first person investigation. By providing a standardized framework in which to communicate discrete dimensions of a conscious experience, psychophysical techniques enable pain scientists to collate the introspective reports of large numbers of subjects and to make inferences about how various manipulations elicit consistent responses in human populations.

Introspective, psychophysical methods are essential for the study of pain, given the highly individual nature of the experience. In sharp contrast with vision, subjective reports of pain evoked by a given stimulus may vary wildly from one individual to the next. Some individuals are highly sensitive while others are remarkably insensitive. In one of our recent studies of seventeen human subjects, subjective reports of pain intensity ranged from 1.05/10 to 8.9/10, despite the fact that each individual received an identical 49°C heat stimulus (Coghill, McHaffie, and Yen 2003). In the face of such widely varying subjective reports of pain, the pain scientist (and often the pain clinician) is forced to wonder if these different psychophysical ratings actually reflect interindividual differences in the subjective experience of pain or if they are simply a confound of individual differences in scale usage or introspective capacity. In order to identify objective correlates of these differing subjective experiences, we imaged brain activation while these different individuals experienced pain. The six most sensitive individuals were directly compared with the six least sensitive individuals (the five in the middle were excluded from analysis). The highly sensitive individuals activated several brain regions significantly more frequently and more robustly than the insensitive individuals. These brain regions included the primary somatosensory cortex, the anterior cingulate cortex, and the prefrontal cortex. All three of these brain regions have been implicated in various aspects of nociceptive pain processing by multiple lines of experimental evidence and are well positioned to contribute to the conscious experience of pain. Given that (1) individuals used introspection to provide a psychophysical report of their subjective experience of pain intensity and that (2) the magnitude and frequency of activation in brain regions important in the processing of pain was significantly related to the magnitude of reported pain across different individuals, we conclude that introspection is a valid method for examining the subjective experience.

The act of introspection, however, is not without its limitations. An experience of pain that is being actively evaluated by an introspective process may be substantially different than an experience of pain that is not being actively assessed for subsequent reporting. Attention and distraction can powerfully modulate the experience of pain and can easily alter its quality or magnitude. However, such changes are some of the many difficult confounds that can occur during any experimental investigation of pain. Although a limitation, having third-person access to even an altered first-person perspective is superior to having no access to that perspective. Without the first-person perspective, one is forced to examine nociception rather than pain.

4 Integrating First-Person Introspection with Third-Person Methodologies

The major strength of the classic psychophysical approach—the emphasis on obtaining reports of a very narrowly defined dimension of a conscious experience—is also its greatest weakness. Subjects may volunteer reports of experiences beyond those being measured, but without being included in the experimental design, such reports are often disregarded. Without a first-person perspective, the investigator is completely ignorant of the rich complexity of a given conscious experience and therefore is distanced from a meaningful interpretation of the underlying neural (or psychological) mechanisms.

The experiential/phenomenological approach advocated by Price and Barrell provides a method to overcome the limitations of such narrowly focused psychophysical studies (Price and Barrell 1980). This approach represents a critical formalization of a strategy that many scientists studying human pain have already employed—using first-person experiences to formulate questions for a third-person experimental phase. Scientists typically participate in some aspect of a preliminary study prior to recruiting (generally) naive subjects for the actual experiment. Yet, in almost all instances, this participation is relatively informal and is directed at a very limited set of practical rather than intellectual questions (i.e., will the subjects get a burn from this stimulus? will my stimulator work in an MRI scanner?). By formalizing the participation of the investigator(s) in a preliminary first-person phase, the complexity of the conscious experiences that may have occurred during the experimental paradigm becomes accessible to the investigator(s). Subsequently in the third-person phase of the study, experimental questions can be directed specifically at conscious experiences identified during the first-person phase of the investigation.

The experiential/phenomenological approach provides a powerful tool, particularly when coupled with vertical paradigms where the neural correlates of differing dimensions of the subjective experience can be examined. By obtaining a more complete view of the first-person perspective, the capacity to explain neural events occurring during the experiment is vastly enhanced. In our own laboratory, we have started using this strategy. Remarkably, after nearly two decades of using (and experiencing) noxious thermal stimuli, we have identified several novel experiential dimensions in first-person experiments that are now being actively investigated in third-person designs.

One significant consideration in using experiential/phenomenological designs is experimenter bias. If an experimenter has an expectation of a certain outcome and a desire for that outcome, the results of the first-person phase of the experiment may be significantly biased. Expectations and desire may actually reshape the perceptual experience of the investigator. Subsequently, if the questions formulated for the third-person phase are based on the reshaped experience of the investigator, the experiences of the subjects (or their reports of their experiences) may be substantially and similarly altered. One possible example of how such biases can occur can be seen in the thermal grill illusion. This illusion is evoked by stimulating the skin with a grill in which alternating bars are innocuously cool and innocuously warm. A complex sensation with thermal qualities is elicited. When subjects are instructed to rate the magnitude of pain evoked during such an experience, pain reports are obtained (Craig et al. 1996). However, when subjects are provided a broad list of verbal descriptors of thermal qualities as well as pain, the pain descriptor is rarely chosen (Fruhstorfer, Harju, and Lindblom 2003). Importantly, if the experiential/phenomenological paradigm is employed by investigators who are aware (or attempt to be aware) of their own potential biases, questions formulated for the third-person phase can be made sufficiently broad to capture a more accurate view of the subjective experience of naive individuals.

5 Conclusion

As a sensory and emotional experience, pain is uniquely private and extremely difficult to appreciate from a third-person perspective. Integrating first-person methodologies into the study of populations of subjects is absolutely essential for relating the complex, multiple dimensions of the experience of pain to their underlying neural mechanisms.

Acknowledgment

R. C. C. is supported by NIH RO1 NS39426.

References

Baxter, D. W., and J. Olsezewski. 1960. "Congenital Universal Insensitivity to Pain." *Brain* 83: 381–393.

Beecher, H. K. 1959. *Measurement of Subjective Responses.* New York: Oxford University Press.

Coghill, R. C., J. G. McHaffie, and Y. F. Yen. 2003. "Neural Correlates of Interindividual Differences in the Subjective Experience of Pain." *Proceedings of the National Academy of Science USA* 100: 8538–8542.

Coghill, R. C., C. N. Sang, J. M. Maisog, and M. J. Iadarola. 1999. "Pain Intensity Processing within the Human Brain: A Bilateral, Distributed Mechanism." *Journal of Neurophysiology* 82: 1934–1943.

Craig, A. D., E. M. Reiman, A. Evans, and M. C. Bushnell. 1996. "Functional Imaging of an Illusion of Pain." *Nature* 384: 258–260.

Dejerine, J., and G. Roussy. 1906. "La Syndrome Thalamique." *Revue Neurologique* 14: 521–532.

Fruhstorfer, H., E. L. Harju, and U. F. Lindblom. 2003. "The Significance of A-Delta and C Fibres for the Perception of Synthetic Heat." *European Journal of Pain* 7: 63–71.

Morgan, M. M., M. M. Heinricher, and H. L. Fields. 1994. "Inhibition and Facilitation of Different Nocifensor Reflexes by Spatially Remote Noxious Stimuli." *Journal of Neurophysiology* 72: 1152–1160.

Price, D. D. 1999. *Psychological Mechanisms of Pain and Analgesia.* Seattle: IASP Press.

Price, D. D., and J. J. Barrell. 1980. "An Experiential Approach with Quantitative Methods: A Research Paradigm." *Journal of Humanistic Psychology* 20: 75–95.

Sherrington, C. S. 1908. *The Integrative Action of the Nervous System.* London: Archibald Constable.

———. 1910. "Flexion-Reflex of the Limb, Crossed Extension-Reflex, and Reflex Stepping and Standing." *Journal of Physiology* 40: 28–121.

17   The Problem of Pain

Eddy Nahmias

When my wife was pregnant, our birthing coach asked the class "What is pain?" I thought I might finally get to display some of my philosophical training, but alas, the correct answer was: "Pain is whatever she says it is." The coach's "sufferer-centric" definition echoes the one offered by the International Association for the Study of Pain (IASP)—"Pain is always subjective"—as well as the definition of pain offered by the philosopher Saul Kripke in his argument against identity theory: "Pain . . . is picked out by the property of being pain itself, by its immediate phenomenological quality" (1972/1980, p. 152).

These subjective conceptions of pain pose problems for the scientific study of pain, as Price and Aydede point out in the introduction of their chapter. If the essence of pain is its phenomenological quality, then it seems the only way to study it *directly* is through introspection and subjects' verbal reports on their conscious experience, but this is often considered to be an unverifiable and unreliable method. Conversely, *indirect* information about the objective properties associated with pain experiences seems inadequate to fully explain the essential phenomenological quality of pain. The problem, which generalizes to many other conscious experiences, is that any materialist theory that attempts to explain subjective experiences in terms of objective properties seems doomed to leave out the very essence of what it is trying to explain—for example, the *feeling* of pain. And any nonmaterialist theory seems doomed to be unscientific. What I will call the "problem of pain" has us taking a materialist approach to the study of pain while conceiving of pain in a way suggestive of metaphysical dualism.[1]

The first step in responding to the problem of pain is to recognize the possibility of *epistemological* dualism. As Price and Aydede explain this idea, certain complex neural processes can be accessed in two fundamentally different ways: first, conscious agents in whom the processes occur can introspect on them, and second, other agents (often with the aid of scientific instruments) can observe the processes as well as their objective causes (e.g., stimuli) and effects (e.g., behavior).[2] The two types of access provide two different kinds of information, experiential and physical, about the same process. The authors do not flesh out this theory here, but they are right to point out that if we cannot make sense of the idea that one and the same physical process can be accessed in these two fundamentally different ways, then we seem stuck having to choose between some form of metaphysical dualism (perhaps epiphenomenalism) and some form of reductive materialism that eliminates conscious experiences from our ontology (i.e., our catalog of what is real and can be studied scientifically).

The second step in responding to the problem of pain is to recognize that, though pain experience is essentially subjective, there is no *single* and *simple* pain experience. As the authors discuss, people can distinguish differences between, for instance, experiences of "first pain" and "second pain" and between the intensity and the unpleasantness of pain experiences, and these differences correlate with specific activity in the central nervous system. Indeed, by the time a subject experiences a peripheral noxious stimulus as pain, the experience has become *subject-relative* in that the incoming neural signals have interacted with complex brain activity specific to the subject. This activity includes emotional and cognitive processes, such as the subject's fears about being in pain and beliefs about the long-term effects of the pain. These subject-relative processes explain the diversity of introspective reports in response to identical stimuli, which encourages the idea that pain cannot be identified with any objective processes and that introspection is unreliable. On the contrary, however, using introspective methodologies in combination with objective observations is required to understand the complexity of pain experience and to test the reliability of introspective reports (see below).

Most pain researchers have been painfully aware of the complexity of factors involved in the subjective experience of pain and introspective reports about them. In contrast, when philosophers present the problem of pain, they often suggest that pain is a simple experience corresponding to a simple type of peripheral neural process (e.g., the C-fibers "identified" with Kripke's arguments). Paradoxically, it is only by recognizing the *complexity* of pain experiences that we can begin to imagine a satisfying response to the problem of pain. Any successful theory explaining how neural processes can have experiential properties will have to refer to the range of emotional and cognitive states involved in pain experiences and to the corresponding range of diverse neural states.[3]

Notice the vague language in the previous sentence regarding the processes and states at issue. The third step in responding to the problem of pain is to recognize how far we are from understanding how to individuate the terms on either side of the equation—both the subjective states, including emotional and cognitive states, and the corresponding neural processes. This point reminds us of William James's description of the state of psychology in 1892: "We don't even know the terms between which the elementary laws would obtain if we had them" (1892/1961, p. 335). So, while one way to describe the problem of pain is to point out that we have no theory to explain how neural states can *be* conscious states, one response is to point out that we do not yet have a theory about which neural states to pick out or how to individuate the conscious states at issue—including, for instance, how to obtain complete and accurate descriptions of pain experiences. What we do have is a lot of inductive evidence that some such correlations exist and that suggests such a theory is precisely what we should be looking for. What we need, in the meantime, is a methodology to arrive at such a theory.

It should be clear by now that I think Price and Aydede gesture toward each of these three steps in responding to the problem of pain. So, those who prefer scathing critiques may prefer to stop reading this commentary. Instead, I will continue to clarify and to extend some of the authors' most important claims, now turning specifically to their proposed methodology.

Epistemological dualism is a philosophical theory in need of empirical support. The best methodology for garnering such support—and for challenging both some versions of dualism and eliminative versions of materialism—requires using introspective reports.[4] The goal is to map out the "phenomenological space" of conscious experience and to map this structure onto the "neurobiological space." The latter mapping, what Price and Aydede call the "vertical phase" of their experiential approach, is a well accepted if nascent method in cognitive psychology and neuroscience. But the idea of mapping the phenomenology of pain (and other conscious experiences), what they call the "horizontal phase" of their approach, is more controversial and underexplored.[5]

The most novel suggestion the authors offer is to encourage pain researchers to become subjects of their own investigations. This idea, as they point out, has been used in some prior pain research. It was also the standard methodology of the introspectionist psychologists. But unlike the introspectionists, Price and Aydede emphasize that the experimenter should *passively* observe their experiences rather than actively attending to particular aspects of it. This helps to avoid introspective reports that interpret experiences in terms of one's theoretical commitments and that unduly elaborate on the experience itself (for instance, some introspectionists obtained twenty-minute reports of two-second experiences). However, I do not think the authors are clear enough about what they think introspection is, and the success of their methodology may vary depending on what model of introspection it employs (see below).

The authors are also not explicit enough about why their methodology *requires* experimenters to test their ideas on themselves rather than others. Their four-step methodology suggests that hypothesis generation and data interpretation will be facilitated by this approach (and surely it is more practical for researchers to do pilot studies on themselves). But it should also be emphasized that introspection, considered analogously to other types of observation, is a *skill* that can be improved with practice and with knowledge of the basic goals of inquiry. Untrained and uninformed subjects are less likely to attend to the subtle aspects of their experiences, especially the distracting and unpleasant experience of pain. And while one of the problems with the classical introspectionist methodology was that training led to theory-laden reports, in fact untrained subjects are also inclined to offer explanations (biased by their lay theories) about *why* they are experiencing what they are rather than just descriptions of *how* things seem to them.

Experimenters using Price and Aydede's methodology can practice introspecting pains produced by the same stimuli numerous times with the aim of producing consistent and clear descriptions of those experiences that are minimally tainted by theory. They can examine the relationships between different experiences produced by numerous different stimuli.[6] And they can compare introspective reports, including the vocabulary used, among themselves to test for inter-rater reliability. Unlike the introspectionists, these reports should be treated as neither definitive nor final data but rather as preliminary data, which can then be used to generate hypotheses to test on untrained subjects. The experimental stage involves correlating objective data in the form of untrained subjects' verbal reports with objective data about both the stimuli and the resulting neural processes. Indeed, as the authors point out, "the independent variable becomes the experiential dimension to be manipulated as opposed to the external conditions used to produce changes in the experiential dimension" (this volume, p. 264).

An interesting recent example of this method is demonstrated in an experiment by Coghill, McHaffie, and Yen (2003). Previous experiments have shown consistent within-subject relationships between brain activity and subjects' pain reports as evoked by different stimuli. This experiment established *between*-subject correlations among different subjects' brain activity and their varying reports of pain intensity in response to the same stimuli. High-sensitivity subjects (who report a high degree of pain on a visual analog scale for pain intensity) showed more activity in anterior cingulate cortex and primary somatosensory cortex than did low-sensitivity subjects (while there were no significant differences in thalamic activity). This suggests that the experienced intensity of pain is mediated by cortical areas associated with emotional and cognitive functions.[7] This experiment provides evidence that subjects' introspective reports are reliable indicators of objectively measured neural activity, as predicted by epistemological dualism. It also demonstrates the second step in responding to the problem of pain discussed above—that the experience of pain (even intensity alone) will be identified with a complex, though consistent, range of neural activity rather than with some simple neural process.

The problem of pain mirrors traditional arguments for dualism. It just doesn't seem like the subjective feeling of pain can be the same thing as electrochemical processes. But the correlations begin to look less contingent and the identity claim less mysterious as researchers uncover more detailed and complex correlations between the phenomenal and the neural. It makes sense that C axons, which fire slowly, correspond to dull, throbbing experiences of second pain, while A-delta axons, which fire more rapidly, correspond to sharp, stinging experiences of first pain. It makes sense that increased activity in anterior cingulate cortex, which is associated with emotions such as fear, increases the experienced intensity of pain. Using better introspective methods in combination with improved tools for measuring brain activity is the best way to move toward an explanation of how epistemological dualism works—or to show that it doesn't.

In fact, Price and Aydede suggest that their methodology works on either physicalist *or* dualist assumptions. They mean that it will work to map the correlations between conscious experiences and brain activity, whether considered as two dimensions of one process (e.g., type identity theory) or as two distinct things or properties connected in lawlike ways (e.g., dualist epiphenomenalism). However, there are some forms of physicalism, such as anomalous monism, that suggest such lawlike correlations will not be found, because there are no psychophysical laws. And there are forms of dualism—namely Cartesian interactionism—that should also predict that such correlations will not be found, because some of the emotions and thoughts affecting the experience of pain occur in an immaterial mind beyond neuroscientific study. That is, first-person reports of pain, if modulated at all by the input of a nonphysical mind, should not be expected to correlate consistently with the same types of neural activity.

Indeed, I think Price and Aydede attempt to be *too* inclusive in their metaphysical claims. Whereas they suggest that their methodology is amenable to various types of dualism and physicalism, they should argue more explicitly that their methodology works only if certain metaphysical theories of mind are true, and conversely, that their methodology offers a way to put pressure on other metaphysical theories by discovering the correlations that those theories predict should not be found.[8] If we can use introspective methodologies to improve the horizontal mapping of the structure of our phenomenology and to bring to fruition the vertical mapping between phenomenology and neurobiological processes, then we will have empirical evidence suggesting type identity between the mental and the physical—especially if the vertical mapping also provides information about *why* the phenomenological states have the properties they have (just as in other scientific reductions where the lower-level properties *explain* the properties at the higher levels).

I'll conclude with a question about the authors' view of introspection. Price and Aydede (perhaps wisely) attempt to avoid an analysis of introspection, and they suggest that their methodology is amenable to various conceptions of introspection, such as the higher-order thought (HOT) model as well as the higher-order perception (HOP) model. But their methodology may work differently depending on what introspection is—and if there are different types of introspection, it may depend on what type the experimenters and subjects use.[9] For instance, does introspecting on a pain involve a distinct process from just consciously experiencing that pain? If the introspection is a distinct process, then we should expect to see different neural activity in subjects who are introspecting on their pain than in those who are simply experiencing pain. If the passively attentive introspection the authors advocate is a different process than the actively attentive introspection used by the introspectionists, then we should expect different neural activity and different reports from subjects who employ these different types of introspection. Perhaps we will be able to examine how the process of introspecting alters the neural activity associated with the pain experience and alters it differently depending on the type of introspection employed. Presumably, the introspective methodology Price and Aydede outline can be adapted to study the nature of introspection itself, though I'll leave it as an exercise for the reader to consider how this might work.

Notes

1. Price and Aydede point out several ways that pain poses different problems (and solutions) than other conscious experiences. The more general problem is posed by several different arguments in the philosophical literature. Kripke's argument against identity theory, roughly, is that pain states are not identical with neural states (e.g., C-fibers firing) because, unlike other cases of discovered identities (e.g., water = H_2O), we have no explanation for the apparent contingency of this purported necessary identity. The essential property of pain states, their phenomenological quality, seems like it could exist without a particular neural state (or any neural state), and

the neural state seems like it could exist without the specific feeling of pain, and we have no explanation for its seeming to be this way. Kripke (1972/1980) sees this as a problem for any type–type identity theory (though not as an argument *for* dualism; see p. 155).

2. See their note 6 for philosophical discussions of epistemological dualism.

3. The complexity and diversity of experiences such as pain may explain why some philosophers have seen token–token identity theory as a more plausible alternative than type–type identity theory. I think the plausibility of type–type identities depends on how liberal we are about the types at issue. We will not discover an identity between the type "pain experience" and any specific type of neural process (e.g., C-fibers firing). But that does not entail that there is no identity between more fine-grained types of pain experience and more fine-grained complexes of neural activity.

4. The authors correctly point out that introspection (sometimes in disguised form) has been used often in experimental psychology since the demise of introspectionism. For further defense of the scientific legitimacy (and necessity) of introspective methodologies, see Goldman 1997; Jack and Shallice 2001; Vermersch 1999; Nahmias 2002; and various articles in the "Trusting the Subject" editions of the *Journal of Consciousness Studies* (2003 and 2004).

5. In addition to the introspective methodologies the authors discuss, other recent approaches include the phenomenological interview (e.g., Pollio, Henley, and Thompson 1997); descriptive experience sampling (e.g., Hulbert and Heavey 2001); and protocol analysis (Ericcson and Simon 1993).

6. Few college sophomores will have had extensive experience with a variety of different pains, but (dedicated!) pain researchers can expose themselves to a range of painful stimuli and examine the relationships between the resulting experiences (though, my wife assures me, unless they experience childbirth, they'll never understand *real* pain).

7. It would be interesting to test the brain activity of a "stoic subject," one whose cortical activity suggests she experiences a high degree of pain intensity though she *reports* a low intensity. Presumably, she would show interesting differences from both high- and low-sensitivity subjects in other cortical areas associated with cognition and language.

8. For instance, Price and Aydede claim (note 8) that nothing about their methodology crucially depends on what sort of relation holds between physical and mental states—for example, type identity versus supervenience. However, as they note, supervenience allows that the same type of mental state (e.g., pain states) may be realized in different types of neural states. Depending on the range of neural states at issue, some forms of supervenience may thus make their methodology ineffective. The first-person introspective reports would not pick out any neural kinds to be studied using third-person methods. On the other hand, the success of their methodology and the case studies they discuss provide tentative support for type-identity theory (though it may be unable to rule out dualist epiphenomenalism).

9. See Prinz 2004 for a discussion of the different processes that fall under the category of introspection.

References

Coghill, R., J. McHaffie, and Y. Yen. 2003. "Neural Correlates of Interindividual Differences in the Subjective Experience of Pain." *PNAS* 100: 8538–8542.

Ericsson, A., and H. Simon. 1993. *Protocol Analysis*. Cambridge, Mass.: MIT Press.

Goldman, A. 1997. "Science, Publicity, and Consciousness." *Philosophy of Science* 64: 525–545.

Hulbert, R., and C. Heavey. 2001. "Telling What We Know: Describing Inner Experience." *Trends in Cognitive Sciences* 5: 400–403.

Jack, A. I., and T. Shallice. 2001. "Introspective Physicalism as an Approach to the Science of Consciousness." *Cognition* 79: 161–196.

James, W. 1892/1961. *Psychology: The Briefer Course*. New York: Harper and Row.

Kripke, S. 1972/1980. *Naming and Necessity*. Cambridge, Mass.: Harvard University Press.

Nahmias, E. 2002. "Verbal Reports on the Contents of Consciousness: Reconsidering Introspectionist Methodology." *Psyche*, 8(21). Available at http://psyche.cs.monash.edu.au/v8/psyche-8-21-nahmias.htlm.

Pollio, H., T. Henley, and C. Thompson. 1997. *The Phenomenology of Everyday Life*. Cambridge: Cambridge University Press.

Prinz, J. 2004. "The Fractionation of Introspection." *Journal of Consciousness Studies* 11(7/8): 40–57.

Vermersch, P. 1999. "Introspection as Practice." *Journal of Consciousness Studies* 6: 17–42.

Murat Aydede and Donald D. Price

We are grateful to our commentators for their useful and insightful comments. We are delighted to see that there is much agreement on key issues between us and the commentators, which might be an indication for a sea change in the scientific study of pain and other types of conscious experiences. This makes the issues addressed all the more important and urgent.

Gallagher and Overgaard

Gallagher and Overgaard's commentary has two parts. The first and shorter part is on introspection in general. The second, longer part contains a very useful survey of positions similar to ours along with a critical internal comparison among them. We have almost nothing to say about the second part except to express our gratitude for such an enormously useful comparison of our proposal to the converging work of others—enormously useful to us, at least, since we were not familiar with many of them. The present form of the view proposed here has its origins in the work of Price and Barrell (1980, 1984; Price, Barrell, and Barrell 1985) as well as Barrell and Barrell (1975). We are of course delighted to learn that there has been such a convergence and are looking forward to learning more about it while savoring the realization that such a growing convergence might be an exciting sign of what is to come in the study of human consciousness.

In the first part of their commentary, however, Gallagher and Overgaard argue, contrary to our view, that not every form of first-person access to one's conscious experience and its features is introspective. Introspection, in other words, is not the only first-person way of knowing our experiences. They criticize us for not making the distinction, even though, as they themselves remark, our proposal demands only the introspection as they characterize it. Namely: "... introspection is understood as a second-order cognitive act that thematizes first-order phenomenal experience, and makes that experience the object of reflection." So, in a sense, their criticism is clearly incidental and not directed against our main proposal. Nevertheless, we want to make two points on this.

First, in the target article we recommended introspection as a passive noticing of what is or has just taken place in experience and reporting the content of experience in a manner that faithfully reflects the experience (first person, present tense). We certainly proposed that the participants reflect on the common factors or elements that were present in multiple experiences of a given phenomenon, such as anger or performance anxiety, *but only after the experiences have taken place and have been reported*. This listing of common elements was then to be followed by an analysis of the necessary and sufficient elements of the type of experience in question. In principle, this is similar to the phenomenologists' search for structural invariances. The quoted definitions listed by Gallagher and Overgaard appear not to clearly reflect the distinction between the act of observing the contents of experience and the act of analyzing their nature. We think this is an important distinction regardless of how introspection is defined.

Second, we are very skeptical as to whether there is any first-person direct access to one's own experiences that doesn't involve a second-order mental state whose content is about the experience (or an aspect of it) thus accessed—as indeed the very notion of access suggests. Gallagher and Overgaard talk about our having ". . . an implicit, nonintrospective, prereflective self-awareness of our own experience." They continue: "At the same time that I see the light, I know that I see the light. This knowledge of seeing the light is not based on reflectively or introspectively turning our attention to our own experience. It is rather built into our experience as an essential part of it, and it is precisely that which defines our experience as conscious experience." It is not clear to us how we should understand this talk of knowledge about our experience being built into the experience itself. Experience is one thing, the knowledge of it is another, especially reportable knowledge. Our awareness (knowledge) of experiences is not mediated by the very same experiences. To be sure, experiences are typically representational mediators through which or with which we become aware of what those experiences sensorially represent. But it is totally mysterious to us how an experience can mediate our awareness of it through the very same experience itself—unless, by this, we just mean to make the rather obvious remark that they are conscious states (but this is a different matter). If we look at the history of philosophy, we see a conception of introspection (especially in Kant, but also in British empiricists like Locke) as internal perception. Just as ordinary external perception has its object and is distinct from it, introspection was meant to have other (immediately available) mental states as *its* objects while being distinct from them. This much, we assume, is essential for introspection and not controversial. But we also think that it is almost analytic that any first-person (direct, noninferential) knowledge that we can report about our experiences is introspective in just this minimal and rather traditional sense (*possibly* including the historical phenomenological tradition). Note that we didn't claim that for an experience to be conscious it has to be introspected (or, even introspectable). But these are deep waters that need extended discussion. Given that their criticism is not directed at how we propose to use "introspection" where this is understood in the sense they and we both agree upon, we may safely put this issue aside.

Gallagher and Overgaard complain that we seem to trivialize the use of introspection when we say that introspection is and has been indispensable for psychophysics (and other similar experimental paradigms). They complain that in the sense in which we use 'introspective report,' any sincere utterance counts as introspective reporting (among other things) since they indicate that the speaker had an intention, say, to communicate. We think that this is not the right way to look at the issue. Although some psychophysical experiments are designed in terms of detecting the presence or absence of stimuli, many are designed in terms of estimating magnitudes and qualities of sensory experiences. In the case of magnitude judgments, this requires that the subject notice and represent sensory magnitude using a rating method or representing the perceived magnitude of one type of sensation by using another sensory modality (e.g., adjusting a sound to represent the intensity of a pain). This task requires subjects to notice what is present in their experience and represent what is present in the form of words or nonverbal responses. The main bulk of psychophysics is not concerned with the relation of stimulus parameters and *any* response on the part of the subjects. Most often and crucially, it is concerned with the relation between the stimuli and *sensations*—conscious sensations. Thus, magnitude judgments inevitably reflect a simple form of introspection.

Importantly, in the case of pain (and other similar bodily sensations), reports of pain *are* introspective reports partly because pain is less easily linked to stimulus properties than are other sensory modalities (cf. IASP definition of 'pain'). Notwithstanding the confused folk practice of locating pain in body parts, pains are *experiences* in the first instance. Reporting pain *is* therefore reporting experiences (among other things). If this is done in a first-person way, genuine pain reports are introspective in our sense. For instance, subjects are often not asked to just detect the noxious *stimuli* in classical psychophysical sense but rather are asked to determine when *feeling* the stimulus becomes painful, that is, pain. Thus, pain thresholds, for example, could not have been established by simply asking the subjects when they detect certain kinds of stimuli.

Finally, at some point in their commentary, Gallagher and Overgaard question our suggestion that scientists should use themselves as subjects, similar to Nahmias's statement that we are not explicit enough about why we think experimenters should first test their ideas on themselves. Clearly, our rationale for this position needs further clarification and justification. Gallagher and Overgaard remind us that investigators have certain hypotheses and results that they hope to find and thus are more likely to be biased as subjects. We fail to see why this should be a concern, particularly for our proposal, especially since evaluations of biases of either investigators or subjects are by no means routinely made. Investigators become susceptible to bias the moment they form their hypotheses or design their experiments and there are multiple ways they can transfer their biases to their subjects. Biases have numerous origins, including the past or existing literature, imagining how something works, and vested interest in supporting one's past results. Far from being a problem for our position, involving the investigators themselves as subjects has the potential of eliminating or minimizing the impact of such biases within the formal framework of our proposal. Indeed in the experiential method we propose, the participation of investigators as subjects offers a means whereby bias can be recognized and dealt with more effectively. If the subject matter of a study is that of a particular type of experience, such as a type of pain or an emotion, then wouldn't it be better for the investigators to encounter these phenomena in their own direct experience as the basis for formulating hypotheses and designing experiments as opposed to only relying on published accounts of others or their imagination? The former approach reflects being objective about the subjective, whereas the latter approach reflects being subjective about the subjective. The experiential approach may be especially effective when there are multiple investigators-subjects. The exploration and results can be compared across participants and checked for possible biases. It is important to note in this regard that judging and bias can be part of what is passively noticed by the participant, as pointed out in our article.

Robert D'Amico

D'Amico thinks that there is a radical epistemic asymmetry between first-person introspective data and third-person empirical data: while the latter are revisable and refineable, the former are not—not even in principle. He thinks that the asymmetry is not marked merely by degree but rather it's the manifestation of a radical difference between two conceptual schemes. Because of this, they cannot be put into service in the way our proposed methodology requires; in other words our methodology is deeply confused by treating these two types of data as if they were on a par—conceptually. He writes: "We cannot imagine, I suggest, that the experimenter could be led by some persistent or even robust brain-imaging results to the conclusion that such and such an experience that he or she is now having is not in fact a pain experience after all; or that the diffuse pain felt in that testing was in fact a sharp pain after all, given the experiment's third-person results." This would be imaginable if our methodological proposal were sound. Evidently, D'Amico thinks that it is logically inconceivable that someone competent with the relevant mental vocabulary and conceptual repertoire and familiar with the normal/ordinary range of sensory experiences, taking his time and upon careful reflection, can be mistaken about whether or not he is having a certain sensation of a familiar kind, or whether he is having this sensation rather than some other ordinarily easily distinguished sensation. Although D'Amico doesn't put the point in modal terms, we believe he would not object to our putting his unrevisability claim as a metaphysical or logical impossibility claim: under optimal conditions, competent subjects cannot be mistaken in their core judgments about their own current sensations above threshold levels. By "core judgment" we mean identifying their own sensations in terms of their general kind and above-threshold spatiotemporal properties. The cases suggested by D'Amico's own examples serve well in this regard. Here is another rather extreme example—to pump the relevant modal intuition. Suppose under optimal psychophysical conditions, I take a good bite of a ripe cantaloupe melon, chew it for ten seconds, and sincerely misidentify the taste as a good-quality dark chocolate taste. The intuition we are invited to share here is that *necessarily* if I indeed sincerely (to my own great surprise) identify the taste as a dark chocolate taste, then it must have tasted so to me. It is unimaginable (thus, metaphysically impossible) that I have the melon taste but sincerely identify it as the taste of dark chocolate. In short, one can certainly make subtle mistakes about one's sensations at threshold levels, but no gross introspective mistakes under optimal conditions are possible—there is simply no logical room here. Call this unrevisability claim the Cartesian Transparency Thesis regarding one's Current Sensations (CTT_{cs}). We think CTT_{cs} is false. Thus we reject the radical unrevisability claim that D'Amico presents as constituting an essential barrier to the kind of methodology we propose.

The issues surrounding this traditional and venerable thesis are deep and many. We cannot hope to address all the questions raised by our rejection. However, we do want to provide some support for why such a rejection is not crazy. First, we want to point out that there is a distinction between *having a sensation or experience* and *making a judgment about an experience* one is currently having. However tight the connection may be, it is clear that they are distinct events that do not metaphysically or logically necessitate each other. Many animals or even young children may plausibly be said to have experiences without being capable of making judgments about their experiences. It is plausible to attribute to some animals sensations without attributing capability of making judgment at all (let alone judgments about their own sensations). Furthermore, the psychoneural mechanisms subserving these capacities are probably functionally distinct. The capacity to have sensations may not necessitate the capacity to have concepts and use them in making judgments—even though having a sensory system may be nomologically necessary for having concepts at all. Thus, insofar as CTT$_{cs}$ is a thesis about the revisability of one's judgments about one's current sensations and insofar as making a judgment is a distinct activity from having a sensation, it should in principle be metaphysically or logically possible that these judgments are revisable in a way that makes CTT$_{cs}$ false. This is a sort of transcendental argument against CTT$_{cs}$, but one might legitimately ask whether we have any *positive conception* of how revisability of sensation judgments is possible. We think we can come close enough to make positive sense of this possibility, but explaining how would go beyond the scope of this reply. However, we don't think that our rejection of CTT$_{cs}$ depends on our offering a positive conception. Moreover, we think that there are very good (principled) reasons why we are typically having difficulty in forming a first-person positive conception of such cases. Nevertheless, we would like to present some empirical cases that it is plausible to describe as cases where a mismatch occurs between one's conscious experience and one's judgments about these in a way that puts the burden on the defenders of CTT$_{cs}$ to show that they don't falsify CTT$_{cs}$.

For example, repeated encounters with the same stimulus object may lead to different sensations depending on several psychological factors. Coghill's example of the thermal grill illusion (see commentary by Coghill, this volume) offers an example of how a stimulus that evokes a complex sensation produces very different types of subjective reports depending on how the experimenter instructs the subjects to attend and the methods used to report sensations. In the case of his example, if the same subjects had been exposed to the same thermal grill in the two experiments described by Coghill, it is very likely they would have revised or refined their own reports of how they experience the thermal grill in a context that is not influenced by biases fostered by the experimenter.

More radical cases may be certain forms of anosognosia—in particular visual anosognosia (Anton–Babinski syndrome) where people suffering from this disorder typically claim, indeed insist, that they can see, offer visual descriptions of what they claim to see, or attempt to move around, all the while being technically blind, that is, not capable of receiving and processing visual *information* around their immediate surrounding in a way that matches their actual behavior. When asked about why they bump into objects or when the accuracy of their reports is challenged, they often confabulate, sometimes wildly, but rarely accept their blindness. There are competing theories that could potentially explain what is happening in visual and other anosognosias. While some postulate as part of their explanation that these patients are in fact having hallucinatory sensations (unrelated to their actual surrounding reality), others attempt to explain them as involving complex disconnection problems among various subsystems of which the disconnection of the conceptual/verbal system from others, including various sensory visual subsystems, plays an important role. Our point here is not that the latter explanation should be preferred, but rather the fact that an explanatory hypothesis of this sort *could* be offered shows that the rejection of CTT_{cs} where radical error of this sort is logically ruled out is not so crazy after all.

Other cases may involve some forms of agnosias, especially forms of associative visual agnosia, where low-level vision of details is often preserved but the patient cannot recognize the visual stimulus. One way to describe such a syndrome is by saying that although the patient can see every low-level visual detail (say, lines, edges, colors, light and color gradations, etc.) more or less just like a normal subject can see, the patient cannot see the object presented in this detail as what it is. What would the visual phenomenology of such patients be like? Despite good visual acuity, the visual world of ordinary objects seems to disappear from the visual phenomenology of such patients. These patients are not blind, their entire visual field preserves some qualitative phenomenology, in fact in some sense exactly like the phenomenology of the normal subjects, but they just can't see the apple, the lighter, match box, the pencil, the cup, the plate, and so on, presented in this phenomenology. These patients often are unable to visually imagine such objects or remember what they looked like— despite the fact that they have near perfect nonvisual concepts of these objects: when they touch, taste, hear, or smell them, oftentimes they can recognize them easily. It is as if almost all the visual information has been erased from the mental files (concepts) they keep for object kinds. Sometimes some patients can figure out what the visually presented objects must be by painstakingly reasoning out from the low-level visual cues they can see.

This case doesn't *directly* bear on the possible falsity of a radical unrevisability thesis about one's current sensations. But it shows how recognitional judgments in a given sensory modality can be impaired while the phenomenology of (early) sensory processing in that modality is preserved. (We are assuming that, for instance, visual object recognition is partly a modality–specific conceptual capacity involving classifying different visual stimuli under a visual concept.) Anosognosia is a deficit that attaches to various neuropsychological deficits, such as blindness, hemianopsia, hemiplegia, amnesia, and others. We are not aware of an actual case of associative visual agnosia with anosognosia. Such a case may well be empirically impossible, but then again, maybe not. If such a case is empirically possible, then we would have a person who cannot visually identify objects but insists that she can—despite having complete (low-level) visual acuity and phenomenology. A person in this condition would judge, for example, "it appears to be a cup" when in fact *nothing* visually *appears* to be a cup *to her*. Again, our point here is not that there may be actual cases of this sort (for all we know, there may be), but rather, certain empirical hypotheses (e.g., disassociation between sensory systems and conceptual information pick up or categorization or verbalization mechanisms) naturally suggest themselves that cannot be put aside a priori. If these hypotheses turn out to be better than others in explaining such deficits, CTT_{cs} would be in serious trouble even as a *nomological* necessity claim—we would have *actual* counterexamples. But we believe that we have already shown that we have *possible* counterexamples. Hence CTT_{cs} is false as ordinarily defended—as a metaphysical–logical necessity claim. And this is enough to rebut D'Amico's main criticism that there is a radical asymmetry *in kind* between first-person and third-person modes of epistemic access to data, which shows, according to D'Amico, that we have two radically different (logically incongruent) conceptual schemes involved in these two kinds of access that cannot be experimentally juxtaposed in the way our methodological proposal requires.

Finally we cannot help but point to the actual practice of scores of experimental neuropsychologists who have been using first-person data fruitfully for decades and now increasingly in brain-imaging studies with great success and amazing results that have sometimes immediate practical and clinical consequences.

Robert Coghill and Eddy Nahmias

We don't really have much to say about Coghill's and Nahmias's supportive commentaries. They expand on some of the things we touch upon in our paper briefly, and so in this regard they are very helpful and complementary—for which we are grateful. We have only a few minor points to make.

In the beginning of his paper, Coghill writes: "In other words, while being burnt with a 51°C stimulus, one is not aware of the thermal qualities of the stimulator, but is simply experiencing pain. In contrast, while one is being stimulated with a 40°C warm stimulus, one attributes the sensation of warmth to the external object (i.e., the stimulator)." In our paper, we emphasize the differences between pain and other more obviously representational perceptual experiences. But we don't mean to deny that pain experiences are representational, they represent actual or potential tissue damage and its various features. Of course, pain experiences do more than just represent. Also, we don't think it is plausible to say that in having a thermal pain caused by, say, 51°C stimulus, one is *not* aware of thermal qualities. It may be that in having the pain experience one is aware of these thermal qualities while registering them as (potentially) damaging the skin area the stimulus is applied to.

We concede Nahmias's point that not all dualist positions require lawlike psychophysical correlations. However, we are puzzled about why he thinks that Cartesian interactionist substance dualism suggests that no such correlations will be found. If psychophysical causation (causal interaction in both directions) requires psychophysical covering laws, then Cartesian dualism entails that there are lawlike correlations between some properties instantiated in a physical body and some properties instantiated in a nonphysical substance (soul) "attached" to that body. But perhaps Nahmias has in mind the high variability of pain experiences under the influence of other mental states. Such a variability would certainly make things difficult, but probably not impossible. Nevertheless, we agree with Nahmias's insightful observation "that [our] methodology offers a way to put pressure on other metaphysical theories by discovering the correlations that those theories predict should not be found."

We also agree with Nahmias's claim that our "methodology may work differently depending on what introspection is—and if there are different types of introspection, it may depend on what type the experimenters and subjects use." In the target article, we didn't have space to discuss the difficult issue of what introspection consists in, and so we left its conception at an intuitive level. But this issue needs to be addressed especially in the light of our positive methodological proposal. One of us (M.A.) has a detailed theory about the basic mechanisms of experiential introspection and its ontogenesis, according to which having an experience of pain is a different activity/event than introspecting it. Introspection requires cognitive (conceptual) capacities—see our reply to D'Amico above, and Aydede and Güzeldere 2005, forthcoming, for more details.

References

Aydede, Murat, and G. Güzeldere. 2005. "Cognitive Architecture, Concepts, and Introspection: An Information-Theoretic Solution to the Problem of Phenomenal Consciousness." *Noûs* 39(2): 197–255.

———. Forthcoming. *Information, Experience, and Concepts: A Naturalistic Theory of Phenomenal Consciousness and Its Introspection.* Oxford: Oxford University Press.

Barrell, J. J., and J. E. Barrell. 1975. "A Self-Directed Approach for a Science of Human Experience." *Journal of Phenomenological Psychology* (fall): 63–73.

Price, D. D., and J. J. Barrell. 1980. "An Experiential Approach with Quantitative Methods: A Research Paradigm." *Journal of Humanistic Psychology* 20(3): 75–95.

———. 1984. "Some General Laws of Human Emotion: Interrelationships between Intensities of Desire, Expectations, and Emotional Feeling." *Journal of Personality* 52: 389–409.

Price, D. D., J. E., Barrell, and J. J. Barrell, 1985. "A Quantitative-Experiential Analysis of Human Emotions." *Motivation and Emotion* 9: 19–38.

19 Closing the Gap on Pain: Mechanism, Theory, and Fit

Thomas W. Polger and Kenneth J. Sufka

A widely accepted theory holds that some emotional experiences are mediated mainly in a part of the human brain called the amygdala. A different theory asserts that color experience is largely determined by a small subpart of the visual cortex called V4. If these theories are correct, or even approximately correct, then they are remarkable advances toward a scientific explanation of human conscious experience. Yet simply understanding the claims of such theories—much less evaluating them—raises some puzzles. Conscious experience does not present itself as a brain process. Indeed experience seems entirely unlike neural activity. For example, to some people it seems that an exact physical duplicate of you could have different sensations than you do, or could have no sensations at all. If so, then how is it even possible that sensations could turn out to be brain processes?

One answer is that it is not possible. According to this line of reasoning, the puzzle of how brain states could be experiences is a genuine contradiction. Facts about brains do not settle questions about conscious experience because experience is of a different sort. Our ability to imagine like-brained but differently experienced critters reveals a deep metaphysical truth.[1]

A more modest but still troubling conclusion is that our imaginative successes are indicative of explanatory failures. It may be that sensations are brain processes. However, that does not mean that we can explain facts about conscious experiences with facts about brains. It may be that emotions occur in the amygdala and color sensations in V4. But there is nothing special about the amygdala that explains why certain emotions occur there rather than in V4 or some other part of the brain, or in things that aren't brains. The problem is not a metaphysical fissure but rather an epistemological one. Between minds and brains there is an *explanatory gap*.[2]

The conclusion that there is an explanatory gap, however modest, is still worrisome. First, it places limits on any possible science of the mind, and on the more general ideal of a natural science of our world. This is particularly frustrating because consciousness could be a unique holdout in a world that has otherwise succumbed to natural explanation. Second, the modest conclusion may be exploited to argue for the more radical conclusion that there is a *metaphysical gap* between mind and brain (e.g., Chalmers 1996; Chalmers and Jackson 2001).

In light of these concerns, it would be an achievement to show that there is no explanatory gap. We don't suppose that we can achieve that much, but we think we can make some progress. This paper has four parts: first, we sketch a general approach to the explanatory gap challenge. Next, we argue that our approach counts as a response to the gap problem. Third, we provide an example of how the approach can be implemented in the case of pain experience. Finally, we argue that the example fits the model provided earlier. If we are right, there is reason to doubt that there is an unbridgeable explanatory gap.

1   The Explanatory Gap and Identity Conditions

The idea that there is an explanatory gap between mind and brain—particularly between conscious experiences and neural processes—can be developed in many ways. The basic idea, however, is simple. It is hard to understand how any physical theory of brain processes can explain how or why a particular brain process must be correlated with a particular sensation. Because this puzzle can be repeated for every brain process and every sensation, it is hard to understand how any brain process could be associated with any sensation at all. Believers in the explanatory gap argue that it is not possible for physical explanation to tell us what we really want to know about conscious experience, namely, why any sensation depends necessarily on any particular neural activity. Why couldn't that sensation depend on different neural activity, or that neural activity mediate a different sensation? As Joseph Levine puts it: "What is left unexplained by the discovery of C-fiber firing is *why pain should feel the way it does!* For there seems to be nothing about C-fiber firing which makes it naturally 'fit' the phenomenal properties of pain, any more than it would fit some other set of phenomenal properties" (1983, p. 357). Because of the lack of natural fit between pains and C-fiber firings—or any particular neural process—no explanation that appeals to neural processes will tell us what we want to know about the nature of pains. Specifically, no account will tell us why pain sensation depends on C-fiber activity in such a way that the sensation is necessitated by the neural activity. The explanatory gappists do not deny that there are correlations between sensations and brain processes, and that neuroscience can discover these correlations. But they doubt that the correlation is other than contingent, and thus that the brain sciences are discovering anything about the nature of conscious experience. To do that, they maintain, we have to know not only that pain is correlated with C-fiber firing but moreover that it must be so.[3] The demand that we explain the "fit" between sensations and brain processes is a demand for an account that ties them together as a matter of necessity. This is the challenge of the explanatory gap.

Here, then, is the situation. Advocates of the explanatory gap (call them *mysterians*, adapting the terminology of Flanagan 1991) argue that it is impossible to explain all the features of conscious experience in entirely natural or neuroscientific terms. There is one sure-fire way to show that something is possible, and that is to show that it is actual. So it goes for explanations of experience. Philosophers, psychologists, and neuroscientists have offered numerous examples wherein some phenomenon of consciousness is given a naturalistic explanation (e.g., Hardin 1988; Damasio 2000; Flanagan 1995, 2000; Gustafson 1998; Ledoux 1998; Price and Aydede, this volume). If some kinds of conscious experience have been explained naturally, this surely demonstrates that conscious experience can be explained naturally.[4] And if so, then the mysterians are wrong.

As a matter of fact, we think that some of the naturalistic responses to the gap should be counted as successful. But for the most part mysterians have been unmoved by the neuroscientific accounts. And why should they be when the explanatory gap reasoning concludes that no such data can close the gap? So despite the availability of naturalistic explanations of conscious experience, in one way the mysterians are right to be unimpressed. What is missing is an account of why some neuroscientific model should count as explaining a kind of consciousness in the way that the mysterian demands. (One option, of course, is to deny that the mysterian is making reasonable explanatory demands. Perhaps that is right. But for now we'll play along.)

The key to satisfying the mysterian is to take seriously the requirement that explanations of conscious experience explain the "fit" between brain processes and conscious mental occurrences. What is demanded is that we show how some brain states and experiential states go together necessarily.[5] Anything less would leave unanswered the question of why some brain activity, for example, C-fiber firing, must feel the way it does. This is why merely making room for the possibility of a connection (Hill 1991; Hill and McLaughlin 1999; Block and Stalnaker 1999; Polger 2004) is not sufficient to close the gap. This is also why merely adding more empirical detail to the explanation of sensations is not sufficient. (After all, mysterians will "spot" the naturalists whatever physical facts they might want.) An account of the "fit" between sensations and brain processes will include empirical data, but the mysterian is correct that establishing the necessary connection requires more. We argue, however, that closing the gap does not require the mysterian's kind of more. Rather, closing the gap requires some explanatory and theoretical resources; but those resources are already available to the naturalists like us.

Nearly everyone agrees that there is at least one kind of explanation of the relationship between experiences and brain processes that can satisfy the mysterians' demands. If conscious experiences are identified with brain processes then there is a necessitating dependency between brain processes and experiences. Identity is a relation that holds with necessity. This is as decisive an answer to the question of fit between sensations and brain processes as one would want. Now we can see the beginning of an answer to why some neuroscientific data should count against the explanatory gap argument. It will count when it is evidence that some sensation can be identified with some neural process. It may be that a less robust relation could close the explanatory gap, but identity will do the job.

Mysterians typically agree that mind–brain identification, if correct, would cut off the explanatory gap reasoning.[6] But they each believe that the mind–brain identity theory is, in fact, false. We are quite a bit more optimistic about the identity theory, as you'll soon see. The more salient point at the moment is that what it takes to close the explanatory gap, on our model, is just whatever it takes to justify mind–brain identity claims for conscious experiences. If we are justified in asserting mind–brain identities, then we are justified in holding that there is a necessary connection between mind and brain, and denying that there is an explanatory gap. This will give us the "fit" we're looking for. But identities are not themselves part of the observational data. To make our proposal a substantial claim we need to provide an answer to the question: How can we justify mind–brain identity claims?

2   A Model for Mind–Brain Identification

These days there is widespread agreement that we can empirically discover the truth of some scientific identities. It has been discovered that temperature in a gas is mean molecular kinetic energy, that the evening star is the morning star, that gold is the substance with atomic number 79, and that water is H_2O. None of these examples is uncontroversial, however the controversies are generally over whether these particular identifications are correct rather than over the very possibility of such identifications. In this way explanatory gap concerns differ from other disputes about identity. The mysterians' argument purports to show that no mind–brain identifications can be correct. Yet mysterians do not reject the possibility of scientific identifications in general.[7] The question, then, is: how can we establish or justify scientific identity claims?

The basic answer is that if x is identical to y then x and y have the same identity conditions. Identity conditions are, as Alan Sidelle puts it,

the sorts of things that are represented by statements saying, for any possible object, what features it must have in order to be, or those which suffice for it to be (identical to) some particular thing (or, for kinds or properties, for something to be a member of that kind, or possess that property). . . . a specification of identity conditions need not state with full precision—need not mention—what the relevant features are; for certain purposes, 'this chemical microstructure' or 'this thing's origin' will do as well as 'H_2O' or 'sperm S and egg O'. (1992, p. 291)

Sameness of identity conditions is at least good evidence for identity and may be sufficient for identity. After all, since identity conditions determine (to wit) the identity of a kind of thing, if two kinds of things have exactly the same identity conditions then they are the same kind of thing. We shall assume for present purposes that establishing that two things have the same identity conditions is sufficient for establishing their identity.[8]

So the general answer is that sensations and brain processes are identical if they have the same identity conditions. The next question is what their respective identity conditions are, and how their sameness can be established. The answer to this question cannot be a platitude about identity. We will have to provide a substantial model for establishing certain scientific claims about sensations. Earlier we suggested that the key idea in the explanatory gap argument is the demand that we be able to explain the necessary "fit" between sensations and brain processes. Meeting that demand is central to establishing identities and closing the gap. Robert Van Gulick writes: "The more we can articulate structure within the phenomenal realm, the greater the chances for physical explanation; without structure we have no place to attach our explanatory 'hooks'. There is indeed residue that continues to escape explanation, but the more we can explain relationally about the phenomenal realm, the more the leftover residue shrinks toward zero" (Van Gulick 1993, p. 565). This tactic is what we call the Structure of Experience strategy (see Polger and Flanagan 1999).[9] The idea is that once we have a thorough description of experiential phenomenology and its structure, and of the correlated structures of neural events and processes, then we will be able to see that the two "fit" together after all. We will be able to see the fit, because we will see that the experiential and neural processes have the same identity conditions (Polger 2004). This still seems right to us. But we have come to see that the Structure of Experience approach needs to be further elaborated if it is to be more than a platitude about identities.

When we talk about phenomenal and physiological structures fitting together, there is a temptation to think of the relation as a kind of flat mapping. But this is too simplistic, for the phenomenal and physiological structures are not one dimensional.

Rather, we need an account of how phenomenology and its neural bases fit into a multilevel mechanistic explanation of activity and experience. Machamer, Darden, and Craver (2000; Craver 2001) have provided a useful model for thinking about mechanistic explanation. Figure 19.1 illustrates their proposal. Providing a full mechanistic explanation of a system involves not only describing what a thing does, but also showing how it fits into a broader context and how its behavior is realized by its constituents. Different phenomena may be explained at different levels—some by explaining how the system contributes to the activity of a containing system, and others by reference to the behavior or its parts.

Moreover, even this sort of multilevel mechanistic explanation does not stand on its own. It must be situated in the context of general background theories that unify scientific explanations (Polger and Flanagan 1999). When a multilevel mechanistic explanation is anchored in a background theory, theoretical identifications are the natural consequence. As Lewis argued, "theoretical identifications *in general* are implied by the theories that make them possible—not posited independently" (1972;

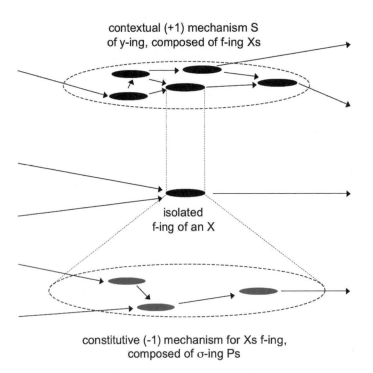

Figure 19.1
The multilevel mechanistic notion of "fit," adapted from Craver 2001.

in Block 1980, p. 207).[10] Take, for example, the atomic theory of chemical elements. Could one adopt the atomic theory and still wonder whether gold is the element with atomic number 79? It does not seem so. Once one adopts the atomic theory the identity of gold falls out. It looks like this is because the atomic theory fixes the identity conditions for chemical elements. Furthermore, we cannot ask whether gold could have different properties than it does, for the atomic theory explains the properties of elements in terms of their identity constituting atomic structure. The periodic table is a systematized representation of the basic identities and property explanations implied by this theory of chemical elements.

The case of sensations and brain processes is no different, in principle. In practice the case is not so simple, for we presently lack anything like a general background theory of neuroscience against which to set our mechanistic accounts of experience. Specifically, there is no general structural theory of neuroscience that tells us how to identify neuroscientific entities and properties. For this reason, mind–brain identities often seem ad hoc and they fail to get the intuitive traction of other scientific identities. We don't see how they "fit." Yet there is a background theory for biological and psychological sciences that can be of some use, namely, evolutionary theory. This theory cannot give us all that we need, but it is a start. Providing an evolutionary account of experience can help us secure the mechanistic hooks in our account. Such an explanation may (but need not) take the form of an adaptation explanation. That is, it need not account for some aspects of experience as products of evolution by natural selection favoring experience because it is fitness enhancing. William James (1890/1981) asserts that conscious experience is necessary for flexible behavior, a claim that is often repeated. But Owen Flanagan (1995, 2000) argues that the phenomenal experience of dreams is a spandrel, and we have argued the same for the phenomenology of chronic pain (Sufka 2000). If experience is a natural feature of biological organisms then it must have a place in the general theories of how organisms come to be as they are. One such theory is evolutionary theory. But it is not the case that every evolutionary explanation will be an adaptation explanation.

Evolutionary theory is a historical theory, rather than a structural theory like the modern atomic theory of chemical elements. As such, it does not imply mind–brain identities on its own, but it is a grounding framework for structural theories that do that job. Thus we may be seen as offering something akin to what Machamer, Darden, and Craver (2000) call a mechanism *sketch* or *schema*, rather than a completed mechanistic explanation. And that should be enough. A mechanism sketch is all that is had in the case of an explanation of life in inorganic terms. Sometimes the problem of explaining life is compared to the problem of explaining consciousness. This comparison is apt in many ways, not least of which being that both share evolutionary theory as part of their background. There is no explanatory gap in the case of life (as even most mysterians have acknowledged, e.g., Chalmers 1996; Chalmers and Jackson 2001), so it must be that mechanism sketches are sufficient for showing that an explanatory gap can be closed. For it is clear that we do not yet have a complete mechanistic explanation of life.[11] This is because the mechanistic explanation and its grounding are what yield the necessitating fit between phenomena, even when the details of exactly what fits with what are still sketchy.

The explanatory model that we are urging, which we argue has the resources to justify identity claims and thereby bridge the explanatory gap, is quite general. Consider the standard case of the identification of water with the molecule H_2O. How does this theoretical identification work? We begin with a qualitative analysis of water. Water is the clear, wet, potable stuff that rains from the sky, is found in lakes and rivers, boils at 100°C, and so forth. Some qualitative facts about water are readily observed and others take more work, for example, facts about the reactions into which water can enter. This gives us a picture of the structure of the phenomenon, in this case, water. These qualitative facts about water are then matched with qualitative facts about molecules. These are harder to ascertain but no less crucial. When enough is known about the structure and activities of water and the structure and activities of certain molecules, we can explain how the molecular structure of water accounts for its qualitative chemical properties.[12] The multilevel mechanistic explanation of water identifies it with a certain molecule type, H_2O, and explains its behavior in terms of the behavior and composition of H_2O molecules. Water is H_2O. But, of course, one doesn't just make the identification of water and H_2O. That identity only makes sense if other substances are also identified with other molecules, and if elements are identified with atomic kinds, and so forth. In short, the identity of water and H_2O works because it is set in the context of molecular and atomic chemistry. Those background theories entail the identifications. This is just the point we made earlier with respect to the atomic number of gold. One could, we suppose, wonder whether water is correctly identified with H_2O rather than some other molecule; but within the context of chemistry as we know it, one cannot sensibly wonder whether water might fail to be any kind of molecular substance at all. There is no explanatory gap about water and H_2O.

This is the model that we urge in general, and for conscious experience in particular. If it seems like a commonsense model, we agree. Yet this model is quite different from those that are widely discussed. On the one hand, we agree with the mysterians that standard naturalistic replies fall short because contingent explanations are not enough to close the explanatory gap. No amount of empirical detail will, by itself, explain how sensations depend necessarily on brain processes. For this reason, what Price and Aydede (this volume) call the "vertical phase" of explanation is not completed by merely describing correlations. On the other hand, we deny that the required necessary dependence must be an analytic connection, or that it is uniquely problematic in the case of conscious experience. This runs against the model assumed by Chalmers and Jackson (2001), for example. Our approach goes well beyond the familiar responses to the explanatory gap because we take seriously the task of showing how a naturalist can account for the necessary "fit" between sensations and brain processes.

In the remainder of this paper we show how to apply the model to the case of pain experience. Pain experience has distinctive sensory and affective structures. There is a well-integrated multilevel account of the physiological mechanisms that explain the qualitative structures of pain. This mechanistic explanation is set against the background of (admittedly nascent) theories of neurons and neuronal systems. There are also evolutionary explanations of pain experience, but these are not always adaptation explanations. Together these explanations show how the mechanisms of pain fit the phenomenology of pain. Once pain mechanisms are identified, there is no room for an explanatory gap.

3   Case Study: Pain

Pain is "an unpleasant sensory and emotional experience associated with actual or potential tissue damage, or described in terms of such damage" (Merskey and Bogduk 1994, pp. 207–213; see also Merskey et al. 1986). Like other sensory systems, the pain sensory system is designed to carry out feature detection processes for a given sensory modality. To accomplish this task, the pain sensory system contains (1) specialized receptors that respond only to certain kinds of noxious stimuli, (2) a multitude of dedicated pathways that carry specific features of this sensory modality (i.e., line-labeling) to the central nervous system, and (3) numerous distinct subcortical and cortical areas that process and coordinate the various sensory and emotional features and behavioral responses of this sense modality. Functionally, the pain sensory system is responsible for detecting stimuli that have the potential for damaging tissue and for causing the organism to engage in escape and avoidance responses.

It will be useful to keep in mind an outline of the neuroanatomy of the pain sensory system (figure 19.2). Noxious stimuli, which have the potential to damage tissue, are detected by a special class of receptors called nociceptors, which are located at the

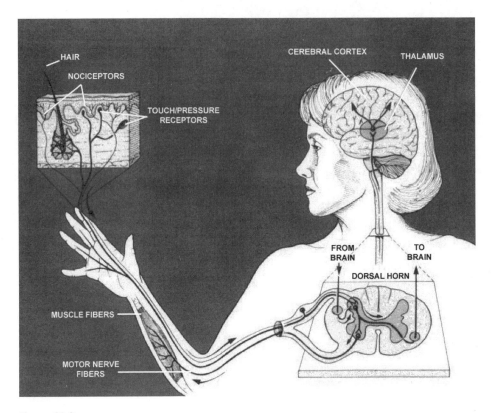

Figure 19.2
The pain sensory system.

distal ends of primary afferent fibers (PAFs) or simply put, first order sensory neurons. Following transduction, the process by which physical energy is converted to a neural code, information in the form of action potentials is carried to the central nervous system via two classes of PAFs: Aδ- and C-fibers. These Aδ- and C-fibers terminate in several of the ten distinct laminae or layers of the spinal cord dorsal horn and synapse onto nociceptive specific, wide dynamic range second-order neurons. The best way to think of these second-order neurons is that their feature detection capabilities are maintained through the principle of line-labeling. Line-labeling preserves receptor individuation throughout a pathway. For example, adjacent receptors project to adjacent locations in the thalamus, and so onward. Thus the whole pathway can be thought of as coding for the feature detected by the receptor, and distinct from that of the adjacent receptor or pathway. The line-labeled second-order neurons project via both monosynaptic and polysynaptic parallel pathways to numerous supraspinal centers—predominately the thalamus, but also the hypothalamus, the amygdala, and the brainstem periaqueductal gray and reticular formation. Projection neurons from these centers extend their axons to distinct regions of the cortex, including primary somatosensory, secondary somatosensory, anterior cingulate, insula, and medial pre-frontal cortical areas. As we will discuss below, electrophysiological, regional cerebral blood flow, positron emission tomography and functional resonance-imaging studies all demonstrate that activity in these areas follows the presentation of noxious or tissue damaging stimuli. These activities can be linked with various conscious sensory and emotional states. The more we are able to describe the phenomenology of pain and map these experiences onto neurobiological processes, the closer we come to eliminating the explanatory gap.

3.1 The Phenomenal Structure of Pain Experience

With the above outline of pain neurophysiology in mind, let us examine the phe-nomenological structure of pain experience.[13] Pain is a multidimensional experience that involves both sensory and affective dimensions. The sensory dimension of a pain experience typically refers to the quality or type, intensity, and spatiotemporal dynam-ics characteristics. Melzack and Torgerson (1971), in an attempt to develop a valid and reliable pain questionnaire, compiled, categorized and scaled a list of descriptors of pain phenomenology (figure 19.3). Interestingly, these verbal descriptors not only clustered in predictable ways around the multidimensional phenomenology, the

				SENSORY						
	Temporal	Spatial	Punctate	Incisive	Constrictive	Traction	Thermal	Brightness	Dullness	
1									Dull	1
2	Quivering		Pricking		Pinching			Smarting		2
3	Pounding	Shooting		Sharp		Tugging	Hot		Heavy	3
4			Stabbing	Lacerating	Crushing		Searing			4
5										5

			AFFECTIVE				EVALUATIVE	
	Tension	Autonomic	Fear	Punishment	Misc.		Anchor Words	
1							Mild	1
2	Tiring						Discomforting	2
3	Exhausting	Sickening	Frightful	Gruelling	Wretched		Distressing	3
4		Suffocating	Terrifying		Blinding		Horrible	4
5				Killing			Excruciating	5

Figure 19.3
The phenomenology of pain. Adapted from Melzack and Torgerson 1971.

individual descriptors within a single dimension revealed a scaling with respect to an intensity dimension (using anchor words: 1 = mild, 2 = discomforting, 3 distressing, 4 = horrible, 5 = excruciating). For example, a thermal pain experience might be described as ranging from hot to burning to searing, while a mechanical (pressure) pain experience might be described as ranging from pinching to cramping to crushing. For the temporal feature of a pain experience, one might describe it in terms ranging from flickering to pulsing to pounding. These sensory characteristics of type, intensity, and location can be mapped to neurobiological processes in the pain sensory system.

Another dimension of a pain experience that can be mapped onto neurobiological processes is its affective–emotional component. The affective dimension of a pain experience has two components: primary and secondary affect (Price 2000; Wade et al. 1996). Primary affect refers to the initial unpleasantness one experiences as a result of a threat. It comprises "the moment by moment unpleasantness, distress and possible annoyance that are often closely linked with the intensity of the painful sensation and the accompanying arousal" (Price 1999, p. 59). Secondary affect refers to one's "elaborate reflection and relates to memories and imagination about the implications of having pain, such as how pain may interfere with different aspects of one's life" (ibid.).

In addition to the dimensions of pain outlined above, there are a number of alterations in pain sensation that can follow an injury or illness and include allodynia and hyperalgesia (see Willis 1992). Allodynia is pain due to a stimulus that does not normally produce pain. Hyperalgesia is an increased response to a stimulus that is normally painful. The difference between allodynia and hyperalgesia is in the relationship between the stimulus and the painful experience. Allodynia involves two modes of processing where a nonnoxious stimulus is perceived as being noxious, and hyperalgesia involves a single mode of processing where a noxious stimulus is perceived as being more noxious. Allodynia and hyperalgesia are hallmark signs of chronic pain and can be mapped onto changes in both peripheral and central nervous system functioning.

3.2   A Multilevel Mechanistic Model of Pain

We claimed that mechanisms that explain pain phenomenology have been identified. Now we need to deliver on this promise. In this section we illustrate how pain phenomenology (outlined in section 3.1) can be explained by specific neuroanatomical and neurophysiological processes (outlined in section 3.0). We will restrict this discussion to coarse-grained phenomenology of pain sensation (type, intensity, and location), pain affect, and the chronic pain symptoms of allodynia and hyperalgesia. Other dimensions of pain phenomenology are open to such explanation, and progress continues to be made in understanding the neurobiological underpinnings of pain.

Type coding   People are quite adept at differentiating the various types of noxious or tissue damaging stimuli. Our pains can be felt as coming from either mechanical stimuli (smashing a finger in a door), thermal stimuli (touching a hot oven), or chemo-inflammatory stimuli (getting lemon juice into an open cut). The pain sensory system is capable of this level of feature detection through three types of nociceptors (Belemonte and Cervero 1996; Reichling and Levine 1999). Although least studied, mechanical nociceptors are probably the simplest in terms of the transduction process and consist of high threshold mechano-gated ion channels; low threshold mechano-receptors are for cutaneous or touch sensory system. Noxious mechanical stimuli open ion channels of this receptor leading to a depolarizing current in the primary afferent fibers. Thermal nociceptors are a bit more complex. High intensity thermal stimuli activate nonspecific cation channels as well as both vanilloid and non-vanilloid-like receptors. The influx of positively charge ions leads to depolarizing current in the PAFs. Chemo-inflammatory nociceptors are perhaps the most complex. A variety of exogenous substances, for example capsaicin, and endogenous neural and nonneural substances, for example bradykinins, prostaglandins, and serotonin can activate our pain sensory system through binding to their respective receptors (Reichling and Levine 1999). Some of these chemo-inflammatory nociceptors depolarize primary afferent neurons through iontropic mechanisms while others do so through metabotropic or second messengers mechanisms. Adding to this kind of feature detection are nonspecific nociceptors that seem to respond to more than one kind of stimulus. The VR1 receptor is a good example in that it responds to thermal stimuli and changes in pH (ibid.). In short, our pain sensory system, like other sensory systems, engages in feature detection processes due to, in part, a principle of receptor specialization. Receptor differences are the primary explanation for why veridical acute pain is experienced as coming in different types that have different responses. Many nonveridical cases involve the mistaken stimulation of those receptors or their projections. But type is only one dimension of the structure of pain phenomenology.

Intensity coding   The intensity of pain experiences seems to be initially coded at the receptor level as well. Elegant neural recording studies in PAFs show that the activity in these fibers parallel psychophysical assessment of pain intensity for both noxious mechanical and thermal stimuli (Handwerker and Kobal 1993; Gybels, Handwerker, and Van Hees 1979; Bromm, Jahnke, and Treede 1984). However, there is not a one-to-one correspondence between stimulus intensity and actual transduction processes. This is due to an ensemble of mechanisms that interact to modulate the eventual pattern of activity in PAFs. Among these mechanisms include multiple transduction avenues for a single stimulus, functional interactions between ion channels due to their spatial distribution, and the ability of second messenger systems to modulate receptors, ion channels and other second messenger systems (Reichling and Levine 1999). Nevertheless, there seems to be a strong relationship between a noxious stimulus's intensity and verbal ratings using the Visual Analogue Scale (VAS); this relationship yields a positive power function with an exponent of 3.0–3.5 (Price and Harkins 1987). Thus, a psychophysical relationship exists between level of the noxious stimulus presented and the perceived intensity of pain that is explained by the response characteristics of the nociceptor and the eventual activity along the PAFs.

Temporal coding   One interesting phenomenon of pain is in its temporal dynamics. Some kinds of tissue damage initially produce a "sharp and highly localized" pain that is followed by a "dull and more diffuse" pain. The temporal ordering of these distinct pain experiences maps well to physiological properties of the A-delta and C PAFs.[14] These two fibers differ in axonal diameter and whether they are covered with myelin. Axonal diameter and myelin sheathing greatly affect signal conduction velocities. A-delta fibers have medium-sized myelinated axons of around 2–6 μm in diameter with conduction velocities of 12–30 m/sec. C-fibers have thin unmyelinated axons of around 0.4–1.2 μm in diameter with conduction velocities of 0.5–2.0 m/sec. (Besson and Chaouch 1987; Laem et al. 1993). While both fibers are responsive to noxious stimulation, A-delta fibers evoke the sharp, first phase of the pain experience whereas C-fibers evoke the dull, second wave of the pain experience (Price and Dubner 1977; Price 1996; Price and Aydede, this volume). For example, stubbing a toe leads to two distinct and sequentially processed pain experiences: the first is a sharp and highly localized pain experience that is followed by a dull, throbbing and diffuse pain experience. These distinct phenomenal qualities nicely map onto the differences in signal processing capabilities found in the A-delta and C PAFs.

Location coding   To understand how an organism is capable of spatially localizing noxious stimulation requires an understanding of receptive fields, dermatomes, and cortical somatotopy. A receptive field refers to an area of skin in which a stimulus will elicit a response in a receptor or, on a larger scale, a single PAF. The receptive fields for C-fibers range in size up to about 10 mm in diameter depending upon the body location. Dermatomes are essentially very large receptive fields for one of the thirty-one pairs of spinal nerves. Cervical nerves 1–8 have dermatomes in head, neck, shoulder, and arm regions; thoracic nerves 1–12 have concentric dermatomes down the midsection, and so forth for the 5 lumbar, 5 sacral, and single coccygeal nerve pairs. Through line-labeling, the integrity of this spatially discriminated information is maintained up to the level of the cortex where in areas of the primary and secondary somatosensory cortex of the parietal lobe, several somatotopic maps are located. The discovery of human somatotopic maps dates back to Wilder Penfield's pioneering research from the 1930s to the 1950s (Penfield and Jasper 1954) and are essentially maps of the body surface projected onto the surface the post-central gyrus of parietal lobe. PET studies have revealed these somatotopic maps indeed map pain loci by showing that noxious stimulation to different body regions (i.e., receptive fields) give rise to increased activity in distinct areas of S1 that yield pain experiences to a particular body region (Andersson, Lilja, and Hartvig 1997). This supports the conclusion that the spatial features of pain experience can be explained by the spatial features of somatosensory cortex and the receptors and pathways that project into it.

Affect coding Numerous studies demonstrate that noxious stimulation activates several distinct regions of the cortex including primary somatosensory (S1), secondary somatosensory (S2), anterior cingulate (ACC), insula, and medial prefrontal cortical areas. Clever studies by Catherine Bushnell and her colleagues have been able to dissociate the sensory and affective dimensions of pain perception and to localize these dissociated processes to activity in distinct brain areas. In one PET study, rCBF levels were obtained during a thermal grill illusion task. This is an illusion in which spatially alternating nonnoxious warm and cool thermal stimuli leads to reports of pain and unpleasantness despite neither stimuli producing pain or unpleasantness when presented alone (Craig et al. 1996). Using a subtraction technique for the two conditions, only the increase in activity in the ACC seemed to account for the pain experience.

In another study, researchers used hypnotic suggestion to manipulate subjects' pain affect while holding their perception of pain intensity constant while subjects' hands were submerged in hot water. Verbal reports confirmed that hypnotic suggestion was successful in dissociating pain unpleasantness and intensity. Again rCBF results after subtraction (hand in warm water) show a significant increase in ACC activity that actually parallels subjects' level of pain unpleasantness ratings (Rainville et al. 1997). A similar study that held subjects' perception of pain unpleasantness constant while manipulating perception of pain intensity produced strikingly different results. Here, the strongest change in pain-intensity-evoked activity was localized to S1 (Rainville et al. 1999).[15]

These findings parallel the case of a 57-year-old male who suffered a stroke that damaged the hand region of his right S1 and S2 cortex (Ploner, Freund, and Schnitzler 1999). Noxious cutaneous laser stimulation presented to his left foot produced a well-localized pain sensation; however, the same stimulus presented to his left hand at three times the intensity failed to evoke a pain sensation. Interestingly, the patient "spontaneously described a 'clearly unpleasant' intensity dependent feeling emerging from an ill-localized and extended area 'somewhere between the fingertips and shoulder' that he wanted to avoid" (ibid., p. 213). Further, this patient was "completely unable to describe quality, localization and intensity of the perceived stimulus." Case studies like this, as well as the PET studies described earlier, provide compelling evidence for the differential involvement of cortical structures in our sensory and affective dimensions of pain experiences.

By now our pattern of argument is familiar. We conclude that the quantitative aspects of veridical pain experience including location, type, and intensity are explained mainly by activity in S1, while the qualitative or affective dimension of pain experience is explained mainly by activity in the ACC (Sufka and Lynch 2000).

Altered sensory coding   Finally, allodynia and hyperalgesia are two primary symptoms of chronic pain and reflect perceptual alterations to nonnoxious and noxious stimuli, respectively. Considerable work over the last decade has elucidated the neurobiological processes that underlie the development of allodynia and hyperalgesia both of which result from neuroplastic changes in the peripheral and central nervous systems. More specifically, allodynia seems to result from a phenotypic switch in A-beta PAF (Neumann et al. 1996). A-beta fibers are normally responsible for processing nonnoxious cutaneous information. However, prolonged tissue trauma can cause a subset of these A-beta fibers to switch phenotypes to one that resembles A-delta and C-fibers. This phenotypic switch is at the level of the neurotransmitter released onto second-order neurons. The result is a cross-talk between the touch system and the pain system and it occurs in spinal cord relay neurons.

Hyperalgesia results from several related processes that ultimately increase the rate of firing in second-order pain neurons in the spinal cord, a phenomenon known as central sensitization (see Sufka 2000). Four neurobiological changes have been identified as being responsible for central sensitization induction. These include activation of silent nociceptors, increased release of glutamate from PAFs, increased numbers of and signal conduction in glutamate receptors on second-order pain transmission neurons, and loss of inhibitory interneurons in the spinal cord dorsal horn (Woolf and Salter 2000).

A brief aside Before we move on, two clarifications are in order. First, we are not asserting that different features of pain experience (pain qualia, if you like) can be individually identified with the individual physiological mechanisms that are associated with them in veridical pain sensation. In some cases that may turn out to be the case; but our argument does not require that outcome. Rather, taken together these explanatory components support the conclusion that the individual mechanisms combine to form an overall system that we call the pain sensory system. Pain sensations are processes occurring in that system. Some of these processes may be relatively localized, others quite global. One way to think of the situation is that the above mechanism sketches tell us about some of the core mechanisms that comprise the total mechanism of pain experience.[16] A complete mechanistic explanation of pain experience will identify pain sensations with processes in and characteristics of the pain sensory system.

Second, we are aware that the coarseness of the phenomena that we are considering leaves ample room for question on each feature. Yet mechanism sketches are sufficient for some purposes (Machamer, Darden, and Craver 2000). And above we argued that, contrary to prevailing sentiment among naturalists, closing the explanatory gap is one such purpose. This is because closing the gap involves showing in general how sensations and brain processes fit, and thereby establishing the necessity of the dependence. Grounding a mechanism sketch or schema of pain sensations in a general background theory can be accomplished even in the absence of all the details that would fill out a complete mechanistic explanation. And that is fortunate, for otherwise we would have no mechanistic explanations or any phenomena and the world would be riddled with explanatory gaps. So we only need to provide enough detail to justify a mechanistic model of pain sensations that identifies them with neural and bodily processes, even if many of the details remain to be discovered.[17]

3.3 Background Theories

As more and more details of pain processing are revealed, more pain phenomenology can be explained in terms of neurobiological processes. With each new discovery the case for closing the explanatory gap grows stronger. But we have also suggested that although empirical progress inevitably proceeds bit by bit, the identification of pain experience with neural processes is not piecemeal. We do not discover one identity at a time and then add them together. Rather, what justifies our claims to have identified the mechanisms of pain is the way the whole multilevel mechanistic story hangs together. We have emphasized that different aspects of pain experience are explained by different parts of the mechanism—some at the level of synapses, some neurons, some neural systems. No one of the identifications makes sense without the others, nor without a background understanding of neurons and organisms. So in order to make good on the claims in the previous section, we must articulate the background theories that are relevant to the case pain. To satisfy this requirement, we provide a multilevel, theoretical consideration of neuronal communication, sensory system organization and functioning, and evolutionary explanations of pain.

To our knowledge there is no widely accepted background theory that applies to all sensory systems in terms of organization, processing, and functioning. This does not mean that nothing can be said. Neural communication is an electrochemical process. Given the appropriate input from a receptor or other neurons, a cell transitions from a resting membrane potential to an action potential. The mathematical model that describes neuronal membrane potentials is the Goldman Equation (Goldman 1943; Hodgkin and Katz 1949). Basically, the determination of the membrane potential is heavily influenced by a given ion concentration and its membrane permeability. Further, when membrane permeability is exceptionally high for a given ion, the Goldman Equation reduces to the Nernst Equation for that particular ion. Communication between neurons occurs via electrical and chemical transmission, although the latter being far more common in the nervous system. There are two competing theories (i.e., *Classical* Heuser–Reese model versus Ceccarelli *Kiss and Run* model) that describe neurotransmitter release, both of which entail a calcium-dependent exocytosis of vesicle contents at the active zone (Holt and Jahn 2004). For one wanting yet a lower-level theoretical explanation, a decade of elegant studies by Roderick MacKinnon detail the structure and necessary requirements for the opening and closing of neuronal ion channels responsible for cellular communication (see MacKinnon 2003).

Moreover, sensory systems, including the pain sensory system, seem to be designed in ways to allow them to carry out feature detection processes. How these sensory systems accomplish this task is, in most ways, quite similar across the various sensory modalities. Each sensory system has specialized receptors that respond only to certain kinds of stimuli. Further, these receptors typically respond in a manner that parallels, up to a point, the intensity of that stimulus. Each sensory system contains a multitude of dedicated, line-labeled pathways that carry specific information to the central nervous system. Finally, each sensory system routes information to various distinct cortical areas that are dedicated to the conscious processing of the various features of that sensory modality. How each sensory system accomplishes the complex processing task of feature detection is not fully understood. One theory of visual processing for form perception (viz., the Multichannel Model) holds that neuronal cells throughout the visual system (from the retina to the cortex), which vary in their receptive field size, naturally extract information by a process analogous to spectral/Fourier analysis of a scene's four grating properties: spatial frequency, contrast, orientation, and spatial phase (Campbell and Robson 1968).

The background framework for the neurosciences may not be as established as the atomic theory is for chemistry, but some pieces are coming into focus. Of course, the neurosciences can rest on the general background theories that anchor biological sciences, for example, the background of organic chemistry. And one general theoretical framework for both the biological and mind–brain sciences is that of evolutionary theory. Evolutionary theory posits that organisms evolved through gradual change over time, subject to the influences of natural selection, which typically favors those biological forms that have an ecological or adaptive advantage in their environments. Pain seems to have ecological value in that it alerts the organism to injury or threat of injury and elicits an escape–avoidance response to terminate tissue trauma or a state of quiescence to prevent further injury and permit haling and recovery processes to ensue (Sufka and Turner in press). Further evidence of the adaptive value of pain comes from a cluster of rare medical conditions known as Hereditary and Congenital Pain Insensitivity Syndromes (Nagasako, Oaklander, and Dworkin 2003). People with these syndromes fail to experience the normal subjective and objective responses to noxious stimuli. Sufferers of these syndromes do not express aversion to noxious stimuli nor attempt to prevent injury from recurring. Further, they fail to notice injuries and illnesses and often die in childhood; rarely do these individuals live beyond their thirties. Syndromes such as these suggest that pain normally serves an adaptive function (Melzack and Wall 1983; Nagasako, Oaklander, and Dworkin 2003). However, it is not necessary that all pains are adaptive. Chronic pain typically leads to states of withdrawal, anhedonia (i.e., loss of pleasure), and environmental indifference—in essence, depression. For some individuals, their chronic pain persists long after resolution of the initial tissue trauma and likely involves possibly permanent maladaptive changes in nervous system functioning (Scholz and Woolf 2002; Sufka and Price 2002). Whereas acute pain plausibly has protective functions, Millan (1999) and Sufka (2000; Sufka and Turner 2005) have argued that chronic pain has no ecological value and is simply a spandrel or byproduct of the highly adaptive phenomenon of neural plasticity.

As we conceded earlier, evolutionary theory does not by itself entail mind–brain identity claims. (Mysterians, after all, do not reject evolutionary theory.) Rather, evolutionary theory provides some background constraints on structural theories of neuroscience. These latter we only have the beginnings of at the moment. And we only have them in a piecemeal way—as with the theories of electrochemical communication and feature detection pointed to above. But these are the kinds of theories that warrant identifications.

Note that this is not the blank check strategy, often lamented of naturalistic philosophy. Although there are details to be filled in, our argument does not hinge on a bet about how those details will turn out. Rather, we already know enough to know what form the yet-to-be-discovered theories will take. The cash value of our proposal is our claim about the basic structure of mechanistic mind–brain explanations: they identify mental occurrences with neural processes. We believe this not because the data alone tell us so, but because that is how the overall mechanistic account of experience identifies its kinds. We have argued that we already have evidence of this, and we have sketched examples. The missing details are just that. An analogous case would be the situation, which we suppose actually occurred, in which one has sufficient evidence that the atomic theory of chemistry is correct but has not yet identified all the elements. In that situation you would already be justified in asserting that, say, uranium is identical to some atomic type that is individuated by its atomic number, even if you did not yet know what the atomic number of uranium is. You would also be justified in asserting that the chemical properties of uranium are explained by its atomic structure. This is what we claim about sensations and brain processes. If we are right, we are already justified in concluding that sensations are identical to brain processes.

4   Closing the Gap on Pain

The explanation that we sketched for pain experience fits the model that we offered in section 2. We showed how features of the phenomenology of pain can be identified with features of the pain sensory system and its activity. These identities do not stand on their own. They are supported by a general view of transduction, coding, and processing that applies not only to pain but to other sensory systems, and in some cases to neural systems in general. These, in turn, can be understood and situated in and constrained by the background assumption that biological systems are evolved systems.

Our model is quite general. Indeed it helps us to see why some identifications are more intuitively compelling than others. The identification of water and H_2O is convincing because we know quite a bit about the general framework from which it derives, that of molecular chemistry. The identification of sensations and brain processes remains somewhat mysterious because the mechanistic explanations and background theories are still under development. We have urged that this is a mundane fact about the status of neuroscientific theories rather than a special fact about conscious experiences. Good empirical theories, after all, are hard to come by. But when we have them, we have all that we need to close the explanatory gap.

As we see it, the explanatory gap worries successfully point to a lacunae in the naturalistic explanations offered to date. But that kind of gap is not distinctive to explanations of consciousness. Nor is it located, as it were, between the phenomenology and the neurophysiology. Rather, the missing information involves enriching the detail of both the phenomenology and neurophysiology and—crucially—situating them together against the background of general theories of perception and sensation. We have indicated how some of this missing information can be provided in the case of pain experience. And we illustrated how the multilevel mechanistic explanation schema of pain fits the model that we endorse. Of course there is always more information that could be added and further detail that could be articulated. This is as much the case for water and H_2O as for pains and neural processes.

Acknowledgments

We would like to thank audiences at the Southern Society for Philosophy and Psychology, the Association for the Scientific Study of Consciousness, and Washington University in St. Louis for useful feedback on versions of this paper. We also benefited from conversations with Murat Aydede, Carl Craver, George Graham, Güven Güzeldere, Michael Lynch, Eddy Nahmias, Rob Skipper, and Derek Turner. Polger's work on this project was supported in part by the Charles P. Taft Fund at the University of Cincinnati.

Notes

1. Though no one probably would endorse this quick sketch as is, versions of this line of reasoning are meant to be familiar, for example, from the works of Descartes (1641/1986), Kripke (1980), and Chalmers (1996).

2. This is the expression made famous by Levine (1983, 2001). Similar reasoning can be found in Nagel 1974 and McGinn 1991. Jackson (1982) and (especially) Chalmers (1996) exploit this line of reasoning to drive the stronger conclusion, mentioned above.

3. As it happens we will discuss the example of pain in great detail, and even say a few things about C-fibers. But we think the best way to understand talk about "C-fibers" in most

philosophical discussions is as a placeholder for some future neuroscientific account. For the most part, where you read *C-fiber* in a philosophical discussion you should interpret that as something like, *neural process F*. This is why it will not defuse the question of the explanatory gap to simply observe that it is in fact false that all pains are C-fiber firings.

4. Strictly speaking, this leaves the possibility that some kinds of consciousness still cannot be explained naturalistically. But the explanatory gap reasoning concludes that no kind of consciousness can be explained naturalistically, and that is the problem we are concerned with at present.

5. This is where we would get traction if we wanted to dispute the mysterians' criteria. We would argue that it is too much to require a necessary connection, or that a contingent connection would be sufficient.

6. Recently, Chalmers and Jackson (2001) have argued that even identity is not enough. They demand not just that the relationship be metaphysically necessary but that it also be "epistemically" necessary. We're not convinced, so we're going to neglect that objection for present purposes. However, if you think that we have to answer that objection in order to close the gap, see Polger and Skipper, unpublished manuscript.

7. The explanatory gap is supposed to be a special problem for consciousness, and not for a posteriori identities in general. (See, for example, Block and Stalnaker 1999; and Chalmers and Jackson 2001.) There may be other problems about such identities. We don't mean to downplay those difficulties, but they are not the explanatory gap.

8. In assuming that sameness of identity conditions is sufficient for identity, we do not take ourselves to be making a move that would not be acceptable to mysterians. The explanatory gap argument is comfortably construed as working from the premises that conscious experiences and brain processes have different conditions of identity. Some may wish to argue that sameness of identity conditions is not sufficient for identity, but this is a different argument. See the previous footnote.

9. We called it the "Structure of Consciousness" reply back then, but we now think that "Structure of Experience" sounds better. This is not meant to be a substantive change.

10. Lewis may have thought, and thereby encouraged others to think, that theories analytically entail identity claims. That is not our position.

11. Here, as above, we emphasize that the explanatory gap is supposed to be a special problem about consciousness. We welcome the comparison of the problem of explaining consciousness to that of explaining life. We are puzzled as to why Chalmers and others think this case favors the explanatory gap reasoning. If the mysterian falls back to general concerns about identities, reference, and natural kinds, then that will be a rather different debate that is not peculiar to consciousness.

12. In fact we believe available explanations to be incomplete, and thus only mechanism schema, even in the case of the properties of molecular substances. For example, as far as we know there

is no complete mechanical explanation of fluid properties (e.g., viscosity and turbulence) in terms of atomic and molecular features.

13. See also Price and Aydede, this volume, particularly concerning the methodology for acquiring this information.

14. Finally, C-fibers!

15. See also Price and Aydede, this volume, for a more detailed discussion.

16. This way of talking about "core" and "total" mechanisms is derived from Shoemaker's (1981) distinction between core and total realizers.

17. This seems only fair: as the mysterian argument does not depend on the particular details, so the solution to it should not depend on those details. The mistake that many naturalists made was to think that the gap would collapse on itself under the weight of empirical data. Not so; an argument is needed.

References

Andersson, J., A. Lilja, and P. H. Hartvig. 1997. "Somatotopic Organization along the Central Sulcus, for Pain Localization in Humans, as Revealed by Positron Emission Tomography." *Experimental Brain Research* 117: 192–199.

Belemonte, C., and F. Cervero. 1996. *Neurobiology of Nociceptors*. Oxford: Oxford University Press.

Besson, J. M., and A. Chaouch. 1987. "Peripheral and Spinal Mechanisms of Nociception." *Physiological Review* 67: 67–186.

Block, N. ed. 1980. *Readings in Philosophy of Psychology*. Cambridge, Mass.: Harvard University Press.

Block, N., O. Flanagan, and G. Güzeldere, eds. 1997. *The Nature of Consciousness: Philosophical Debates*. Cambridge, Mass.: MIT Press.

Block, N., and R. Stalnaker. 1999. "Conceptual Analysis, Dualism, and the Explanatory Gap." *Philosophical Review* 108(1): 1–46.

Bromm, B., M. T. Jahnke, and R. D. Treede. 1984. "Responses of Human Cutaneous Afferents to CO_2 Laser Stimuli Causing Pain." *Experimental Brain Research* 55: 158–166.

Campbell, F. W., and J. G. Robson. 1968. "Application of Fourier Analysis to the Visibility of Gratings." *Journal of Physiology* 197: 551–566.

Chalmers, D. 1996. *The Conscious Mind: In Search of a Fundamental Theory*. New York: Oxford University Press.

Chalmers, D., and F. Jackson. 2001. "Conceptual Analysis and Reductive Explanation." *Philosophical Review* 110(3): 315–361.

Craig, A. D., E. M., Reiman, A. C. Evans, and M. C. Bushnell. 1996. "Functional Imaging of an Illusion of Pain." *Nature* 384: 258–260.

Craver, C. 2001. "Role Functions, Mechanisms, and Hierarchy." *Philosophy of Science* 68: 53–74.

Damasio, A. 2000. *The Feeling of What Happens: Body and Emotion in the Making of Consciousness*. New York: Harvest Books.

Descartes, R. 1641/1986. *Meditations on First Philosophy*. Ed. J. Cottingham. Cambridge: Cambridge University Press.

Flanagan, O. 1991. *Consciousness Reconsidered*. Cambridge, Mass.: MIT Press.

———. 1995. "Deconstructing Dreams: The Spandrels of Sleep." *Journal of Philosophy* 92: 5–27.

———. 2000. *Dreaming Souls*. New York: Oxford University Press.

Goldman, D. E. 1943. "Potential, Impedance, and Rectification in Membranes." *Journal of General Physiology* 27: 37–60.

Gustafson, D. 1998. "Pain, Qualia, and the Explanatory Gap." *Philosophical Psychology* 11: 371–387.

Gybels, J. M., H. O. Handwerker, and J. Van Hees. 1979. "A Comparison between the Discharges of Human Nociceptive Nerve Fibers and the Subject's Ratings of His Sensations." *Journal of Physiology* 292: 193–206.

Handwerker, H. O., and G. Kobal. 1993. "Psychophysiology of Experimentally Induced Pain." *Physiology Review* 73: 639–671.

Hardin, C. 1988. *Color for Philosophers*. Indianapolis, Ind.: Hackett.

Hill, C. 1991. *Sensations: A Defense of Type Materialism*. Cambridge: Cambridge University Press.

Hill, C., and B. McLaughlin. 1999. "There Are Fewer Things in Reality than Are Dreamt of in Chalmers' Philosophy." *Philosophy and Phenomenological Research* 59: 445–454.

Hodgkin, A. L., and B. Katz. 1949. "The Effect of Sodium Ions on the Electrical Activity of the Giant Axon of the Squid." *Journal of Physiology* (London) 108: 37–77.

Holt, M., and R. Jahn. 2004. "Synaptic Vesicles in the Fast Lane." *Science* 303: 1986–1987.

Jackson, F. 1982. "Epiphenomenal Qualia." *Philosophical Quarterly* 32(127): 127–136.

James, W. 1890/1981. *The Principles of Psychology*. Cambridge, Mass.: Harvard University Press.

Kripke, S. 1980. *Naming and Necessity*. Cambridge, Mass.: Harvard University Press.

Laem, W., W. D. Willis, S. C. Weller, and J. M. Chung. 1993. "Differential Activation and Classification of Cutaneous Afferents in the Rats." *Journal of Neurophysiology* 70: 2411–2424.

Ledoux, J. 1998. *The Emotional Brain: The Mysterious Underpinnings of Emotional Life*. New York: Simon and Schuster.

Levine, J. 1983. "Materialism and Qualia: The Explanatory Gap." *Pacific Philosophical Quarterly* 64: 354–361.

———. 2001. *Purple Haze: The Puzzle of Consciousness.* New York: Oxford University Press.

Lewis, D. 1972. "Psychophysical and Theoretical Identifications." *Australasian Journal of Philosophy* 50: 249–258. Reprinted in Lewis 1999.

Machamer, P., L. Darden, and C. Craver. 2000. "Thinking about Mechanisms." *Philosophy of Science* 67: 1–25.

MacKinnon, R. 2003. "Potassium Channels." *FEBS Letters* 555: 62–65.

McGinn, C. 1991. *The Problem of Consciousness.* Oxford: Basil Blackwell.

Melzack, R., and W. S. Torgerson. 1971. "On the Language of Pain." *Anesthesiology* 34: 50–59.

Melzack, R., and P. D. Wall. 1983. *The Challenge of Pain.* New York: Basic Books.

Merskey H., and N. Bogduk. 1994. *Classification of Chronic Pain. Descriptions and Chronic Pain Syndromes and Definitions of Pain Terms.* 2d ed. Seattle: IASP Press.

Merskey, H., et al. 1986. "Pain Terms: A Current List with Definitions and Notes on Usage." *Pain* 3 (suppl.): S217.

Millan, M. J. 1999. "The Induction of Pain: An Integrative Review." *Progress in Neurobiology* 57: 1–164.

Nagasako, E. M., A. N. Oaklander, and R. H. Dworkin. 2003. "Congenital Insensitivity to Pain: An Update." *Pain* 101: 213–219.

Nagel, T. 1974. "What Is It Like to Be a Bat?" *Philosophical Review* 83(4): 435–450.

Neumann, N. R., T. P. Doubell, T. Leslie, and C. J. Woolf. 1996. "Inflammatory Pain Hypersensitivity Mediated by Phenotypic Switch in Myelinated Sensory Neurons." *Nature* 384: 360–364.

Penfield, W., and H. Jasper. 1954. *Epilepsy and the Functional Anatomy of the Human Brain.* Boston: Little, Brown.

Ploner, M., H. J. Freund, and A. Schnitzler. 1999. "Pain Affect without Pain Sensation in a Patient with a Postcentral Lesion." *Pain* 81: 211–214.

Ploner, M., F. Schmitz, H. J. Freund, and A. Schnitzler. 2000. "Differential Organization of Touch and Pain in Human Primary Somatosensory Cortex." *Journal of Neurophysiology* 83: 1770–1776.

Polger, T. 2004. *Natural Minds.* Cambridge, Mass.: MIT Press.

Polger, T., and O. Flanagan. 1999. "Natural Answers to Natural Questions." In *Where Biology Meets Psychology*, ed. Valerie Hardcastle. Cambridge, Mass.: MIT Press.

Polger, T., and R. Skipper. Unpublished manuscript. "Naturalism, Explanation, and Identity."

Price, D. D. 1996. "Selective Activation of A-Delta and C Nociceptive Afferents by Different Parameters of Nociceptive Heat Stimulation: A Tool for Analysis of Central Pain Mechanisms." *Pain* 68: 1–3.

————. 1999. *Psychological Mechanisms of Pain and Analgesia.* Seattle: IASP Press.

————. 2000. "Psychological and Neural Mechanisms of the Affective Dimension of Pain." *Science* 288: 1769–1772.

Price, D. D., and R. Dubner. 1977. "Neurons that Subserve Sensory-Discriminative Aspects of Pain." *Pain* 3: 307–338.

Price, D. D., and S. W. Harkins. 1987. "The Combined Use of Experimental Pain and Visual Analogue Scales in Providing Standardized Measurement of Clinical Pain." *Clinical Journal of Pain* 3: 1–8.

Rainville, P., B. Carrier, R. K. Hofbauer, M. C. Bushnell, and G. H. Duncan. 1999. "Dissociation of Sensory and Affective Dimensions of Pain Using Hypnotic Induction." *Pain* 82: 159–171.

Rainville, P., G. H. Duncan, D. D. Price, B. Carrier, and M. C. Bushnell. 1997. "Pain Affect Encoded in Human Anterior Cingulated but Not Somatosensory Cortex." *Science* 277: 968–971.

Reichling, D. B., and J. D. Levine. 1999. "The Primary Afferent Nociceptor as a Pattern Generator." *Pain* 69 (suppl.): S103–S109.

Scholz J., and C. J. Woolf. 2002. "Can We Conquer Pain?" *Nature Neuroscience* 5 (suppl.): 1062–1067.

Shoemaker, S. 1981. "Some Varieties of Functionalism." *Philosophical Topics* 12(1): 83–118.

Sidelle, Alan. 1992. "Identity and Identity-like." *Philosophical Topics* 20(1): 269–292.

Sufka, K. J. 2000. "Chronic Pain Explained." *Brain and Mind* 1: 155–179.

Sufka, K. J., and M. P. Lynch. 2000. "Sensations and Pain Processes." *Philosophical Psychology* 13: 299–311.

Sufka, K. J., and D. D. Price. 2002. "Gate Control Theory Reconsidered." *Brain and Mind* 3: 277–290.

Sufka, K. J., and D. D. Turner. 2005. "An Evolutionary Account of Chronic Pain: Integrating the Natural Method in Evolutionary Psychology." *Philosophical Psychology* 18: 243–257.

Van Gulick, R. 1993. "Understanding the Phenomenal Mind: Are We All Just Armadillos?" In Block, Flanagan, and Güzeldere 1997.

Wade, J. B., L. M. Dougherty, C. R. Archer, and D. D. Price. 1996. "Assessing the Stages of Pain Processing: A Multivariate Approach." *Pain* 68: 157–167.

Willis, W. D. 1992. *Hyperalgesia and Allodynia.* New York: Raven Press.

Woolf, C. J., and M. W. Salter. 2000. "Neuronal Plasticity: Increasing the Gain in Pain." *Science* 288: 1765–1768.

20 Deciphering Animal Pain

Colin Allen, Perry N. Fuchs, Adam Shriver, and Hilary D. Wilson

1 Introduction: Pain by Analogy

In this paper we[1] assess the potential for research on nonhuman animals to address questions about the phenomenology of painful experiences. Nociception, the basic capacity for sensing noxious stimuli, is widespread in the animal kingdom. Even relatively primitive animals such as leeches and sea slugs possess nociceptors, neurons that are functionally specialized for sensing noxious stimuli (Walters 1996). Vertebrate spinal cords play a sophisticated role in processing and modulating nociceptive signals, providing direct control of some motor responses to noxious stimuli, and a basic capacity for Pavlovian and instrumental conditioning (Grau et al. 1990; Grau 2002). Higher brain systems provide additional layers of association, top-down control, and cognition. In humans, at least, these higher brain systems also give rise to the conscious experiences that are characteristic of pain. What can be said about the experiences of other animals who possess nervous systems that are similar but not identical to humans?

Much of our knowledge of the mechanisms underlying pain in humans is derived from research on nonhuman animals. Yet the extent to which animals (e.g., rats) provide a model of conscious pain experiences remains a matter of uncertainty among most pain researchers, and controversy among others. Experimenters who apply treatments such as capsaicin, heat, and electric shocks to rats frequently self-apply the stimuli to reassure themselves that the degree of pain that is caused is tolerable (e.g., Grau 2002). The supposition behind these self-tests is that the phenomenology of pain in rats is similar to the phenomenology of pain in humans. But this supposition is regarded by most as relatively insecure, founded only on broad analogies between the anatomy, physiology, and behavior of humans and other animals.

Analogical arguments are widely exploited in the animal welfare literature (see Allen 2004 for discussion). Anatomical similarities including the presence of nociceptors connected to a central nervous system, physiological similarities including the existence of endogenous opioids, and behavioral similarities such as withdrawal, vocalization, and "nursing" responses to injury, have all been cited to support the view that many nonhuman animals suffer from pain and thus deserve moral consideration and legal protection. Some of the authors working in this area acknowledge that there is room for doubt about the force of the argument by analogy, but they apply the precautionary principle that it is better to err on the side of too much protection rather than too little. Other authors, however, place considerably more weight on the analogy argument, considering it to be firmly established scientifically that there are no significant differences between humans and many other animals in the capacity to feel pain. This conclusion is often bolstered by appeal to evolutionary continuity between the species.

Lists of shared features may be adequate to reassure those who are predisposed to accept the conclusion that there are no significant differences between pain experiences in humans and (at least some) other animals. But this approach is rather weak in the face of certain skeptical challenges that are commonly raised. Such challenges have at least two independent sources. Some skeptics are doctrinally convinced that conscious experiences of any sort are beyond the reach of empirical investigation. On this view, no scientific procedures can tell us about something that is (allegedly) essentially subjective and private. Other skeptics allow that empirical results are relevant, but they are convinced that there exist significant disanalogies between humans and other animals, which make it unlikely that the experiences of nonhuman animals are anything like the conscious experiences of humans. The capacity for speech and possession of "theory of mind" are two features of human cognition that are frequently presumed to mark a distinction between human consciousness and that of other animals (e.g., Carruthers 1996, 2000).

The standard analogy arguments that have been advanced by many philosophers are not sufficient to overcome either of these two variants of skepticism. This is because the providers of these lists of similarities generally do not provide any theoretical reasons for connecting them to attributions of conscious pain. Specialized nociceptors are found in such relatively primitive organisms as sea slugs and leeches, and as such do not provide strong grounds for attributing conscious pain to these organisms. Opioid systems are also widespread among animals. Many withdrawal responses, and even some forms of learning about noxious stimuli, can be accomplished by spinal cords without mediation by higher brain systems (Grau 2002). If items on the list do not individually entail conscious experience of pain, it is not clear why satisfying multiple criteria should add up to conscious experience. Analogy arguments are vulnerable because for all the similarities between humans and other nonhuman animals, there are dissimilarities that can be used to deny the inference to conscious pain in nonhumans (cf. Nelkin 1986, p. 137). While human brains may be similar to animal brains at the level of gross anatomy and physiology, more fine-grained analysis reveals numerous differences. It is also open to critics to point out the many ways in which human behavior is not identical to the behavior of other animal species. Consequently, without an adequate framework for understanding the connection between the observed similarities and conscious pain, analogy arguments remain essentially weak.

In this paper we pursue the suggestion made by Allen (2004) that a functional understanding of pain in the context of learning would provide a framework for assessing comparisons of anatomy, physiology, and behavior. Fuchs's work (described below) on the sensory and emotional aspects of pain experiences in rats provides a context in which the functional roles of different components of the phenomenology of pain could be investigated with respect to anatomy (particularly the role of the anterior cingulate cortex), physiology (the effect of opioid substances), and behavior (avoidance of aversive contexts). While the development of such a framework may not ultimately convince either type of skeptic, it may help to preempt skeptical and antiskeptical arguments that are based on overly simplistic ideas about the functions of pain. Our aim is to chart a middle course between the excessively skeptical view that animal pain cannot be studied empirically and the overly credulous view that scientific investigation has already revealed that other animals (other mammals, at least) feel pain much as we do. We do not intend to show that rats experience pain consciously in this paper, but we do intend to outline an empirical research program that allows sophisticated comparisons to be drawn between the pain experiences of humans and those of other animals.

2 Functional Anatomy: The Role of the Anterior Cingulate Cortex

The link between pain and learning appears to go, at least in mammals, through the anterior cingulate cortex (ACC). Before going into detail about the role of the ACC in learning, it is useful to describe the phenomenological properties commonly associated with ACC activity in humans. The "classical" view of pain, as described elsewhere (Hardcastle 1997; Melzack and Casey 1968), generally holds that there are at least two key pathways involved in the experience of pain in humans. One pathway is responsible for the sensory or descriptive component of pain sensations (such as the intensity of the pain, the location of the pain, whether the pain is "sharp" or "dull," etc.), and the other is responsible for the affective or motivational dimension of pain (i.e., how much one "minds" the pain or how unpleasant the pain is). Activity in the ACC is generally correlated with the latter of these two pathways and, in fact, research on the ACC has played a central role in establishing the existence of different pathways.

One of the earliest experimental indications that there may be multiple pathways involved in the processing of pain came from reports of patients who were administered morphine (Price et al. 1985). Patients who were given morphine while in pain stated that they still felt the pain but that it did not bother them as much. Interestingly, the subjects of cingulotomies (a procedure that ablates the ACC) reported the same thing (Foltz and White 1962; Gybels and Sweet 1989). These results seem to suggest a dissociation between sensations of pain and the unpleasantness that usually accompanies such sensations.

Several other recent discoveries have supported the dual pathway hypothesis. A study by Rainville et al. (1997) used hypnotic suggestion to increase or decrease the amount of unpleasantness subjects felt while holding their hands under 47°C water. Subjects rated the intensity of the pain sensation the same regardless of whether or not they underwent hypnotic suggestion, but their ratings of the unpleasantness of the sensation varied accordingly to the suggestion (i.e., they ranked pains as "more unpleasant" when given a suggestion that the pain would be more unpleasant, and "less unpleasant" when given the suggestion that it would be more unpleasant). Positron emission tomography conducted during the study revealed that the anterior cingulate cortex but not other areas (such as the somatosensory cortex) was increasingly active relative to "more unpleasant" rankings given by subjects and decreasingly active during "less unpleasant" rankings.

A later experiment by Rainville et al. (1999) used hypnotic suggestion to increase or decrease the intensity of the pain intensity in the same experimental setup. In this case, both the ACC and the somatosensory cortex were more active during suggestions of increased intensity and less active during suggestions of decreased intensity. This, according to Price (2002), suggests that there is a "direction of causation" that flows from the sensory qualities of pain to the perceived unpleasantness of the pain. Price accepts that there are dual pathways, but he rejects the view that consciously felt affect is processed in a parallel system that is entirely independent from the sensory processing stream. He favors a model in which the pain affect processing stream receives input from the pain sensation stream. Such a model allows for a dissociation between the affective and sensory aspects of pain, and Price provides evidence based on the psychophysical relationship between pain intensity and heat stimuli and the psychophysical relationship between pain unpleasantness and heat stimuli. Both are positively accelerating power functions, but, according to Price, "pain unpleasantness ratings are less than those of pain sensation intensity ratings in response to 45–50°C temperatures" and consequently the exponents in the power functions must be different. The fact that pain sensation and pain unpleasantness bear different relationships to the pain-causing stimuli further supports the idea that there are at least two functionally distinct elements of the experience of pain.

If, as the evidence suggests, the ACC does play a central role in the experience of the unpleasantness of pain in humans, then it is clear that more detailed knowledge of the ACC will have important implications for both ethical and practical questions regarding the treatment of pain in humans. Likewise, because unpleasantness is such an important part of how humans conceptualize our own experience of pain, it also seems likely that ACC research on nonhuman animals could provide important information in relation to the question of whether animals consciously feel pain. However, as was indicated earlier, it is not enough to simply point out the presence or absence of the ACC in other animals as the deciding factor in an argument by analogy or disanalogy. It is important to place its function in an explanatory framework, and as such there are a couple more features of the ACC that are worth mentioning before moving on to a discussion of how animal research has and can aid our understanding of pain.

In addition to its role in the affective dimension of pain, recent research has suggested that the ACC plays a role in learning to avoid noxious stimuli. One of the ways this takes place is in the detection of errors. A study by Kiehl, Liddle, and Hopfinger (2000) found that the rostral region of the ACC, the same region that is active during pain, was activated during a task only after the subject made inappropriate responses and not when the subject made appropriate responses. (This connection to endogenous error signals is especially interesting in light of the suggestion by Allen and Bekoff [1997, ch. 8] that autonomous error correction could provide a significant source of evidence about animal consciousness.)

Eisenberger and Lieberman (2004) found that the dorsal ACC was active for subjects in situations of perceived social exclusion. And Carter et al. (1998) have shown that the conditions with strong competition between different response drives caused strong activation in the ACC. Though these results cover a wide range of situations, all of them involve circumstances in which the subject would potentially benefit by learning to avoid specific behaviors in the future. Thus, at least a large part of the function of the ACC appears to be critically linked with the ability of the organism to learn in various contexts.

Allen (2004) identifies learning as a key area for investigating the functions of conscious pain experiences. It is important, however, to realize that various forms of learning about noxious stimuli may take place in the spinal cord, with no implications for consciousness. Some kinds of learning, however, seem to be closely correlated to conscious awareness, for example, trace conditioning as opposed to delay conditioning (see Clark and Squire 1998) and to depend on higher brain mechanisms, for example, more sophisticated kinds of operant learning as opposed to simpler forms of instrumental learning (Grau 2000, 2002; Grau, Barstow, and Joynes 1998). It is these that are of most interest for our present purposes (see Allen 2004 for discussion).

While the ACC appears to be of considerable interest in understanding the connection of pain processing to higher cognitive functions, it is important not to place too much importance on any single chunk of neural tissue. Because the "classical" account pictures the pain system as having two tracks, comprised of a sensory-discriminative system that projects to somatosensory cortex and an affective-motivational system that projects to frontal lobes, some philosophers have been tempted to make bold claims about the functional isolation of these tracks in specific parts of the brain. Hardcastle, for example, writes: "The frontal lobe (and its connections) process our actual suffering" (1997, p. 391). But Coghill warns that while two subsystems of the classical account can be experimentally separated, "current views that sensory-discriminative and affective-motivational dimensions of pain are processed in parallel by distinct neural structures cannot account for a number of aspects of pain" (1999, p. 67). Coghill discusses brain-imaging studies that show that pain is processed by multiple regions in a highly distributed system, that there is no single brain region whose destruction completely abolishes the experience of pain, that sensory processing occurs in areas classically associated with affect, and affect itself is at least a three-layer process with the prefrontal cortex being mainly important for the latest stage of processing.

The ACC is thus just one part of a very complex system. Its particular interest to us lies in its functional significance for linking painful experiences in humans and other mammals to learning and motivation, but it is important to note that these functions may belong to other structures in nonmammalian animals with independently evolved neural architectures. In the remainder of this paper, however, we focus on organisms that do possess ACCs, namely rodents, and we seek to show how the investigation of functional connections between nociception, motivation, and learning in rodents has the potential to advance our understanding of the nature of painful experiences in nonhuman animals.

3   Testing the Functions of Pain

We are now in a position to discuss what research on certain nonhuman animals has taught us about pain. An examination of the literature reveals that the classical hypothesis of separate, interacting systems that process sensory-discriminative and emotional-motivational determinants of pain has not been rigorously examined in animals. A major factor for lack of direct testing is likely related to limitations in the commonly used behavioral testing methods in rodent models of nociception. For instance, following nerve injury, animals display changes in withdrawal behaviors in response to noxious stimulation that are thought to reflect clinical conditions of neuropathic pain. Quantification of these stimulus-evoked nociceptive behaviors is based almost exclusively on withdrawal thresholds—that is, measures of the strength or duration of the stimulus needed to provoke a response. A lowered threshold for responding (plotted graphically as a leftward shift in the stimulus–response function) following peripheral nerve injury or the induction of an inflammatory condition is thought to be the analogue for clinical symptoms of allodynia (displaced or inappropriate pain) and hyperalgesia (oversensitivity to pain). Treatments that have an antinociceptive effect are represented graphically as an attenuation of the leftward shift in the stimulus–response function. Manipulations (e.g., pharmacological) that fail to attenuate the leftward shift in the stimulus–response function are assumed to reflect a lack of antinociception. The withdrawal method has been useful for understanding certain aspects of the neural system and allowing for a fast initial screen for potentially useful analgesic compounds. Most of the studies that have examined supraspinal focal brain stimulation or microinjection of drugs into discrete nuclei have also utilized such measurements of reflexive behavioral response to acute noxious stimulation. Indeed, activation of various subcortical, brainstem, and spinal cord systems produce antinociception as revealed by an increase in the threshold or latency to respond to noxious stimulation (Basbaum and Fields 1978).

There is little doubt that the knowledge gained from behavioral testing based on withdrawal responses is of extreme importance. Nonetheless, the methodology has significant limitations (Fuchs 2000). First, recall that reflexive measures such as tail flick and leg withdrawal show the same response profiles in spinally transected rats as intact rats, and that various kinds of learning also take place in spinalized rats (Grau 2000, 2002; Grau, Barstow, and Joynes 1998). The absence of any connection to the brain in the spinalized condition indicates that any apparent hyper- or hypoalgesia need not reflect a change in conscious experience even in intact animals. Second, King et al. (1996) showed that manipulations in rats (exposure to either heat or shock) that increased tail withdrawal latencies (apparent hypoalgesia) simultaneously lowered the latency to vocalize (apparent hyperalgesia). Thus, two different reflexive measures yield results that point in opposite directions if both are taken as simple measures of the intensity of experience. These results caution against simple reliance on reflexive measures when trying to determine the aversiveness of a stimulus experienced by an animal. Hence, analogical arguments about consciousness of pain in animals that are based on similarities between withdrawal behaviors or vocalizations elicited by noxious stimuli are also precarious. Specifically, the reflexive behavioral paradigms do not provide definitive information about the aversive and unpleasant qualities of a persistent pain condition (Fuchs 2000).

It has proven difficult to dissociate the affective-motivational and sensory-discriminative components of pain processing in rodents using reflexive behavioral measures alone. It becomes even more problematic when it is assumed that decreases in stimulus-evoked behavior, following manipulations to neural structures thought to be involved in emotional aspects of behavior, reflect changes in the affective-motivational dimension of pain processing (Fuchs and Melzack 1995; Donahue, LaGraize, and Fuchs 2001). This assumption has not been rigorously tested in animal models. To directly test the role of different neural substrates on the emotional aspect of nociceptive behavior, additional measures are needed.

Additional measures based on operant principles have been recently developed and are being used to evaluate the sensory and discriminative aspects separate from the motivational and affective components of pain, and to explore potential higher order brain processing in animals (Johansen, Fields, and Manning 2001; LaBuda and Fuchs 2000a; Vierck et al. 2002). One such approach is the use of operant techniques that measure escape and avoidance of animals (LaBuda and Fuchs 2000a; Mauderli, Acosta-Rua, and Vierck 2000; Yeomans, Cooper, and Vierck 1995). The assumption is that flexible exercise of choice between alternative behaviors to achieve escape or avoidance of a noxious stimulus is a clear indication that the stimulus's aversiveness is being registered at higher cognitive levels (see also Grau, Barstow, and Joynes 1998; Grau 2002). A behavioral test method based on this assumption has been recently reported (LaBuda and Fuchs 2000a). In this paradigm, the animals are allowed to "choose" where the mechanical stimulus is applied—an injured or uninjured region of the body. The basic paradigm requires the use of a chamber that is equally divided into a dark side and a light side. Under normal conditions, rats naturally prefer the dark area of the environment. During behavioral testing, a mechanical stimulus (476 mN von Frey monofilament—a flexible plastic wire used to simulate a pinprick of a fixed intensity) is applied to an injured hindpaw when the animal is within the dark area or to the noninjured contralateral hindpaw when the animal is within the light area of the chamber. The amount of time that animals spend within the light side of the chamber is recorded. Control animals spend about 20 to 40 percent of the time in the light side of the chamber. However, the animals that have an injury demonstrate escape–avoidance behavior toward the dark side of the chamber and a shift in preference toward the light side of the chamber (see LaBuda and Fuchs 2000a for details of controls and quantitative results). The escape–avoidance behavior supports the notion that mechanical stimulation of the treated paw during a neuropathic or inflammatory pain condition is aversive and when given a choice, animals will perform purposeful behavior to minimize stimulation of the afflicted body region.

To further test the validity of the escape–avoidance test paradigm the effect of two commonly used analgesic agents, morphine and gabapentin have been tested (LaBuda and Fuchs 2000b). The rationale for these series of experiments is that analgesic compounds that decrease nociception should also decrease the aversive nature of stimulating an injured body region. Decreased mechanical sensitivity was measured using the traditional stimulus-evoked threshold method with von Frey monofilaments following nerve injury immediately prior to and then thirty minutes following the administration of different doses of morphine (1 mg/kg or 10 mg/kg) or gabapentin (30 mg/kg or 90 mg/kg). Compared to injured animal controls treated with the drug-delivery vehicle only, both morphine and gabapentin attenuated the enhanced response to mechanical stimuli in a dose-dependent manner (i.e., caused a leftward shift in the stimulus–response function). Of primary interest, however, is the additional finding that both morphine and gabapentin also attenuated the escape–avoidance behavior. Morphine (1 mg/kg) and gabapentin (30 mg/kg and 90 mg/kg) treated nerve-injured animals displayed less time in the light area of the chamber relative to vehicle-treated nerve-injured control animals. The findings indicate that analgesic compounds administered at a dose sufficient to decrease the sensory component of pain also decreased the affective component of the pain condition. In other words, the sensory experience was less intense, so the affective response was also reduced, in accordance with the "direction of causation" mentioned above.

Psychophysical evidence from humans indicates that morphine can selectively modulate the affective dimension of pain without significantly altering the sensory aspect (Price et al. 1985). As we noted previously, patients receiving morphine for pain control sometimes report that the pain is still present, but that it does not bother them as much. It is exactly this aspect of the experience that cannot be reliably measured with stimulus-evoked withdrawal tests but can be quantified using additional behavioral test paradigms. When morphine is administered at a lower dose (0.5 mg/kg), there appears to be a selective attenuation of pain affect with no alteration in mechanical paw withdrawal threshold (LaGraize et al., in press), a finding that is analogous to the clinical reports of morphine in patients with chronic pain conditions. Based on these results, it is concluded that manipulations that decrease hyperalgesia also attenuate escape–avoidance behavior. In other words, the behavioral test paradigm reveals that if the experimental condition is less "painful," then mechanical stimulation of the afflicted region is less aversive.

These behavioral paradigms can also reveal aspects of nociceptive processing that have not previously been accessible in animals. For example, the hypothesis that limbic system structures are involved in the aversiveness of stimulus-evoked pain has also been examined using this test paradigm. As indicated above, the anterior cingulate cortex (ACC) is a critical structure in processing pain in mammals, especially with regard to affect and motivation. This conclusion has been based on three primary lines of evidence. First, electrophysiological studies have revealed nociceptive neurons in the rat and rabbit ACC that have very large receptive fields, consistent with the notion that this region is involved in encoding intensity rather than spatial discrimination (Sikes and Vogt 1992). Second, as mentioned in our previous section, human imaging studies have correlated the unpleasantness of pain with ACC activation (Rainville et al. 1997; Tölle et al. 1999). Third, also mentioned above, cingulotomy provides significant pain relief in humans (Ballintine et al. 1967; Pillay and Hassenbusch 1992; Wong et al. 1997). The effectiveness of this procedure has been attributed to the interruption of the cingulate gyrus concerned with the emotional component of pain (Martinez et al. 1975). Indeed, patients with ACC lesions report that the intensity of their pain remains, but that it is less bothersome (Corkin and Hebben 1981).

The escape–avoidance paradigm has been used to determine if the same pattern of behavioral results could be obtained from the rat. It was predicted that lesions of the rat ACC would decrease the aversiveness of pain without altering the mechanical hyperalgesia. Indeed, results from Fuchs's lab confirm this hypothesis (LaGraize et al. 2004). First, bilateral ACC electrolytic lesions failed to reverse the decrease in mechanical withdrawal threshold following nerve injury. However, the ACC lesion caused a significant attenuation in the shift from the dark side of the test chamber relative to the control group which was not subject to nerve injury. These results are a reflection of the clinical reports of patients who indicate following ACC lesions the intensity of the pain remains, but that it is less bothersome. Although alternative explanations of the data have yet to be excluded (i.e., ACC lesions interrupt memory processes that are critical for the acquisition of the association), this finding provides preliminary evidence to directly support the original proposition by Melzack and Casey (1968) that limbic system structures have different modulatory effects on sensory-discriminative and emotional-motivational dimensions of pain.

In summary, operant behavioral techniques can be used as a method to explore the complex nature of pain in rat models of nociception. Unlike traditional threshold measurement techniques, the general paradigm outlined above provides animals the opportunity to perform purposeful behavior to escape or avoid mechanical stimulation by a von Frey monofilament to the hyperalgesic body region. The assumption is that escape and avoidance behavior indicate that the aversiveness of the mechanical stimulus is playing a role in cognitive decision-making processes of the animal. This is analogous to a patient who might demonstrate guarding behavior and avoidance of environmental conditions that might exacerbate the severe discomfort associated with causalgia (a burning sensation along a peripheral nerve) or other clinical pain syndromes. Although the present approach is by all accounts not the only paradigm that can be used (i.e., Johansen, Fields, and Manning 2001; Mauderli, Acosta-Rua, and Vierck 2000), it is a very important initial step in developing operant behavioral test paradigms to unravel the complex processing of nociceptive information in supraspinal neural substrates.

4 Future Directions and Implications

The move from simple behavioral measures (stimulus–response) to more sophisticated operant behavioral techniques suggests additional methods for investigating the roles that different dimensions of painful experiences might play in higher order forms of learning. For example, we are interested in the question of whether long-term conditioning is differentially affected by blocking the sensory and affective components of pain processing. Does treatment with morphine affect the ability of rats to learn about noxious stimuli? Would treated rats fail to learn associations between contextual cues and noxious stimuli, or is sensory awareness sufficient? Would the effect vary for different types of pain conditions (i.e., is sensory awareness sufficient for acute conditions but not chronic conditions)? Additionally, if given the choice, would rats learn to self-administer sensory and affective pain relief differentially? A second topic of interest is motivational drives. Do animals experiencing food deprivation and pain simultaneously choose to eat, or does the pain drive supersede the hunger drive? Is their choice differentially affected by blocking sensory and affective components of pain processing? Furthermore, does loss of pain affect correlate with loss of affect in other behaviors (i.e., mating, predator–prey, and maternal behaviors)? Do losses of pain affect versus sensory pain experience differentially modify these behaviors? The ability to investigate such questions opens the door to much more detailed analyses of the functions of these different aspects of painful experiences. It is also worth noting that the utility of these measures depends to a large degree on animals exercising choices in conditions where they are not in so much pain as to be rendered immobile or dysfunctional. While the deliberate infliction of pain on another organism is always a matter of moral concern, the experiments we propose generally involve a degree of pain that would be consistent with good overall welfare.

We began this paper with questions about the phenomenology of animal pain, to which we now return. We believe that animal pain studies are on the brink of some exciting advances that will make it possible to describe more precisely the roles played by different aspects of painful experiences. An understanding of how the unpleasantness of pain connects to the complex cognitive capacities of organisms would provide an explanatory framework that would allow behavioral evidence from a variety of species to be assessed. Of course, it is open to the more ideological kind of skeptic that we identified in the introduction to maintain that none of this tells us anything about the conscious nature of animal experiences, because all the anatomical, physiological, and behavioral evidence in the world is compatible with the complete absence of conscious experience. But this view applies just as much to our ability to investigate human experiences of pain as it does to nonhuman animals, and as such provides no special barrier to our understanding of animal pain. The other kind of skeptic we identified thinks that outstanding differences in higher cognitive abilities such as language processing or theory of mind abilities are the crucial elements for understanding the nature of conscious experience. They may be right, but no empirical method has been provided for testing the hypothesis that consciousness serves those functions. In contrast, the techniques we have described make it possible to test an alternative class of hypotheses linking the phenomenology of painful experiences to specific motivational and learning functions. By manipulating dimensions of the painful experience we stand to gain a more detailed view of the complex relationship between behavior, mechanism, and experience, which in turn strengthens the basis for analogical comparisons of animals and humans.

Both philosophers and scientists have, until now, tended to focus only on the most basic responses to painful stimuli, such as withdrawal and nursing behaviors. This has fostered some rather simplistic views about the functions of pain sensations (see Allen 2004 for criticism) that have, in turn, supported a polarized debate. On the one side are those who point to withdrawal and nursing behaviors in nonhuman animals as evidence that their pain systems are essentially no different from the human pain system. On the other side are those who point out that these simple responses can be implemented with simple mechanisms that provide little confidence for the attribution of conscious experiences. If one limits oneself to these kinds of behaviors, it is indeed hard to think of empirical studies that would depolarize this debate, and philosophers, in particular, have an unfortunate tendency to think that if they can't imagine any relevant experiments to address a particular question, then there can't be any. With this paper we hope to have shown that there is much potential for investigating functional aspects of the experience of pain, providing, we hope, a fertile middle ground in which sophisticated comparisons of different species can grow.

Acknowledgments

We thank Murat Aydede, Verne Cox, Jim Grau, Gary Lucas, and Bill Timberlake for comments.

Notes

1. Address for correspondence: Colin Allen, Department of History and Philosophy of Science, 1011 East Third Street, Goodbody Hall 130, Indiana University, Bloomington, IN 47405; phone: (812) 855-8916; fax: (812) 855-3631; email: colallen@indiana.edu.

References

Allen, C. 2004. "Animal Pain." *Noûs* 38: 617–643.

Allen, C., and M. Bekoff. 1997. *Species of Mind: The Philosophy and Biology of Cognitive Ethology*. Cambridge, Mass.: MIT Press.

Ballintine, H. T., W. L. Cassidy, N. B. Flanagan, and R. Marino. 1967. "Stereotaxic Anterior Cingulotomy for Neuropsychiatric Illness and Intractable Pain." *Journal of Neurosurgery* 26: 488–495.

Basbaum, A. I., and H. L. Fields. 1978. "Endogenous Pain Control Mechanisms: Review and Hypothesis." *Annals of Neurology* 4: 451–462.

Carruthers, P. 1996. *Language, Thought, and Consciousness*. Cambridge: Cambridge University Press.

———. 2000. *Phenomenal Consciousness*. Cambridge: Cambridge University Press.

Carter, C., T. Braver, D. Barch, M. Botvinick, D. Noll, and J. Cohen. 1998. "Anterior Cingulate Cortex, Error Detection, and the Online Monitoring of Performance." *Science* 280: 747–749.

Clark, R. E., and L. R. Squire. 1998. "Classical Conditioning and Brain Systems: The Role of Awareness." *Science* 280: 77–81.

Coghill, R. C. 1999. "Brain Mechanisms Supporting the Pain Experience." Pp. 67–76, in *Pain 1999—An Updated Review*, ed. M. Max. Seattle: IASP Press.

Corkin, S., and N. Hebben. 1981. "Subjective Estimates of Chronic Pain before and after Psychosurgery or Treatment in a Pain Unit." *Pain* 1 (suppl.): S150.

Donahue, R. R., S. C. LaGraize, and P. N. Fuchs. 2001. "Electrolytic Lesion of the Anterior Cingulate Cortex Decreases Inflammatory, but Not Neuropathic Nociceptive Behavior in Rats." *Brain Research* 897: 131138.

Eisenberger, N. I., and M. D. Lieberman. 2004. "Why Rejection Hurts: A Common Neural Alarm System for Physical and Social Pain." *Trends in Cognitive Sciences* 8: 294–300.

Foltz, E. L., and L. E. White. 1962. "Pain 'Relief' by Frontal Cingulotomy." *Journal of Neurosurgery* 19: 89–100.

Fuchs, P. N. 2000. "Beyond Reflexive Measures to Examine Higher Order Pain Processing in Rats." *Pain Research and Management* 5: 215–219.

Fuchs, P. N., and R. Melzack. 1995. "Analgesia Induced by Morphine Microinjection into the Lateral Hypothalamus of the Rat." *Experimental Neurology* 134: 277–280.

Grau, J. W. 2000. "Instrumental Conditioning." Pp. 767–769, in *The Corsini Encyclopedia of Psychology and Behavioral Science*, 3rd edition, ed. W. E. Craighead and Nemeroff, C. B. New York: John Wiley and Sons.

———. 2002. "Learning and Memory without a Brain." Pp. 77–88, in *The Cognitive Animal*. ed. M. Bekoff, C. Allen, and Burghardt, G. M. Cambridge, Mass.: MIT Press.

Grau, J. W., D. G. Barstow, and R. L. Joynes. 1998. "Instrumental Learning within the Spinal Cord: I. Behavioral Properties." *Behavioral Neuroscience* 112: 1366–1386.

Grau, J. W., J. A. Salinas, P. A. Illich, and M. W. Meagher. 1990. "Associative Learning and Memory for an Antinociceptive Response in the Spinalized Rat." *Behavioral Neuroscience* 104: 489–494.

Gybels, J. M., and W. H. Sweet. 1989. *Neurosurgical Treatment of Persistent Pain*. Basel: Karger.

Hardcastle, V. G. 1997. "When a Pain Is Not." *Journal of Philosophy* 94: 381–409.

Johansen, J. P., H. L. Fields, and B. H. Manning. 2001. "The Affective Component of Pain in Rodents: Direct Evidence for a Contribution of the Anterior Cingulate Cortex." *Proceedings of the National Academy of Sciences of the USA* 98: 8077–8082.

Kiehl, K. A., P. F. Liddle, and J. B. Hopfinger. 2000. "Error Processing and the Rostral Anterior Cingulate: an Event-Related fMRI Study." *Psychophysiology* 37: 216–223.

King, T. E., R. L. Joynes, M. W. Meagher, and J. W. Grau. 1996. "The Impact of Shock on Pain Reactivity II: Evidence for Enhanced Pain." *Journal of Experimental Psychology: Animal Behavior Processes* 22: 265–278.

LaBuda, C. J., and P. N. Fuchs. 2000a. "A Behavioral Test Paradigm to Measure the Aversive Quality of Inflammatory and Neuropathic Pain in Rats." *Experimental Neurology* 163: 490–494.

———. 2000b. "Morphine and Gabapentin Decrease Mechanical Hyperalgesia and Escape/ Avoidance Behavior in a Rat Model of Neuropathic Pain." *Neuroscience Letters* 290: 137–140.

LaGraize, S. C., C. J. LaBuda, M. A. Rutledge, R. L. Jackson, and P. N. Fuchs. 2004. "Differential Effect of Anterior Cingulate Cortex Lesion on Mechanical Hyperalgesia and Escape/Avoidance Behavior in an Animal Model of Neuropathic Pain." *Experimental Neurology* 188: 139–148.

LaGraize, S. C., J. Borzan, Y. B. Peng, and P. N. Fuchs. In press. "Selective Regulation of Pain Affect Following Activation of the Opioid Anterior Cingulate Cortex System." *Experimental Neurology*.

Martinez, S. N., C. Bertrand, P. Molina, and J. M. Perez-Colon. 1975. "Alteration of Pain Perception by Stereotaxic Lesions of Fronto-Thalamic Pathways." *Confinia Neurologica* 37: 113.

Mauderli, A. P., A. Acosta-Rua, and C. J. Vierck. 2000. "An Operant Assay of Thermal Pain Conscious, Unrestrained Rats." *Journal of Neuroscience Methods* 97: 19–29.

Melzack, R., and K. L. Casey. 1968. "Sensory, Motivational, and Central-Control Determinants of Pain." Pp. 423–443, in *The Skin Senses*, ed. D. R. Kenshalo. Springfield, Ill.: Charles C. Thomas.

Nelkin, N. 1986. "Pains and Pain Sensations." *Journal of Philosophy* 83: 129–147.

Pillay, P. K., and S. J. Hassenbusch. 1992. "Bilateral MRI-Guided Stereotactic Cingulotomy for Intractable Pain." *Stereotactic and Functional Neurosurgery* 59: 33–38.

Price, D. D. 2002. "Central Neural Mechanisms That Interrelate Sensory and Affective Dimensions of Pain." *Molecular Interventions* 2: 392–403.

Price, D. D., A. Von Der Gruen, J. Miller, A. Rafii, and C. A. Price. 1985. "Psychophysical Analysis of Morphine Analgesia." *Pain* 22: 261–269.

Rainville, P., B. Carrier, R. K. Hofbauer, M. C. Bushnell, and G. H. Duncan. 1999. "Dissociation of Sensory and Affective Dimensions of Pain Using Hypnotic Modulation." *Pain* 82: 159–171.

Rainville, P., G. H. Duncan, D. D. Price, B. Carrier, and M. C. Bushnell. 1997. "Pain Affect Encoded in Human Anterior Cingulate but Not Somatosensory Cortex." *Science* 277: 968–971.

Sikes, R. W., and B. A. Vogt. 1992. "Nociceptive Neurons in Area 24 of Rabbit Cingulate Cortex." *Journal of Neurophysiology* 68: 1720–1732.

Tölle, T. R., T. Kaufmann, T. Siessmeier, S. Lautenbacher, A. Berthele, F. Munz, W. Zieglgansberger, F. Willoch, M. Schwaiger, B. Conrad, and P. Bartenstein. 1999. "Region-Specific Encoding of Sensory and Affective Components of Pain in the Human Brain: A Positron Emission Tomography Correlation Analysis." *Annals of Neurology* 45: 40–47.

Vierck, C. J., A. Acosta-Rua, R. Nelligan, N. Tester, and A. Mauderli. 2002. "Low Dose Systemic Morphine Attenuates Operant Escape but Facilitates Innate Reflex Responses to Thermal Stimulation." *Journal of Pain* 3: 309–319.

Walters, E. T. 1996. "Comparative and Evolutionary Aspects of Nociceptor Function." Pp. 92–114, in *Neurobiology of Nociceptors*, ed. C. Belmonte and Cervero, F. New York: Oxford University Press.

Wong, E. T., S. Gunes, E. Gaughan, R. B. Patt, L. E. Ginsberg, S. J. Hassenbuch, and R. Payne. 1997. "Palliation of Intractable Cancer Pain by MRI-Guided Cingulotomy." *Clinical Journal of Pain* 13: 260–263.

Yeomans, D. C., B. Y. Cooper, and C. J. Vierck. 1995. "Comparisons of Dose-Dependent Effects of Systemic Morphine on Flexion Reflex Components and Operant Avoidance Responses of Awake Non-Human Primates." *Brain Research* 670: 297–302.

21   On the Neuro-Evolutionary Nature of Social Pain, Support, and Empathy

Jaak Panksepp

1   Prelude: Pain and Affective Consciousness

One of the great mysteries of the brain is how it creates affective experiences—the many unconditional feelings of goodness and badness (with dimensions of valence, arousal, and urgency) that neural systems generate because of the way our minds were designed through evolution. For instance, experiences of pain, as so many other intrinsic capacities of consciousness, are firmly ingrained in our brain organization as evolutionary birthrights that emerge spontaneously during certain interactions with environmental stimuli. Such automatic, instinctual processes of the nervous system, albeit epigenetically refined, generate a primitive, primary-process form of affective consciousness that may be foundational for secondary forms of consciousness (thoughts), which are essential for tertiary variants (thoughts about thoughts). When we hit our thumb with a hammer, even though practically all our initial behavioral responses are reflexive, they are not long unconscious. In other words, affective consciousness is an automatic process of the human brain, and to all indications, that of many other animals as well.

Although we do not yet understand the evolution of any of the variants of consciousness, it is reasonable to assume that the most ancient, affective form emerged with feelings of pain. Granted this, if we could decipher how the brain constructs the primordial feeling of pain in other animals, we might obtain a solid understanding of the ancestral sources of human consciousness. This strategy is important because we cannot obtain detailed information about the underlying brain systems through ethically acceptable human research. However, the biggest stumbling block in pursuing such projects is the seemingly endless debates we (contentious primates that we are) can sustain about whether pain, and other forms of consciousness, exist in the brains of other animals as well as young infants that cannot speak (e.g., Derbyshire 2001; Benatar and Benatar 2001; Rose 2002). Even when we obtain as comprehensive a list of external indicators of pain as is scientifically feasible through animal research (e.g., all the sounds, movements, autonomic and hormonal responses to pain; see Bateson 1991), there will still be skeptics who encourage agnosticism on experiential issues. However, if we place such data on the scales of reasoned judgment, the "weight of evidence" supports the likelihood that other animals, with relevant brain circuits similar to ours, also experience pain when they exhibit the symphony of behavioral and other bodily changes that characterize pain in our own species.

Despite modern functional imaging of the human brain (e.g., Coghill et al. 1999; Jones, Kulkarni, and Derbyshire 2002), there is no definitive mindscope for monitoring the many natural psychological functions of the brain that we may share, homologously, with other mammals. Thus, our capacity to link subjective experiences with brain processes, especially in other animals, is completely dependent on our willingness to accept behavioral and other bodily indices of emotional states as indices of their affective feelings (Panksepp 1998). It is primarily through the detailed study of the brains of other animals that any deep knowledge about the evolutionary underpinnings of the human mind can be experimentally obtained (Knutson, Burgdorf, and Panskepp 2002; Panksepp et al. 2002). Since detailed neuroscientific work on the brain underpinnings of raw affective experiences cannot be ethically conducted on humans, we must explore the fertility of animal models thoroughly and openmindedly. Without our willingness to respect indirect behavioral evidence from other animals as indices of the operations of their mental apparatus, little scientific progress can be made on various basic cross-species neuropsychological processes (e.g., basic emotions and motivations) that may be foundational for human consciousness (Panksepp 1998; Watt and Pincus 2004).

Of course, if comparable emotional feelings do exist in the minds of other animals, it becomes debatable whether such work can be ethically conducted on those other creatures. At best, such work would entail an ethical compromise. A justification might be that that kind of work could help clarify the deep evolutionary nature of human affective experience, and thereby promote new somatic and psychiatric therapies. In any event, abundant evidence is available that affective consciousness is largely a function of subneocortical brain systems we share homologously with other mammals (Liotti and Panksepp 2004; Panksepp 1985, 1998; Shewmon, Holmes, and Byrne 1999). However, this is a position on which there is little agreement among neuroscientists.

If it turned out to be the case that affective consciousness is dependent on higher neocortical expansions that are unique to humans, then work on other animals has little chance of clarifying how affective experiences are created in the human brain. However, the weight of evidence currently indicates otherwise—that raw affective experience is critically dependent on subneocortical brain areas (Liotti and Panksepp 2004; Panksepp 1998); our highest regions of the brain (e.g., neocortex) serve more in an emotion–regulatory and cognitive–parsing role, rather than in affect generation (Ciompi and Panksepp 2005). Here I proceed with the empirically based assumption that the raw experience of pain does not require higher neuromental processes mediated largely by the neocortex (i.e., brain areas clearly needed for secondary and tertiary forms of consciousness), even though such brain areas can regulate affective experiences and provide sophisticated cognitive strategies to minimize harm. Further, I will argue that primary-process pain was also the evolutionary precursor for many other variants of psychological distress, such as the pain of loneliness, which may have emerged evolutionarily from preexisting brain functions that originally mediated bodily pain, especially visceral kinds.

In sum, since the neural details of how affective psychological processes—the pleasantness and unpleasantness of experiences—are generated by the interplay and convergence of diverse brain processes must come from the study of animal models (Craig 2003b), it would be wise to suspend our natural scientific skepticism about whether *other animals* actually have affective experiences. Only through the judicious blending of human and animal research can scientific strategies emerge that will allow us to understand many subtle processes, such as the nature of affective experiences, that have long been deemed beyond the pale of scientific analysis (Panksepp 1998). However, since there are still such strong and skeptical opinions about the existence of painful experiences in other animals, or lack thereof, let me consider this contentious issue in some greater detail.

2 The Issue of Animal Pain

When other mammals are injured, they exhibit most of the behavioral indicators of pain exhibited by humans—they scream, they focus on the injured part of the body, they limp and avoid use of the injured limb, and they tend to hide and exhibit recuperative behaviors, which all suggest significant and aversive internal experiences accompany the pain behaviors. Animal investigators have traditionally been reticent to use terms such as *pain* in their empirical inquiries, preferring less value laden terms such as *nociceptive* responses, which is often accompanied by assertions that pain behaviors are reflexive, with the implicit assumption such instinctual responses need not, perhaps could not, be accompanied by experiential states. In fact, behavioral neuroscientists are accustomed to hide a great deal of their ignorance about the nature of the neuromental apparatus under terminologies such as reflexive and instinctual responses. It is just as likely that the very nature of raw affective experience *is* an instinctual response of the nervous system. There is very little to suggest that our basic affective capacities are learned, even though they are surely molded and modulated by learning.

The issue of subjective affective experience in other animals has been a problematic bone of contention among investigators, with most preferring the "agnostic solution"—since we cannot know with positivistic certainty about such issues, it is best to ignore such topics. Of course, to ignore them may be tantamount to discarding some of the key neuronal processes that we need to clarify how brains really operate. My own advocacy of the "naturalistic" folk-psychological perspective—the *dual process monism* assumption that the full patterns of many emotional instinctual behaviors should be provisionally accepted as likely indicators of internal affective states— remains the minority position. It should become the majority position if such views continue to lead to novel neuropsychological predictions at the human level (Panksepp 2004a; Panksepp and Harro 2004). Our inability to see into the minds of other animals does not preclude the possibility that we can make predictions about changes in the affective tendencies of the human mind from studying the neural basis of instinctual emotional patterns in other animals (Panksepp 1999). Indeed, recently our seminal work on the neurochemistry of separation distress, which high- lighted the role of endogenous opioids in controlling this response, is now supported by evidence that human sadness is characterized by low brain opioid activity (Zubieta et al. 2003).

However, even independently of cross-species neuroscientific predictions, there are a variety of behavioral predictions in animals that should coax us to consider the reality of affective feelings in animals (Panksepp 2004a,b,c). With regard to tissue injury, even fish exhibit behavioral symptoms indicative of the experience of an accompanying pain. For instance, Sneddon, Braithwaite, and Gentle (2003) have demonstrated a variety of behavioral changes indicative of the discomfort of experi- enced pain in trout, including latency to resume feeding after acetic acid irritation of the lips. Similar effects have been observed in all warm-blooded vertebrates that have been tested. The fact that similar behavioral effects are evident even in the larvae of flies, and in our experience other invertebrates such as crayfish (unpublished data), suggests that pain perception, and by inference feelings of pain, may be a very ancient experiential state.

The time-honored resistance in the scientific community to accept such conclusions is based partly on the fact that they are inferential. Also, acceptance of pain in lower animals would pose ethical difficulties for common research practices, and why go there, if there are simple ways to argue around the issue? The traditional tack is, of course, to claim that nociceptive reflexes cannot be assumed to reflect pain. Just consider the fact that we all exhibit withdrawal reflexes when we step on a thorn before we experience the full intensity of pain. The comparative slowness with which pain experience is recruited compared to the speed of our behavioral reflexes encourages us to consider that the affective feeling components may be a consequence of some type of higher cerebral activity. Thus, considering our massive neocortical expansions, which are obviously essential for all higher cognitive activities, it is easy to assert that most other animals do not have the requisite higher brain functions.

To the present day, prominent investigators claim that most other animals do not experience pain. One expert recently asserted that the anatomical "evidence indicates that they cannot, because the phylogenetically new pathway that conveys primary homeostatic afferent activity directly to thalamocortical levels in primates . . . is either rudimentary or absent in non-primates" (Craig 2003a). Although anatomical evidence certainly suggests that most other species do not have the sophisticated, neocortically mediated forms of reflective *self-awareness* that humans do, such views ignore the vast amount of psychobehavioral data concerning the role of many primitive brain systems in the generation of raw emotional experiences (Panksepp 1998a). In fact, the behavioral data is quite compelling that all mammals do experience pain and that, just as in humans, certain brain areas extending from the periaqueductal gray of the midbrain to the anterior cingulate cortex are involved in the genesis of painful affect and separation distress (Johansen, Fields, and Manning 2001; Panksepp 2003a,b; Rainville et al. 1997).

Indeed, if we were to extend the logic of scientific skeptics to the logical limits, we might not be willing to accept the existence of pain in other humans. It was not uncommon to hold such views about human infants, leading to circumcisions and other potentially painful medical maneuvers without anesthesia soon after birth. This bioethical debate continues to the present day (e.g., Derbyshire 2001 vs. Benatar and Benatar 2001). As long as skeptics are only willing to accept human verbal self-report as a measure of affect, as opposed to the neural modulation of behavioral choices in all relevant species (i.e., where striking neural homologies do exist) this debate is bound to continue in sterile ways as opposed to being based on the power of scientific predictions. Certainly the neural mechanisms for pain behavior go far back in brain evolution. For instance, "recently, Seymour Benzer's group identified sensory neurons in Drosophila larvae that mediate aversive responses to noxious heat and mechanical stimuli. Thresholds for behavioral and nerve responses are elevated by mutations in the painless gene, which encodes a TRP ion channel protein. Painless thus joins an elite group of TRPs implicated in sensory transduction in insects, nematodes, mammals and fish" (Goodman 2003, p. 643). How shall we know whether pain experience obeys the same physiology?

Let me again briefly consider the case of fish: as noted, Sneddon and colleagues have recently provided compelling behavioral evidence that trout exhibit all the behavioral symptoms, from breathing changes to delayed feeding responses, that one might expect if they are experiencing discomfort or distress after injections of acetic acid into their lip-line (Sneddon, Braithwaite, and Gentle 2003; Sneddon 2003). In counterpoint, it has been vigorously argued by others that fish do not have the neural complexity to experience pain because their brains do not contain enough of the higher neural architecture that is recruited in humans during their exposure to painful stimuli (Rose 2002). However, perhaps those higher neocortical areas only generate thoughts about our feelings, not the raw affect that accompanies the many discriminative distinctions we can cognitively make about our feelings. All the results are presently consistent with the proposition that animals do have many distinct affective experiences, even though it is highly unlikely they have deep thoughts about those feelings. My provisional conclusion is that "The other animals may not *know* that they are conscious. However, their uncomprehending minds may have many experiences, which can help us understand our own" (Panksepp 2004a, p. 38).

The rest of this essay will highlight the empirical riches about painful human emotions that we can harvest if we abide by the working assumption that pain experiences may be a fundamental part of affective consciousness in all mammals, perhaps all vertebrates. Such raw affective experiences may have provided an evolutionary platform for the emergence of more subtle kinds of distress, as in the internalized brain emotional systems that mediate separation distress in animals and feelings of grief in humans.

3   The Pain of Separation Distress

The key idea that guided our initial analysis of the separation-distress and social-bonding systems of the vertebrate brain was that "painful" experiences of social loss, ranging from a perceived lack of social support to the profound internal "ache" resulting from abandonment or death of a loved one, achieves part of its emotional intensity from the brain systems that mediate physical pain (Panksepp 1998). Evidence for such a relationship emerged through the identification of neural systems in mammalian brains that generate "crying" behaviors and the corresponding emotional states. Behavioral brain research on such separation-distress systems of other animals has long suggested evolutionary anatomical and neurochemical relationships between the sources of social isolation-induced distress and physical pain (Panksepp 1981, 2003a,b). Given the dependence of mammalian young on their caregivers, it is easy to understand the strong survival value conferred by neural systems that elaborate both social attachments and the affective qualities of physical pain. As evidence continues to accumulate and support this idea (Eisenberger and Lieberman 2004; MacDonald and Leary 2005; Panksepp 2005), there are intriguing implications for the clinical management of pain, including the power of social influence and placebos on the felt intensity of grief and physical pain. Indeed, the existence of such brain–mind dynamics may help us better appreciate the unique nature and importance of affective processes not only in human lives but those of other animals as well (Panksepp 2003a, 2004c,d).

With the emergence of modern neuropharmacology and neurophysiology, along with functional brain imaging to monitor the dynamics of the human nervous system, the nature of pain has attracted an enormous amount of neuroscientific inquiry (Craig 2003a,b; Rainville 2002). The classic idea that the brain makes a distinction between the cognitive–sensory aspects of pain and its affective intensity and unpleasantness is garnering ever more support from brain imaging studies (Singer et al. 2004; Tölle et al. 1999). It is presumably from the neural substrates that generated the affective components of physical pain that emergent emotional pains, such as the feelings of sadness and grief, evolved.

4   How Does the Brain Feel the Ache of a Broken Heart?

The rest of this essay, adapted from a recent perspective piece in *Science* (Panksepp 2003b) and an essay in *Cancer Pain* (Panksepp 2004b), summarizes some of the evidence for the claim that emotional pain, such as that which accompanies grief and intense loneliness, does share some of the same neural pathways that generate the affective sting of physical pain. It has long been common in diverse human cultures for people to talk about the loss of a loved one in terms of painful feelings. This seems to be more than a semantic metaphor.

Although we have no time machine to return to the evolutionary origins of brain emotional systems, close analyses of living brains do offer important clues. So far the strongest evidence for the evolutionary relations between social pain and physical pain comes from animal brain research. Localized electrical stimulation of many subcortical brain areas that have been implicated in the regulation of pain can provoke separation cries (Herman and Panksepp 1981; Panksepp et al. 1988; Panksepp 1998).

In addition to anterior cingulate circuits, they include the bed-nucleus of the stria ter-minalis, the ventral septal and dorsal preoptic areas, the dorsomedial thalamus and the pariaqueductal central gray (PAG) of the brain stem. Some of these areas, espe-cially the last two, are known to control feelings of physical pain. Indeed, the PAG is the brain area from which emotional distress can be most easily evoked in humans and animals with the lowest levels of brain stimulation. The close proximity of pain systems and emotional distress systems, along with their shared chemistries such as endogenous opioids (Herman and Panksepp 1981; Zubieta et al. 2003), provides a com-pelling link between the distress of physical pain and that engendered by social loss. Thus, it seems likely that certain kinds of psychological pain, especially grief and intense loneliness, have emerged as exaptations from systems that only controlled physical pain at some remote time in our ancestral past.

A growing psychosocial literature is beginning to indicate that the social environ-ment can modulate the affective intensity of pain (e.g., Brown et al. 2003; Leary and Springer 2000), and some recent brain imaging is remarkably consistent with the idea that certain brain areas such as the anterior cingulate cortex, long implicated in the experience of pain (Hurt and Ballantine 1974; Hutchison et al. 1999; Rainville 2002), as well as the anticipation of pain (Koyama, Tanaka, and Mikami 1998) also partici-pate in the genesis of distressing social feelings (Eisenberger, Lieberman, and Williams 2003). Brain research also indicates that this brain area helps mediate social processes such as maternal behavior, social bonding, and separation calls (MacLean 1990; Panksepp 1998), which helps make sense of why this brain area may participate in the distress that arises from being socially ostracized (Eisenberger and Liberson 2004; Williams 2001).

The Eisenberger, Lieberman, and Williams (2003) study capitalized on a behavioral task that allowed feelings of social exclusion to be evoked easily in the solitary con-fines of a modern brain imaging device (Williams, Cheung, and Choi 2000). "Cyberos-tracism" was imposed on thirteen participants during a simulated ball-tossing video game while their cerebral blood oxygenation was monitored via fMRI. Social distress was experienced by these subjects when the two other "players" (actually computer-ized stooges) stopped throwing the ball to the experimental subjects. As the experi-mental subjects were being shunned, two brain regions exhibited substantial arousals. One area was the anterior cingulate cortex. Indeed, the greater the evoked feelings of social distress, the more the anterior cingulate became aroused. The other brain area, in the ventral prefrontal cortex, known to participate in emotional evaluations, exhib-ited opposite patterns. Instead of becoming more and more aroused with escalating distress, it became less and less active as the distress increased. In other words, the two aroused brain areas were moving in opposite directions with reference to the degree of experienced affect.

These patterns suggest that the anterior cingulate cortex (ACC) helps mediate feelings of distress from being ostracized, while prefrontal cortex, a brain region long implicated in emotional regulation (for summary see Devinsky, Morrell, and Vogt 1995), may help counteract the painful feeling of being shunned. Of course, the feelings that can be induced in the context of human brain imaging in laboratory settings are a pale shadow of the real-life feelings that humans and other animals experience in response to sudden social loss. It will be important for more intense feelings of social loss to be studied with fMRI imaging focused on even deeper regions of the mammalian brain that control separation distress, such as the dorsosmedial thalamus and pariaqueductal gray (PAG) (Panksepp et al. 1988). Such regions have already been highlighted with PET scans during intense human sadness (Damasio et al. 2000; Liotti and Panksepp 2004).

Although the Eisenberger, Lieberman, and Williams (2003) findings are consistent with the idea that aversive feelings of social exclusion and physical pain arise, in part, from the same brain regions, there are, of course, many other explanations for such effects. Pain certainly promotes reallocation of attentional resources (Eccleston and Crombez 1999). Thus, the changes in cingulate activity could have reflected many brain–mind effects other than those related to the experienced distress, especially since it is aroused in many other attention-grabbing tasks and situations. The ACC has also long been implicated in autonomic nervous system, attentional, and various cognitive functions (Bush, Luu, and Posner 2000; Paus 2001), and recent work has highlighted how heart rate during various cognitive and motor tasks is highly correlated with ACC arousal, leading to the suggestion "that a principal function of the ACC is the regulation of bodily states of arousal to meet concurrent behavioral demand" (Critchley et al. 2003, p. 2149). Of course, all such brain-imaging studies only provide correlates of psychological functions and do not constitute compelling evidence for causal issues. Indeed, current brain-imaging methods are not optimal for monitoring activity in compact regions where there are many distinct fiber pathways and functionally antagonistic systems, yielding many false negatives, which can skew our appreciation of the full extent of emotional systems.

Although much work remains to be done before the neural causes of experienced pain are adequately understood, the above findings dovetail nicely with what we know about the neural substrates of social separation-distress in other animals, which first highlighted potential relations of separation-distress mechanisms of the brain to the core brain systems for the affective intensity of pain (see Panksepp et al. 1980;

Panksepp 1981; and initial reviews). The obvious fact that crying occurs in response to pain as well as social loss was not our only reason for considering such relationships. Our initial work into the neurochemistry of social attachments a quarter of a century ago was also motivated by the possibility that the loss of a loved one—the painful grief occasioned by sudden loss of social support—could be alleviated by the same brain chemistries that regulate our feelings of physical pain. Opioid alkaloids of plant origin (e.g., morphine), as well as the endogenous opioids of the brain (especially endorphins), proved to be quintessentially effective in alleviating behavioral measures of separation distress (as monitored by the frequency of isolation–cries), in various species including young dogs, guinea pigs, chicks, rats, and primates (Panksepp, Herman, et al. 1980; Panskepp, Siviy, and Normansell 1985). Not only did these findings affirm our conjecture that there may be fundamental similarities between the brain dynamics of narcotic dependence and social dependence–bonding, but also they forged an empirical link between neural mechanisms that regulate feelings of social loss and those that mediate feelings of pain.

Correlative findings such as those of Eisenberger, Lieberman, and Williams (2003) concerning relationships between physical pain and social exclusion distress systems of the brain need to be further evaluated with causal studies. Perhaps such higher brain functions could be deconstructed using causal manipulations such as transcranial magnetic stimulation (Schutter, van Honk, and Panksepp 2004). A straightforward neuropharmacological maneuver would be to monitor the consequences of opioid manipulations. Participants given opiates should feel less distress at being shunned, and the brain areas implicated in elaborating such feelings should not become as aroused. The reverse would be anticipated from the administration of opioid receptor antagonists such as naloxone and naltrexone—intensification of both effects. Of course, one should also concurrently monitor psychological distress and various autonomic effects to evaluate whether dissociations could be demonstrated between the distressed feelings and other bodily changes such as heart rate changes (e.g., Critchley et al. 2003). Since several other brain chemistries have been identified that powerfully and specifically regulate separation–distress, especially oxytocin (Panksepp 1992), a neuropeptide that can also attenuate pain (Lundeberg et al. 1994), additional critical causal tests could be envisioned as soon as clinically useable drugs to modify such neuropeptidergic systems are developed. Many of the needed functional dissections cannot be achieved with human studies, and progress may require investigators to exhibit a new willingness to entertain the idea that psychological processes do exist in the minds of other animals.

Perhaps the biggest stumbling block to future progress is the continuing paucity of studies analyzing brain–affect linkages in animal models where the neural foundations of emotional experiences can be studied in some detail. As noted earlier, biases against accepting animal emotional behaviors as being veridical readouts of their experiences, still lead prominent investigators to argue that whereas emotional behaviors exist in other animals, affective feelings may only exist in humans and other higher primates (Damasio 2003, with critique by Panksepp 2003c). But the behavioral data are quite compelling that all mammals do experience pain, and that just as in humans, the ACC helps create those experiences (Johansen, Fields, and Manning 2001). To all appearances, separation–distress in other animals is a highly aversive state with many bodily consequences (Panksepp 2003a,b), and it may be scientifically wiser to invest in the working hypothesis that other animals do have many affective experiences than to sustain the Cartesian bias that they do not. Animal models, pursued with new ethical sensitivities, may be the only efficient way to work out the underlying causal details (see sections 1–2 above).

5 Placebo Effects and Empathy

Now that more and more human investigators recognize that the affective sting of social pain may share evolutionary relations to physical pain, many new avenues of understanding and clinically useful intervention may open up. The quality of social environments may contribute much not only to our emotional feelings but to our ability to cope with pain (Gatchel and Turk 2000). There has long been evidence that the pain of childbirth is eased with social support (Klaus et al. 1986), and substantial evidence that both cardiac and post-operative pain can be alleviated likewise (King et al. 1993; Kulik and Mahler 1989). These effects may have underlying similarities to how placebos work to control pain.

Indeed, the placebo effect could be conceptualized as the cerebral representation of the "healing-touch," which may be partly mediated by opioid release in the brain. Contact-comfort and the pleasures of touch are partly mediated by release of opioids (Keverne, Martenz, and Tuite 1989; Panksepp, Herman et al. 1980; Panksepp and Bishop 1981). Likewise, placebos in humans operate partly through opioid release (Grevert, Albert, and Goldstein 1983; Petrovic et al. 2002). Indeed, placebos can reduce arousal of anterior cingulate regions of the brain, which are commonly overactive when people are depressed or in distress (Mayberg, Silva, and Brannan 2002). As already noted, damage to this region of the brain reduces the intensity of pain and emotional distress (Foltz and White 1962; Hurt and Ballantine 1974). It is becoming ever clearer that the higher reaches of distress integrating systems, such as the ACC, participate in our capacity to feel empathy for others (Singer et al. 2004). It seems likely that mirror neurons in frontal regions of the brain (Gallese 2003; Rizzolatti et al. 1996) are able to resonate with the distress of others, perhaps sending messages directly to nearby distress systems of the ACC. Thereby, the apparent emotional distress of others may arouse similar emotions in sensitive perceivers, without much need for cognitive deliberations (Dimberg and Thunberg 1998), even though we cannot help but to also cognitively reflect on the resulting affective states.

Recently I had just such an experience, as I arrived in my hometown of Tartu, Estonia, on Sunday, August 29, 2004. As I disembarked from the bus from Tallinn, to be greeted by a kind relative who had come to meet me, my attention was promptly magnetized by a young girl sitting alone on a bench about twenty yards in front of us. This waif of a child was the epitome of despair, head down, face enveloped in profound sadness. She held her hands between her clenched knees, slumped and alone. My cousin once removed and I both felt her despair, resonated spontaneously with her profoundly painful appearance. We looked at each other and realized we had to get involved in a supportive way. We had to offer a helping hand, gently, without scaring her. We decided such a lost child should respond better to the feminine touch, so my cousin went to her and put a hand gently on her knee and asked what was the matter.

This seven-year-old had been put on a bus in the countryside by a grandmother, with instructions that her mother would be waiting at the bus station. The child had now been waiting for several hours, but her mother had not come. Why? We never found out. We asked her if she knew where she lived, and she described the general region of our small university city of Tartu. We coaxed her to come with us; we would take her home. After some initial reluctance, she warmed up, and came with us. Her mood brightened dramatically when we neared her neighborhood. She pointed to familiar sights, and then her own apartment complex. She seemed composed now, and we let her out at her front door. We chose not to confront her family. We wished her well.

Without that dramatic bodily expression of despair by this waif, our empathetic "juices" would not have been engaged. I believe the key ingredients of those empathic juices are the same as those that lead all young children who are lost to feel despair. Perhaps we feel deep empathy because our mirror neurons can be so readily engaged by the emotional gestures of others (Gallese 2003). When we are confronted by joy and sadness, perhaps we resonate with those basic feelings because the chemistries of joy and sadness (ones we surely share with other animals) become engaged in our brains, providing an immediate form of emotional understanding.

The overall message seems obvious. There is still abundant room for human empathy and the healing touch in the management of pain and distress. This can include music, which often quintessentially captures social feelings (Panksepp and Bernatzky 2002). In addition to development of ever better medications and other biological interventions to control pain, psychosocial approaches need to be recruited for effective pain management. Animals exhibit diminished vocal indices of emotional pain to nociceptive stimulation when they have abundant social companionship (Panksepp 1980). Even rats, under the rigorous testing conditions, are adept at reducing the distress of others (Rice and Gainer 1962). Partly we are satisfied by such actions because antipain, antistress chemistries, such as opioids and oxytocin, are recruited in our brains.

One reason opioids are so commonly used to treat the pain of terminal disease is surely not just because the resulting neurochemical changes simply alleviate the physical pain. They also alleviate the emotional pain of dying. This is both humane and appropriate. However, it would be a shame if humans did not continue to be the ultimate carriers of the healing touch. There is now abundant room for evaluating whether analgesics and other psychotropic drugs are more effective in supportive psychosocial contexts (see Panksepp and Harro 2004). Indeed, there is abundant data to argue that brain opioids may be more important in regulating distressing emotional responses than physical pain (e.g., Grevert and Goldstein 1977; Panksepp et al. 1978).

6 Coda

In sum, one of the most difficult forms of emotional pain that we humans, as well as other mammals, can experience is that arising from social loss and ostracism. We are finally in a position in our scientific understanding where we can coherently address emotional aspects of the problem that transcend the purely behavioral and biological realm. Affective states are emergents of brain functions, even in other animals. I would close this essay with some personal and poetic remarks.

Since time immemorial, poets have focused on the pain of a broken heart. It is moving that such poetic insights into the human condition now find some support from neurophysiological studies. When I lost my daughter fourteen years ago in a horrendous traffic accident, among her papers I found a poem that is now carved on her tombstone. The last stanza affirms that

When your days are full of pain,

And you don't know what to do,

Recall these words I tell you now—

I will always care for you

This remains a treasured memento for me—an image that can help melt my own painful social-emotional distress when my mind drifts to that agony of the past.

Ultimately, those aspects of the environment, both external and internal, that sooth the pain of separation-distress deserve to be seen as an integral part of love. Our capacity to love is built partly upon our ability to experience loss. Indeed, a reasonable scientific hypothesis is that social feelings such as love are constructed partly from chemistries that can alleviate feelings of social isolation, as well as those that directly promote prosocial behaviors such as play, sexuality, and desires for friendly togetherness (Panksepp 1998). Poetic words such as:

My love is like a wave upon the greatest sea:

it's home to whales, floats continents,

and fills the clouds with rain.

It warms the air and sets the wind

to blow away our pain

(Anesa Miller 1995, quoted with permission)

convey a symbolic structure of basic truth about the human and animal condition. Our higher emotional feelings are constructed partly out of our more primitive feelings, and our cognitions often "float" upon a sea of basic emotions and motivations.

Unlike most metaphors that are only psychologically moving, there are some that capture relations that may help guide our brain–mind investigations. Will we eventually find that the feeling of a broken heart arises partly from the rich autonomic circuits of the limbic system that control our cardiac dynamics? Will we find that people we consider "cold" or "warm" are perceived in those terms partly because different people influence the abundant thermoregulatory aspects of our emotional brain systems in different ways? A wider recognition of such commonalities, if supported by additional scientific studies, may help promote real emotional education and a better understanding of the many emotional gifts that evolution has constructed within our brains.

Of course, scientists must continue to be skeptical about such hypotheses until they are shown to be productive in the laboratory (Panksepp 1998), as in the recent work of Eisenberger, Lieberman, and Williams (2003) and MacDonald and Leary (2005). Although there are many controversies and species differences in the emotional systems we share as ancestral gifts with other mammals (Panksepp 1998, 2003b), a basic neuroscience that remains open to the shared nature of human and animal affective experiences may be more productive than one that is not. This is not to say that there is nothing unique about the human experience of pain. As Emily Dickinson declared: "After great pain, a formal feeling comes" (Goodman 2003), and we may never know whether other animals experience pain in more subtle ways than the pure aversiveness of pain. At present there is little realistic possibility of understanding the fundamental nature of human *cognitive consciousness* by studying animals, but it seems likely that we can understand the foundations of our own *affective consciousness* by deciphering selected emotional tendencies in other animals (Panksepp 2003a,b,d).

In sum, the felt intensity of affective experiences appears to emerge largely from the neurodynamics of quite ancient regions of the brain, inherited at the very least, by all mammals, but probably many other animals as well. Although higher forms of reflective consciousness may elaborate interesting transformations—reducing the felt intensity of pain in some contexts and providing opportunities for greater suffering under others—most fundamental forms of pain may have emerged much closer to earliest glimmers of primary process consciousness than to our recently acquired ability to cognitively reflect on our experiences. Being the courier of such urgent survival concerns—the need to stay out of harm's way—various other kinds of emotional pain, such as feelings of separation distress, can now be conceptualized as a substantial components of primary-process affective consciousness.

As I noted over a dozen years ago (Panksepp 1989), we now need to answer several critical neurobehavioral questions: (1) *"Does the activation of distress circuits in young and relatively helpless animals evoke resonant activity in the same circuits of nearby adults?"* Lorberbuam and colleagues (2002) have recently provided fMRI images of women listening to the cries of babies, and the brain areas aroused are the same as those clarified separation-distress work in animal brain research. (2) *"If such perceptually induced resonance does exist, is the evoked activity especially strong between bonded individuals?"* Bartels and Zeki (2004) have now demonstrated how brain reward systems clarified through animal brain research are aroused when viewing one's own infant than a strange infant. And finally I asked (3) *"Does such brain activity arouse caregiving in adults?"* Such causal studies can only be pursued "easily" in animals, but detailed neuroscientific work on such topics has barely begun. There is still considerable resistance to granting other animals the capacity to have experiences. Clearly a great deal more research needs to be done on such important issues concerning the cross-species emotional basis of our social brains.

References

Bartels, A., and S. Zeki. 2004. "The Neural Correlates of Maternal and Romantic Love." *NeuroImage* 21: 1155–1166.

Bateson, P. 1991. "Assessment of Pain in Animals." *Animal Behavior* 42: 827–839.

Benatar, D., and M. Benatar. 2001. "A Pain in the Fetus: Toward Ending Confusion about Fetal Pain." *Bioethics* 15: 57–76.

Brown, J. L., D. Sheffield, M. R. Leary, and M. E. Robinson. 2003. "Social Support and Experimental Pain." *Psychosomatic Medicine* 65: 276–283.

Bush, G., P. Luu, and M. I. Posner. 2000. "Cognitive and Emotional Influences in Anterior Cingulate Cortex." *Trends in Cognitive Sciences* 4: 215–222.

Ciompi, L., and J. Panksepp. 2005. "Energetic Effects of Emotions on Cognitions—Complementary Psychobiological and Psychosocial Findings." In *Consciousness and Emotion: Agency, Conscious Choice, and Selective Perception*, ed. R. Ellis and N. Newton. Amsterdam: John Benjamins.

Coghill, R. C., C. N. Sang, J. M. Maisog, and M. J. Iadarola. 1999. "Pain Intensity Processing within the Human Brain: A Bilateral Distributed Mechanism." *Journal of Neurophysiology* 82: 1934–1943.

Craig, A. D. 2003a. "Interoception: The Sense of the Physiological Condition of the Body." *Current Opinion in Neurobiology* 13: 500–505.

———. 2003b. "Pain Mechanisms: Labeled Lines versus Convergence in Central Processing." *Annual Review of Neuroscience* 26: 1–30.

Critchley, H. D., C. J. Mathias, O. Josephs, J. O'Doherty, S. Zanini, B. K. Dewar, L. Cipolotti, T. Shallice, and R. J. Dolan. 2003. "Human Cingulate Cortex and Autonomic Control: Converging Neuroimaging and Clinical Evidence." *Brain* 126: 2139–2152.

Damasio, A. R. 2003. *Looking for Spinoza*. Orlando, Fl.: Harcourt.

Damasio, A. R., T. J. Grabowski, A. Bechara, H. Damasio, L. L. B. Ponto, J. Parvizi, and R. D. Hichwa. 2000. "Subcortical and Cortical Brain Activity During the Feeling of Self-Generated Emotions." *Nature Neuroscience* 3: 1049–1056.

Derbyshire, S. W. G. 2001. "Fetal Pain: An Infantile Debate." *Bioethics* 15: 77–84.

Devinsky, O., M. J. Morrell, and B. A. Vogt. 1995. "Contributions of Anterior Cingulate Cortex to Behaviour." *Brain* 118: 279–306.

Dimberg, U., and M. Thunberg. 1998. "Rapid Facila Reactions to Emotional Facial Expressions." *Scandinavian Journal of Psychology* 39: 39–45.

Eccleston, C., and G. Crombez. 1999. "Pain Demands Attention: A Cognitive-Affective Model of the Interruptive Function of Pain." *Psychological Bulletin* 125: 356–366.

Eisenberger, N. I., and M. D. Lieberman. 2004. "Why Rejection Hurts: A Common Neural Alarm System for Physical and Social Pain." *Trends in Cognitive Sciences* 8: 294–300.

Eisenberger, N. I., M. D. Lieberman, and K. D. Williams. 2003. "Does Rejection Hurt? An fMRI Study of Social Exclusion." *Science* 302: 290–292.

Foltz, E. L., and L. E. White. 1962. "Pain Relief by Frontal Cingulotomy." *Journal of Neurosurgery* 19: 89–100.

Gallese, V. 2003. "The Roots of Empathy: The Shared Manifold Hypothesis and the Neural Basis of Intersubjectivity." *Psychopathology* 36: 171–80.

Gatchel, R., and D. Turk. eds. 2000. *Psychosocial Factors in Pain: Critical Perspectives*. New York: The Guilford Press.

Goodman, M. B. 2003. "Sensation Is Painless." *Trends in Neurosciences* 26: 643–645.

Grevert, P., L. H. Albert, and A. Goldstein. 1983. "Partial Antagonism of Placebo Analgesia by Naloxone." *Pain* 16: 129–143.

Grevert, P., and A. Goldstein. 1977. "Effects of Naloxone on Experimentally Induced Ischemic Pain and on Mood in Human Subjects." *Proceedings of the National Academy of Sciences* 74: 1291–1294.

Herman, B. H., and J. Panksepp. 1981. "Ascending Endorphinergic Inhibition of Distress Vocalization." *Science* 211: 1060.

Hurt, R. W., and H. T. Ballantine, Jr. 1974. "*Stereotactic* Anterior Cingulate Lesions for Persistent Pain: A Report on 68 Cases." *Clinical Neurosurgery* 21: 334–351.

Hutchison, W. D., K. D. Davis, A. M. Lozano, R. R. Tasker, and J. O. Dostrovsky. 1999. "Pain Related Neurons in the Human Cingulate Cortex." *Nature Neuroscience* 2: 403–405.

Johansen, J. P., H. L. Fields, and B. H. Manning. 2001. "The Affective Component of Pain in Rodents: Direct Evidence for a Contribution of the Anterior Cingulate Cortex." *Proceedings of the National Academy of Sciences* 98: 8077–8082.

Jones, A. K. P., B. Kulkarni, and S. W. G. Derbyshire. 2002. "Functional Imaging of Pain Perception." *Current Rheumatology Reports* 4: 329–333.

Keverne, E. B., N. Martenz, and B. Tuite. 1989. "B-endorphin Concentrations in CSF of Monkeys are Influenced by Grooming Relationships." *Psychoneuroendocrinology* 14: 155–161.

King, K. B., H. T. Reis, L. A. Porter, and L. H. Norsen. 1993. "Social Support and Long-Term Recovery from Coronary Artery Surgery: Effects on Patients and Spouses." *Health Psychology* 12: 56–63.

Klaus, M., J. Kennel, S. Robertson, and R. Rosa. 1986. "Effects of Social Support During Parturition on Maternal and Infant Mortality." *British Medical Journal* 293: 585–597.

Knutson, B., J. Burgdorf, and J. Panksepp. 2002. "Ultrasonic Vocalizations as Indices of Affective States in Rats." *Psychological Bulletin* 128: 961–977.

Koyama, T., Y. Z. Tanaka, and A. Mikami. 1998. "Nociceptive Neurons in the Macaque Anterior Cingulate Activate during Anticipation of Pain." *Neuroreport* 9: 2663–2667.

Kulik, J. A., and H. I. Mahler. 1989. "Social Support and Recovery from Surgery." *Health Psychology* 8: 221–238.

Leary, M. R., and C. A. Springer. 2000. "Hurt Feelings: The Neglected Emotion." In *Aversive Behaviors and Interpersonal Transgression*, ed. R. Kowalski. Washington D.C.: American Psychological Association.

Liotti, M., and J. Panksepp. 2004. "Imaging Human Emotions and Affective Feelings: Implications for Biological Psychiatry." Pp. 33–74, in *Textbook of Biological Psychiatry*, ed. J. Panksepp. New York: Wiley.

Lorberbaum, J. P., J. D. Newman, A. R. Horwitz, J. R. Dubno, R. B. Lydiard, M. B. Hamner, D. E. Bohning, and M. S. George. 2002. "A Potential Role for Thalamocingulate Circuitry in Human Maternal Behavior." *Biological Psychiatry* 51: 431–445.

Lundeberg, T., K. Uvnäs-Moberg, G. Ågren, and G. Bruzelius. 1994. "Antinociceptive Effects of Oxytocin in Rats and Mice." *Neuroscience Letters* 170: 153–157.

MacDonald, G., and M. R. Leary. 2005. "Why Does Social Exclusion Hurt? The Relationship between Social and Somatic Pain." *Psychological Review*. In press.

MacLean, P. D. 1990. *The Triune Brain in Evolution*. New York: Plenum.

Mayberg, H. S., J. A. Silva, S. K. Brannan, J. L. Tekell, R. K. Mahurin, S. McGinnis, and P. A. Jerabek. 2002. "The Functional Neuroanatomy of the Placebo Effect." *American Journal of Psychiatry* 159: 728–737.

Miller, A. 1995. *A Road beyond Loss*. Bowling Green, Ohio: Memorial Foundation for Lost Children.

Panksepp, J. 1980. "Brief Social Isolation, Pain Responsivity, and Morphine Analgesia in Young Rats." *Psychopharmacology* 72: 111–112.

———. 1981. "Brain Opioids: A Neurochemical Substrate for Narcotic and Social Dependence." Pp. 149–175, in *Progress in Theory in Psychopharmacology*, ed. S. Cooper. London: Academic Press.

———. 1985. "Mood Changes." Pp. 271–285, in *Handbook of Clinical Neurology* 45(1), ed. J. A. M. Frederiks. Amsterdam: Elsevier Science Publishers.

———. 1989. "Altruism, Neurobiology." Pp. 7–8, in *The Encyclopedia of Neuroscience*, ed. G. Adelman. Boston: Birkhauser Boston, Inc.

———. 1992. "Oxytocin Effects on Emotional Processes: Separation Distress, Social Bonding, and Relationships to Psychiatric Disorders." *Annals of the N.Y. Academy of Science* 652: 243–252.

————. 1998. *Affective Neuroscience: The Foundations of Human and Animal Emotions.* London: Oxford University Press.

————. 2003a. "Can the Anthropomorphic Analysis of 'Separation Calls' in Other Animals Inform Us about the Emotional Nature of Social Loss in Humans?" *Psychological Review* 110: 376–388.

————. 2003b. "Feeling the Pain of Social Loss." *Science* 302: 237–239.

————. 2003c. "Damasio's Error?" *Consciousness and Emotion* 4: 111–134.

————. 2003b. "At the Interface between the Affective, Behavioral, and Cognitive Neurosciences: Decoding the Emotional Feelings of the Brain." *Brain and Cognition* 52: 4–14.

————. 2004a. "Affective Consciousness and the Origins of Human Mind: A Critical Role Brain Research on Animal Emotions." *Impuls* 3: 47–60. Available at www.psykologi.uio.no/impuls.

————. 2004b. "Feelings of Social Loss: The Evolution of Pain and the Ache of a Broken Heart." *Cancer Pain.*

————. 2004c. "Affective and Social Neuroscience Approaches to Understanding Core Emotional Feelings." In *Mental Health and Well-Being in Animals*, ed. F. McMillan. In press. Ames, Iowa: Iowa State University Press.

————. 2004d. "Basic Affects and the Instinctual Emotional Systems of the Brain: The Primordial Sources of Sadness, Joy, and Seeking." Pp. 174–193, in *Feelings and Emotions: The Amsterdam Symposium*, ed. A. S. R. Manstead, N. Frijda, and A. Fischer. New York: Cambridge University Press.

————. 2005. "Why Does Separation-Distress Hurt?: A Comment on MacDonald and Leary." *Psychological Bulletin* 131: 224–230.

Panksepp, J., N. J. Bean, P. Bishop, T. Vilberg, and T. L. Sahley. 1980. "Opioid Blockade and Social Comfort in Chicks." *Pharmacology, Biochemistry, and Behavior* 13: 673–683.

Panksepp, J., and G. Bernatzky. 2002. "Emotional Sounds and the Brain: The Neuro-Affective Foundations of Musical Appreciation." *Behavioural Processes* 60: 133–155.

Panksepp, J., and P. Bishop. 1981. "An Autoradiographic Map of 3H Diprenorphine Binding in the Rat Brain: Effects of Social Interaction." *Brain Research Bulletin* 7: 405–410.

Panksepp, J., and J. Harro. 2004. "Future of Neuropeptides in Biological Psychiatry and Emotional Psychopharmacology: Goals and Strategies." Pp. 627–659, in *Textbook of Biological Psychiatry*, ed. J. Panksepp. New York: Wiley.

Panksepp, J., B. Herman, R. Conner, P. Bishop, and J. P. Scott. 1978. "The Biology of Social Attachments: Opiates Alleviate Separation Distress." *Biological Psychiatry* 9: 213–220.

Panksepp, J., B. H. Herman, T. Villberg, P. Bishop, and F. G. DeEskinazi. 1980. "Endogenous Opioids and Social Behavior." *Neuroscience and Biobehavioral Reviews* 4: 473–487.

Panksepp, J., J. R. Moskal, J. B. Panksepp, and R. A. Kroes. 2002. "Comparative Approaches in Evolutionary Psychology: Molecular Neuroscience Meets the Mind." *Neuroendocrinology Letters*, special issue 23 (suppl. 4): 105–115.

Panksepp, J., L. A. Normansell, B. Herman, P. Bishop, and L. Crepeau. 1988. "Neural and Neurochemical Control of the Separation Distress Call." Pp. 263–299, in *The Physiological Control of Mammalian Vocalizations*, ed. J. D. Newman. New York: Plenum Press.

Panksepp, J., S. M. Siviy, and L. A. Normansell. 1985. "Brain Opioids and Social Emotions." Pp. 3–49, in *The Psychobiology of Attachment and Separation*, ed. M. Reite and T. Fields. New York: Academic Press.

Paus, T. 2001. "Primate Anterior Cingulate Cortex: Where Motor Control, Drive and Cognition Interface." *Nature Reviews* 2: 417–424.

Petrovic, P., E. Kalso, K. M. Petersson, and M. Ingvar. 2002. "Placebo and Opioid Analgesia: Imaging a Shared Neuronal Network." *Science* 295: 1737–1740.

Price, D. D. 2000. "Psychological and Neural Mechanisms of the Affective Dimension of Pain." *Science* 288: 1769–1772.

Rainville, P. 2002. "Brain Mechanisms of Pain Affect and Pain Modulation." *Current Opinion in Neurobiology* 12: 195–204.

Rainville, P., G. H. Duncan, D. D. Price, B. Carrier, and M. C. Bushnell. 1997. "Pain Affect Encoded in Human Anterior Cingulate but Not Somatosensory Cortex." *Science* 277: 968–971.

Rice, G. F., and P. Gainer. 1962. "'Altruism' in the Albino Rat." *Journal of Comparative and Physiological Psychology* 55: 123–125.

Rizzolatti, G., L. Fadiga, V. Gallese, and L. Fogassi. 1996. "Premotor Cortex and the Recognition of Motor Actions." *Brain Research* 3: 131–141.

Rose, J. D. 2002. "The Neurobehavioral Nature of Fishes and the Questions of Awareness and Pain." *Reviews in Fisheries Science* 10: 1–38.

Schutter, D. J. L. G., J. Van Honk, and J. Panksepp. 2004. "Introducing Transcranial Magnetic Stimulation and Its Property of Causal Inference in Investigating the Brain–Function Relationship." *Synthese* 141(2): 155–173.

Shewmon, D. A., D. A. Holmes, and P. A. Byrne. 1999. "Consciousness in Congenitally Decorticate Children: Developmental Vegetative State as Self-Fulfilling Prophecy." *Developmental Medicine and Child Neurology* 41: 364–374.

Singer, T., B. Seymour, J. O'Doherty, H. Kaube, R. J. Dolan, and C. D. Frith. 2004. "Empathy for Pain Involves the Affective but Not Sensory Components of Pain." *Science* 303: 1157–1162.

Sneddon, L. U. 2003. "Trigeminal Somatosensory Innervation of the Head of a Teleost Fish with Particular Reference to Nociception." *Brain Research* 972: 44–52.

Sneddon, L. U., V. A. Braithwaite, and M. J. Gentle. 2003. "Do Fishes Have Nociceptors? Evidence for the Evolution of a Vertebrate Sensory System." *Proceedings of the Royal Society of London Series B Biological Sciences* 270: 1115–1121.

Tölle, T. R., T. Kaufmann, T. Siessmeier, S. Lautenbacher, A. Berthele, F. Munz, W. Zieglgänsberger, F. Willoch, M. Schwaiger, B. Conrad, and P. Bartenstein. 1999. "Region-Specific Encoding of Sensory and Affective Components of Pain in the Human Brain: A Positron Emission Tomography Correlation Analysis." *Annals of Neurology* 45: 40–47.

Watt, D. F., and D. I. Pincus. 2004. "Neural Substrates of Consciousness: Implications for Clinical Psychiatry." Pp. 75–110, in *Textbook of Biological Psychiatry*, ed. J. Panksepp. New York: Wiley.

Williams, K. D. 2001. *Ostracism: The Power of Silence*. New York: Guilford Press.

Williams, K. D., C. K. T. Cheung, and W. Choi. 2000. "Cyberostracism: Effects of Being Ignored over the Internet." *Journal of Personality and Social Psychology* 79: 748–762.

Zubieta, J. K., T. A. Ketter, J. A. Bueller, Y. Xu, M. R. Kilbourn, E. A. Young, and R. A. Koeppe. 2003. "Regulation of Human Affective Responses by Anterior Cingulate and Limbic Mu-Opioid Neurotransmission." *Archives of General Psychiatry* 60: 1145–1153.

Extended Bibliography

The following bibliography is provided for the interested reader who wants to investigate further some of the issues discussed in this volume. It is loosely thematized under the following categories:

pain, access

pain, adverbialism

pain, affect

pain, animal

pain, asymbolia

pain, choice

pain, concept

pain, Dennett

pain, disassociation

pain, general

pain, history

pain, imaging

pain, infant

pain, insensitivity

pain, language

pain, location

pain, nature

pain, perception

pain, phenomenology

pain, privacy

pain, science

pain, sense data

pain, surgery

pain, value

adverbialism

pleasure

sense data

The coverage of most essays here naturally spans more areas than any one of these categories. I have tried to capture this by using multiple, ordered categories under which the essays are sorted alphabetically by authors' last names. Because this endeavor is bound to be somewhat vague and impressionistic, there are, I am sure, inaccuracies. There are cases where a single essay can, or even should, appear under more than one set of categories—but I have avoided multiple listings in order to preserve space and prevent further confusion. There is a more comprehensive bibliography available online at:

http://www.clas.ufl.edu/users/maydede/pain/PainBib.htm

The majority of its entries are annotated. The reader will find more information there that is regularly updated (and is searchable through any browser's "find" command).

The bibliography of philosophical works provided below is not meant to be complete. The listed scientific works were chosen either because they are historically important or of particular interest to philosophers, or simply because they are useful for a general audience. In addition, some of the entries are marked by an asterisk to highlight their historical importance, or because they contain especially clear and intriguing discussion or a comprehensive and accessible up-to-date review of their coverage area. There are others I could have easily marked, but I have tried to keep these to minimum, so as not to overwhelm the reader.

As of the publication date of this volume, the full text of more than half of the works listed below is already electronically accessible on the Internet through the library subscriptions of most research universities and colleges. Where the full text is not available, one can usually find an informative abstract (sometimes along with bibliographic references). Most libraries have their own search facilities for online resources that can easily locate the electronic database through which these are available.

Adverbialism; Pain, Adverbialism; Pain, Perception; Sense Data

Aune, B. 1967. *Knowledge, Mind, and Nature: An Introduction to Theory of Knowledge and the Philosophy of Mind*. New York: Random House.

*Broad, C. D. 1959. *Scientific Thought*. Patterson, N.J.: Littlefield Adams.

Caruso, G. 1999. "A Defense of the Adverbial Theory." *Philosophical Writings* 10: 51–65.

Chisholm, R. M. 1957. *Perceiving: A Philosophical Study*. Ithaca: Cornell University Press.

Douglas, G. 1998. "Why Pains Are Not Mental Objects." *Philosophical Studies* 91(2): 127–148.

Ducasse, C. J. 1952. "Moore's Refutation of Idealism." In *The Philosophy of G. E. Moore*, ed. P. A. Schilpp. New York: Tudor.

*Jackson, F. 1975. "On the Adverbial Analysis of Visual Experience." *Metaphilosophy* 6: 127–135.

Jackson, F. 1977. *Perception: A Representative Theory*. Cambridge: Cambridge University Press.

*Kraut, R. 1982. "Sensory States and Sensory Objects." *Noûs* 16(2): 277–293.

Lycan, W. G. 1987. *Consciousness*. Cambridge, Mass.: MIT Press.

Robinson, H. 1994. *Perception*. London: Routledge.

*Sellars, W. 1975. "The Adverbial Theory of the Objects of Sensation." *Metaphilosophy* 6: 144–160.

*Tye, M. 1984. "Pain and the Adverbial Theory." *American Philosophical Quarterly* 21: 319–328.

Tye, M. 1984. "The Adverbial Approach to Visual Experience." *Philosophical Review* 93(2): 195–225.

Pain, Access; Pain, Concept

Blum, A., and R. Carasso. 1988. "Pain Corrigibility." *Manuscrito* 11: 127–128.

Canfield, J. V. 1975. " 'I Know That I Am in Pain' Is Senseless." Pp. 129–144, in *Analysis and Metaphysics*, ed. K. Lehrer. Dordrecht: Reidel.

Dalrymple, H. 1980. "Can a Person Know That He Is in Pain?" *Southwest Philosophical Studies* 5: 55–63.

Dartnall, T. 2001. "The Pain Problem." *Philosophical Psychology* 14(1) 95–102.

Dartnall, T. 2001. "The Pain Problem: Reply to Garfield." *Philosophical Psychology* 14(1): 109–112.

Garfield, J. L. 2001. "Pain Deproblematized." *Philosophical Psychology* 14(1): 103–107.

Goldstein, I. 2000. "Intersubjective Properties by Which We Specify Pain, Pleasure, and Other Kinds of Mental States." *Philosophy* 75: 89–104.

Hinton, R. T. 1975. "Is the Existence of Pain a Scientific Hypothesis?" *Philosophy* 50: 97–100.

Hodges, M., and W. R. Carter. 1969. "Nelson on Dreaming a Pain." *Philosophical Studies* 20: 43–46.

Malcolm, N. 1958. "Knowledge of Other Minds." *Journal of Philosophy* 55: 969–978.

Nelson, J. O. 1966. "Can One Tell That He Is Awake by Pinching Himself?" *Philosophical Studies* 27: 81–84.

Radford, C. 1972. "Pain and Pain Behaviour." *Philosophy* 47: 189–205.

Pain, Access; Pain, Privacy

Hudson, H. 1961. "Why Are Our Feelings of Pain Perceptually Unobservable?" *Analysis* 21: 97–100.

Leighton, S. R. 1986. "Unfelt Feelings in Pain and Emotion." *Southern Journal of Philosophy* 24: 69–79.

Morris, K. J. 1996. "Pain, Injury, and First/Third-Person Asymmetry." *Philosophy and Phenomenological Research* 56(1): 125–136.

*Palmer, D. 1975. "Unfelt Pains." *American Philosophical Quarterly* 12: 289–298.

Wadia, P. S. 1973. "Multi-Person Pains." *Mind* 82(327): 450–451.

Pain, Affect

Grahek, N. 2001. *Feeling Pain and Being in Pain*. Oldenburg, Denmark: BIS-Verlag, University of Oldenburg.

Noren, S. J. 1974. "Pitcher on the Awfulness of Pain." *Philosophical Studies* 25: 117–122.

*Pitcher, G. 1970. "The Awfulness of Pain." *Journal of Philosophy* 67(14): 481–492.

Pitcher, G. 1976. "Pain and Unpleasantness." In *Philosophical Dimensions of the Neuro-Medical Sciences*, ed. S. F. Spicker and H. T. Engelhardt. Proceedings of the Second Trans-Disciplinary Symposium on Philosophy and Medicine. Dordrecht, Holland: D. Reidel.

Sufka, K. J., and M. P. Lynch. 2000. "Sensations and Pain Processes." *Philosophical Psychology* 13(3): 299–311.

Pain, Affect; Pain, Value

*Goldstein, I. 1980. "Why People Prefer Pleasure to Pain." *Philosophy* 55: 349–362.

Goldstein, I. 1983. "Pain and Masochism." *Journal of Value Inquiry* 17: 219–224.

Goldstein, I. 1988. "The Rationality of Pleasure-Seeking Animals." Pp. 131–136, in *Inquiries into Values*, ed. S. H. Lee. Lewiston: Mellen Press.

Hall, R. J. 1989. "Are Pains Necessarily Unpleasant?" *Philosophy and Phenomenological Research* 49(4): 643–659.

*Puccetti, R. 1975. "Is Pain Necessary?" *Philosophy* 50: 259–269.

Rachels, S. 2000. "Is Unpleasantness Intrinsic to Unpleasant Experiences?" *Philosophical Studies* 99: 187–210.

Pain, Affect; Pleasure; Pain, Value; Emotion

Aydede, M. 2000. "An Analysis of Pleasure vis-à-vis Pain." *Philosophy and Phenomenological Research* 61(3): 537–570.

Edwards, R. B. 1975. "Do Pleasures and Pains Differ Qualitatively?" *Journal of Value Inquiry* 9: 270–281.

Edwards, R. B. 1979. *Pleasures and Pains: A Theory of Qualitative Hedonism*. Ithaca: Cornell University Press.

*Feldman, Fred. 1988. "Two Questions about Pleasure." Pp. 59–81, in *Philosophical Analysis*, ed. D. F. Austin. Norwell: Kluwer.

Feldman, Fred. 2004. *Pleasure and the Good Life*. Oxford: Oxford University Press. (Contains an extended appendix on pain and pleasure.)

Goldstein, I. 1989. "Pleasure and Pain: Unconditional, Intrinsic Values." *Philosophy and Phenomenological Research* 50(2): 255–276.

*McCloskey, M. A. 1971. "Pleasure." *Mind* 80(320): 542–551.

Marshall, H. R. 1892. "Pleasure–Pain and Sensation." *Philosophical Review* 1(6): 625–648.

Marshall, H. R. 1894. *Pain, Pleasure, and Aesthetics: An Essay Concerning the Psychology of Pain and Pleasure*. London and New York: Macmillan.

Marshall, H. R. 1894. "Pleasure–Pain." *Mind* 3(12): 533–535.

Marshall, H. R. 1895. "Emotions versus Pleasure–Pain." *Mind* 4(14): 180–194.

*Penelhum, T. 1957. "The Logic of Pleasure." *Philosophy and Phenomenological Research* 17(4): 488–503.

Plochmann, G. K. 1950. "Some Neglected Considerations on Pleasure and Pain." *Ethics* 61(1): 51–55.

Weiss, P. 1942. "Pain and Pleasure." *Philosophy and Phenomenological Research* 3(2): 137–144.

Pain, Animal

*Allen, C. 2004. "Animal Pain." *Noûs* 38: 617–643.

Betty, L. S. 1992. "Making Sense of Animal Pain: An Environmental Theodicy." *Faith and Philosophy* 9(1): 65–82.

Boonin Vail, D. 1993. "Response—Parsimony Made Simple: Rosenfeld on Harrison and Animal Pain." *Between the Species* 9(3): 137–140.

*Carruthers, P. 1989. "Brute Experience." *Journal of Philosophy* 86(5): 258–269.

Carruthers, P. 1992. *The Animals Issue: Moral Theory in Practice*. Cambridge: Cambridge University Press.

De Grazia, D. 1991. "Pain, Suffering, and Anxiety in Animals and Humans." *Theoretical Medicine* 193–211.

Duran, J. 1994. "Commentary on 'Is Animal Pain Conscious?'" *Between the Species* 10(1/2): 8–9.

Gennaro, R. J. 1993. "Brute Experience and the Higher-Order Thought Theory of Consciousness." *Philosophical Papers* 22(1): 51–69.

Harrison, P. 1989. "Theodicy and Animal Pain." *Philosophy* 64: 79–92.

*Harrison, P. 1991. "Do Animals Feel Pain?" *Philosophy* 66: 25–40.

House, I. 1991. "Harrison on Animal Pain." *Philosophy* 66: 376–379.

Jennings, Jr., D. 1991. "Why Animal Pain? Considerations in Theodicies." *Between the Species* 7(4): 217–221.

Lynch, J. J. 1994. "Harrison and Hick on God and Animal Pain." *Sophia* 33(3): 62–73.

Lynch, J. J. 1994. "Is Animal Pain Conscious?" *Between the Species* 10(1/2): 1–7.

Nollman, J. I. M. 1987. "To Judge the Pain of Whales." *Between the Species* 3: 133–137.

Perrett, R. W. 1997. "The Analogical Argument for Animal Pain." *Journal of Applied Philosophy* 14(1): 49–58.

Robinson, W. S. 1997. "Some Nonhuman Animals Can Have Pains in a Morally Relevant Sense." *Biology and Philosophy* 12(1): 51–71.

Rosenfeld, R. P. 1993. "Parsimony, Evolution, and Animal Pain." *Between the Species* 9(3): 133–137.

Squire, A. N. N. 1985. "On Animals and Pain." *Between the Species* 1: 19–20.

Pain, Choice

*Beardman, S. 2000. "The Choice between Current and Retrospective Evaluations of Pain." *Philosophical Psychology* 13(1): 97–110.

Belshaw, C. 2000. "Death, Pain, and Time." *Philosophical Studies* 97(3): 317–341.

Broome, J. 1996. "More Pain or Less?" *Analysis* 56(2): 116–118.

Gustafson, D. 2000. "Our Choice between Actual and Remembered Pain and Our Flawed Preferences." *Philosophical Psychology* 13(1): 111–119.

Perrett, R. W. 1999. "Preferring More Pain to Less." *Philosophical Studies* 93(2): 213–226.

Silverstein, H. S. 1998. "More Pain or Less? Comments on Broome." *Analysis* 58(2): 146–151.

Pain, Concept; Pain, Access

Margolis, J. 1966. "After-Images and Pains." *Philosophy* 41: 333–340.

Sharpe, R. A. 1983. "How Having the Concept of Pain Depends on Experiencing It." *Philosophical Investigations* 6: 142–144.

Pain, Concept; Pain, Nature

Edwards, J. 1993. "Following Rules, Grasping Concepts, and Feeling Pains." *European Journal of Philosophy* 1(3): 268–284.

Gandhi, R. 1973. "Injury, Harm, Damage, Pain, Etc." *Philosophy and Phenomenological Research* 34(2): 266–269.

Kelly, M. L. 1991. "Wittgenstein and 'Mad Pain,' " *Synthese* 87: 285–294.

Long, T. A. 1965. "The Problem of Pain and Contextual Implication." *Philosophy and Phenomenological Research* 26(1): 106–111.

Long, T. A. 1965. "Strawson and the Pains of Others." *Australasian Journal of Philosophy* 43: 73–77.

Margolis, J. 1965. "The Problem of Criteria of Pain." *Dialogue* 4: 62–71.

Miller, D. S. 1929. "The Pleasure–Quality and the Pain–Quality Analyzable, Not Ultimate." *Mind* 38(150): 215–218.

Ochs, C. R. 1966. "The Sensitive Term 'Pain.' " *Philosophy and Phenomenological Research* 27(2): 255–260.

Sullivan, M. D. 1995. "Key Concepts: 'Pain.' " *Philosophy, Psychiatry, and Psychology* 2(3): 277–280.

Pain, Dennett

Conee, E. 1984. "A Defense of Pain." *Philosophical Studies* 46: 239–248.

*Dennett, D. C. 1978. "Why You Can't Make a Computer That Feels Pain." In *Brainstorms*. Cambridge, Mass.: MIT Press.

Guirguis, M. M. 1998. "Robotoid Arthritis or How Humans Feel Pain." *Philosophical Writings* 7: 3–12.

Kaufman, R. 1985. "Is the Concept of Pain Incoherent?" *Southern Journal of Philosophy* 23: 279–284.

Nikolinakos, D. D. 2000. "Dennett on Qualia: The Case of Pain, Smell, and Taste." *Philosophical Psychology* 13(4): 505–522.

Pain, General; Pain, Science

Fields, H. L., and D. D. Price. 1944. "Pain." In *A Companion to the Philosophy of Mind*, ed. S. Guttenplan. Oxford: Basil Blackwell.

Hardcastle, V. G. 1999. *The Myth of Pain*. Cambridge, Mass.: MIT Press.

Melzack, R. 1961. "The Perception of Pain." *Scientific American* 204(2): 41–49.

Pain, History; Pain, Medicine

Bonica, J. J., and J. D. Loeser. 2001. "History of Pain Concepts and Therapies." In *Bonica's Management of Pain*, ed. J. J. Bonica and J. D. Loeser. Philadelphia: Lippincott, Williams, and Wilkins.

Botterell, E. H., J. C. Callaghan, and A. T. Jousse. 1942. *Sensation and Perception in the History of Experimental Psychology*. New York: Appleton-Century-Crofts.

Brykman, G. 1985. "Pleasure and Pain versus Ideas in Berkeley." *Hermathena* 139: 127–137.

Carter, W. R. 1972. "Locke on Feeling Another's Pain." *Philosophical Studies* 23: 280–285.

Chisholm, R. M. 1987. "Brentano's Theory of Pleasure and Pain." *Topoi* 6: 59–64.

*Dallenbach, K. M. 1939. "Pain: History and Present Status." *American Journal of Psychology* 52: 331–347.

Duncan, G. 2000. "Mind–Body Dualism and the Biopsychosocial Model of Pain: What Did Descartes Really Say?" *Journal of Medicine and Philosophy* 25(4): 485–513.

Keele, K. D. 1957. *Anatomies of Pain*. Springfield, Ill.: C. Thomas.

Raj, P. P. 1996. "History of Pain Medicine." In *Pain Medicine*, ed. P. P. Raj. New York: Mosby.

Rey, R. 1995. *The History of Pain*. Cambridge, Mass.: Harvard University Press.

Pain, Infant

Benatar, D., and M. Benatar. 2001. "A Pain in the Fetus: Toward Ending Confusion about Fetal Pain." *Bioethics* 15(1): 57–76.

Campbell, N. 1989. "A Response to Cunningham Butler's 'Infants, Pain, and What Health Care Professionals Should Want to Know.' " *Bioethics* 3: 200–210.

Cunningham Butler, N. 1989. "Infants, Pain, and What Health Care Professionals Should Want to Know—An Issue of Epistemology and Ethics." *Bioethics* 3: 181–199.

*Derbyshire, S. W. G. 1999. "Locating the Beginnings of Pain." *Bioethics* 13(1): 1–31.

Derbyshire, S. W. G. 2001. "Fetal Pain: An Infantile Debate." *Bioethics* 5(1): 77–84.

Field, T. 1995. "Infancy Is Not without Pain." *Annals of Child Development: A Research Annual* 10: 1–26.

Franck, L., and L. Lefrak. 2001. "For Crying Out Loud: The Ethical Treatment of Infants' Pain." *Journal of Clinical Ethics* 12(3): 275–281.

Griffith, S. 1995. "Fetal Death, Fetal Pain, and the Moral Standing of a Fetus." *Public Affairs Quarterly* 9(2): 115–126.

Kaufman, R. 1985. "Fetal Pain." *Southern Journal of Philosophy* 23: 305–311.

Pain, Language; Pain, Science

Ehlich, K. 1985. "The Language of Pain." *Theoretical Medicine* 6: 177–188.

*Melzack, R., and W. S. Torgerson. 1971. "On the Language of Pain." *Anesthesiology* 34: 50–59.

Pain, Location; Pain, Perception; Bodily Sensations; Sense Data

Addis, L. 1986. "Pains and Other Secondary Mental Entities." *Philosophy and Phenomenological Research* 47(1): 59–74.

Armstrong, D. M. 1964. "Vesey on Bodily Sensations." *Australasian Journal of Philosophy* 42: 247–248.

*Baier, K. 1964. "The Place of a Pain." *Philosophical Quarterly* 14(55): 138–150.

Coburn, R. C. 1966. "Pains and Space." *Journal of Philosophy* 63(13): 381–396.

Holborow, L. C. 1966. "Taylor on Pain Location." *Philosophical Quarterly* 16(63): 151–158.

Holly, W. J. 1986. "The Spatial Coordinates of Pain." *Philosophical Quarterly* 36(144): 343–356.

Hyman, J. 2003. "Pain and Places." *Philosophy* 73(1): 5–24.

Noordhof, P. 2001. "In Pain." *Analysis* 61(2): 95–97.

Noordhof, P. 2002. "More in Pain." *Analysis* 62(2): 153–154.

Taylor, D. M. 1965. "The Location of Pain." *Philosophical Quarterly* 15(58): 53–62.

Taylor, D. M. 1966. "The Location of Pain: A Reply to Mr. Holborow." *Philosophical Quarterly* 16(65): 359–360.

Tye, M. 2002. "On the Location of a Pain." *Analysis* 62(2): 150–153.

Vesey, G. N. A. 1964. "Armstrong on Bodily Sensations." *Philosophy* 39: 177–181.

*Vesey, G. N. A. 1964. "Bodily Sensations." *Australasian Journal of Philosophy* 42: 232–247.

Vesey, G. N. A. 1965. "Baier on Vesey on the Place of a Pain." *Philosophical Quarterly* 15(58): 63–64.

Vesey, G. N. A. 1967. "Margolis on the Location of Bodily Sensations." *Analysis* 27: 174–176.

Pain, Nature; Pain, Concept

*Baier, K. 1962. "Pains." *Australasian Journal of Philosophy* 40: 1–23.

Bain, A. 1892. "Pleasure and Pain." *Mind* 1(2): 161–187.

Bieri, P. 1995. "Pain: A Case Study for the Mind–Body Problem." In *Pain and the Brain: From Nociception to Cognition*, ed. B. Bromm and J. E. Desmedt. New York: Raven Press.

Blum, A. 1996. "The Agony of Pain." *Philosophical Inquiry* 18(3/4): 117–120.

Cowan, J. L. 1968. *Pleasure and Pain: A Study in Philosophical Psychology*. London: Macmillan.

Daniels, C. B. 1967. "Colors and Sensations, or How to Define a Pain Ostensively." *American Philosophical Quarterly* 4: 231–237.

Gert, B. 1967. "Can a Brain Have a Pain?" *Philosophy and Phenomenological Research* 27(3): 432–436.

Gillett, G. R. 1991. "The Neurophilosophy of Pain." *Philosophy* 66: 191–206.

Grahek, N. 1995. "The Sensory Dimension of Pain." *Philosophical Studies* 79(2): 167–184.

Gustafson, D. 1979. "Pain, Grammar, and Physicalism." Pp. 149–166, in *Body, Mind, and Method: Essays in Honor of Virgil C. Aldrich*, ed. D. F. Gustafson and B. L. Tapscott. Dordrecht: Reidel.

Gustafson, D. 1995. "Belief in Pain." *Consciousness and Cognition* 4: 323–345.

Gustafson, D. 1998. "Pain, Qualia, and the Explanatory Gap." *Philosophical Psychology* 11(3): 371–387.

*Lewis, D. 1980. "Mad Pain and Martian Pain." In *Readings in Philosophy of Psychology*, vol. I, ed. N. Block. Cambridge: Harvard University Press.

Natsoulas, T. 1988. "On the Radical Behaviorist Conception of Pain Experience." *Journal of Mind and Behavior* 9: 29–56.

Noren, S. J. 1976. "The Efficacy of Pain." *Journal of Critical Analysis* 6: 71–76.

O'Shaughnessy, B. 1955. "The Origin of Pain." *Analysis* 15: 121–130.

Putnam, H. 1973. "Psychological Predicates." In *Art, Mind, and Religion*, ed. W. H. Capitan and D. D. Merrill. Pittsburgh: University of Pittsburgh. (Reprinted as "The Nature of Mental States" in *Philosophy of Mind*, ed. David Chalmers, Oxford University Press, 2002.)

Singer, E. A., Jr. 1924. "On Pain and Dreams." *Journal of Philosophy* 21(22): 589–601.

Slater, B. H. 2001. "Seeing Pains." *Grazer Philosophische Studien* 62: 65–81.

*Trigg, R. 1970. *Pain and Emotion*. Oxford: Clarendon Press.

Pain, Nature; Pain, Value; Pleasure

Buytendijk, F. J. J. 1957. "The Meaning of Pain." *Philosophy Today* 1: 180–185.

Buytendijk, F. J. J. 1961. *Pain*. London: Hutchinson.

Plochmann, G. K. 1950. "Some Neglected Considerations on Pleasure and Pain." *Ethics* 61(1): 51–55.

Szasz, T. 1988. *Pain and Pleasure: A Study of Bodily Feelings*. New York: Syracuse University Press.

Pain, Perception

*Armstrong, D. M. 1962. *Bodily Sensations*. London: Routledge and Kegan Paul.

*Armstrong, D. M. 1968. *A Materialist Theory of the Mind*. New York: Humanities Press.

Aydede, M. 2001. "Naturalism, Introspection, and Direct Realism about Pain." *Consciousness and Emotion* 2(1): 29–73.

Bain, D. 2003. "Intentionalism and Pain." *Philosophical Quarterly* 53(213): 502–522.

Dretske, F. 1995. *Naturalizing the Mind*. Cambridge, Mass.: MIT Press.

*Dretske, F. 1999. "The Mind's Awareness of Itself." *Philosophical Studies* 95(1/2): 103–124.

Dretske, F. 2003. "How Do You Know You Are Not a Zombie?" in *Privileged Access: Philosophical Accounts of Self-Knowledge*, ed. B. Gertler. Hampshire, U.K.: Ashgate.

Everitt, N. 1988. "Pain and Perception." *Proceedings of the Aristotelian Society* 89: 113–124.

Fleming, N. 1976. "The Objectivity of Pain." *Mind* 85(340): 522–541.

Graham, G., and G. L. Stephens. 1985. "Are Qualia a Pain in the Neck for Functionalists?" *American Philosophical Quarterly* 22: 73–80.

Grahek, N. 1991. "Objective and Subjective Aspects of Pain." *Philosophical Psychology* 4: 249–266.

Grice, H. P. 1962. "Some Remarks about the Senses." Pp. 133–151, in *Analytical Philosophy*, ed. R. J. Butler. Oxford, UK: Blackwell.

Hardcastle, V. G. 1997. "When a Pain Is Not." *Journal of Philosophy* 94(8): 381–409.

Holborow, L. C. 1969. "Against Projecting Pains." *Analysis* 29: 105–108.

Lycan, W. G. 1996. *Consciousness and Experience*. Cambridge, Mass.: MIT Press.

Margolis, J. 1976. "Pain and Perception." *International Studies in Philosophy* 8: 3–12.

Mayberry, T. C. 1978. "The Perceptual Theory of Pain." *Philosophical Investigations* 1: 31–40.

Mayberry, T. C. 1979. "The Perceptual Theory of Pain: Another Look." *Philosophical Investigations* 2: 53–55.

McKenzie, J. C. 1968. "The Externalization of Pains." *Analysis* 28: 189–193.

*Nelkin, N. 1986. "Pains and Pain Sensations." *Journal of Philosophy* 83(3): 129–148.

Nelkin, N. 1994. "Reconsidering Pain." *Philosophical Psychology* 7(3): 325–343.

*Newton, N. 1989. "On Viewing Pain as a Secondary Quality." *Noûs* 23(5): 569–598.

*Perkins, M. 1983. *Sensing the World*. Indianapolis, Ind.: Hackett.

Pitcher, G. 1969. "McKenzie on Pains." *Analysis* 29: 103–105.

*Pitcher, G. 1970. "Pain Perception." *Philosophical Review* 79(3): 368–393.

Pitcher, G. 1971. *A Theory of Perception*. Princeton, N.J.: Princeton University Press.

Pitcher, G. 1978. "The Perceptual Theory of Pain: A Response to Thomas Mayberry's 'The Perceptual Theory of Pain.' " *Philosophical Investigations* 1: 44–46.

Seager, W. 2002. "Emotional Introspection." *Consciousness and Cognition* 11(4): 666–687.

*Stephens, G. L., and G. Graham. 1987. "Minding Your P's and Q's: Pain and Sensible Qualities." *Noûs* 21(3): 395–405.

Tye, M. 1996. *Ten Problems of Consciousness: A Representational Theory of the Phenomenal Mind*. Cambridge, Mass.: MIT Press.

*Tye, M. 1997. "A Representational Theory of Pains and Their Phenomenal Character." In *The Nature of Consciousness: Philosophical Debates*, ed. N. Block, O. Flanagan, and G. Güzeldere. Cambridge, Mass.: MIT Press.

Wilkes, K. V. 1977. *Physicalism*. London: Routledge.

Pain, Phenomenology

Leder, D. 1984. "Toward a Phenomenology of Pain." *Review of Existential Psychology and Psychiatry* 19: 255–266.

Pain, Privacy; Pain, Access; Pain, Concept

Anderson, J. 1985. "Pain, Private Language, and the Mind–Body Problem." *Auslegung* 12: 53–69.

*Averill, E. 1978. "Explaining the Privacy of Afterimages and Pains." *Philosophy and Phenomenological Research* 38(3): 299–314.

Fiser, K. B. 1986. "Privacy and Pain." *Philosophical Investigations* 9: 1–17.

Langsam, H. 1995. "Why Pains Are Mental Objects." *Journal of Philosophy* 92(6): 303–313.

Locke, D. 1964. "The Privacy of Pains." *Analysis* 24: 147–152.

Senchuk, D. M. 1986. "Privacy Regained." *Philosophical Investigations* 9(1): 18–35.

Sprigge, T. L. S. 1969. "The Privacy of Experience." *Mind* 78(312): 512–521.

Taylor, D. M. 1970. "The Logical Privacy of Pains." *Mind* 79(313): 78–91.

Tibbetts, P. 1973. "On Making a Pain Public." *Philosophical Studies (Ireland)* 21: 96–99.

Zemach, E. M. 1971. "Pains and Pain-Feelings." *Ratio* 13: 150–157.

Pain, Science; Pain, General

Aydede, M., and G. Güzeldere. 2002. "Some Foundational Issues in the Scientific Study of Pain." *Philosophy of Science* 69 (suppl.): 265–283.

Barber, T. 1964. "The Effects of 'Hypnosis' on Pain: A Critical Review of Experimental and Clinical Findings." *Journal of the American Society of Psychosomatic Dentistry and Medicine* 11(2).

Beecher, H. K. 1966. "Pain: One Mystery Solved." *Science* 151(3712): 840–841.

Bonica, J. J., and J. D. Loeser. eds. 2001. *Bonica's Management of Pain*. Philadelphia: Lippincott, Williams, and Wilkins.

Bushnell, M. C., G. H. Duncan, R. K. Hofbauer, B. Ha, J. I. Chen, and B. Carrier. 1999. "Pain Perception: Is There a Role for Primary Somatosensory Cortex?" *Proceedings of the National Academy of Sciences of the United States of America* 96(14): 7705–7709.

Chapman, C. R., G. W. Donaldson, Y. Nakamura, R. C. Jacobson, D. H. Bradshaw, and J. Gavrin. 2002. "A Psychophysiological Causal Model of Pain Report Validity." *Journal of Pain* 3(2): 143–155.

Chapman, C. R., and Y. Nakamura. 1999. "A Passion of the Soul: An Introduction to Pain for Consciousness Researchers." *Consciousness and Cognition* 8(4): 391–422.

Chudler, E. H., and W. K. Dong. 1995. "The Role of the Basal Ganglia in Nociception and Pain." *Pain* 60(1): 3–38.

Fields, H. L., ed. 1990. *Pain Syndromes in Neurology*. Butterworths International Medical Reviews, Neurology 10. London, Boston: Butterworths.

Hilgard, E. R. 1967. "A Quantitative Study of Pain and Its Reduction through Hypnotic Suggestion." *Proceedings of the National Academy of Sciences of the United States of America* 57(6): 1581–1586.

Hosobuchi, Y., J. E. Adams, and R. Linchitz. 1977. "Pain Relief by Electrical Stimulation of the Central Gray Matter in Humans and Its Reversal by Naloxone." *Science* 197(4299): 183–186.

Melzack, R. 1992. "Phantom Limbs." *Scientific American* (April): 120–126.

*Melzack, R., and P. D. Wall. 1965. "Pain Mechanisms: A New Theory." *Science* 150(3699): 971–979.

*Melzack, R., and P. D. Wall. 1988. *The Challenge of Pain*. London, New York: Penguin Books.

Melzack, R., and P. D. Wall, eds. 1999. *Textbook of Pain*. Edinburgh: Churchill Livingstone.

*Millan, M. J. 1999. "The Induction of Pain: An Integrative Review." *Progress in Neurobiology* 57: 1–164.

Nakamura, Y., and C. R. Chapman. 2002. "Measuring Pain: An Introspective Look at Introspection." *Consciousness and Cognition* 11(4): 582–592.

*Price, D. D. 1999. *Psychological Mechanisms of Pain and Analgesia*. Seattle: IASP Press.

*Price, Donald D., and M. C. Bushnell, eds. 2004. *Psychological Methods of Pain Control: Basic Science and Clinical Perspective*. Progress in Pain Research and Management, vol. 29. Seattle: IASP Press, 2004. (Contains updated and accessible reviews of recent scientific and clinical research on pain.)

Sufka, K. J. 2000. "Chronic Pain Explained." *Brain and Mind* 1: 155–179.

Turk, D. C., and R. Melzack, eds. 2001. *Handbook of Pain Assessment*. New York: Guilford Press.

Wall, P. D. 2000. *Pain: The Science of Suffering*. New York: Columbia University Press.

Pain, Science; Pain, Access

Craig, A. D., and M. C. Bushnell. 1994. "The Thermal Grill Illusion: Unmasking the Burn of Cold Pain." *Science* 265(5169): 252–255.

Tinnin, L. 1994. "Conscious Forgetting and Subconscious Remembering of Pain." *Journal of Clinical Ethics* 5(2): 151–152.

Vogel, G. 1996. "Illusion Reveals Pain Locus in Brain." *Science* 274(5291): 1301.

Pain, Science; Pain, Affect

Fernandez, E., T. S. Clark, and D. Rudick-Davis. 1999. "A Framework for Conceptualization and Assessment of Affective Disturbance in Pain." In *Handbook of Pain Syndromes: Biopsychosocial Perspectives*, eds. A. R. Block, E. F. Kremer, and E. Fernandez. Mahwah, N.J.: Lawrence Erlbaum.

*Fields, H. L. 1999. "Pain: An Unpleasant Topic." *Pain* (suppl.) 6: 61–69.

*Melzack, R., and K. L. Casey. 1968. "Sensory, Motivational, and Central Control Determinants of Pain: A New Conceptual Model." Pp. 223–243, in *The Skin Senses*, ed. D. Kenshalo. Springfield, Ill.: Charles C. Thomas.

*Price, D. D. 2000. "Psychological and Neural Mechanisms of the Affective Dimension of Pain." *Science* 288(9): 1769–1772.

Price, D. D., S. W. Harkins, and C. Baker. 1987. "Sensory-Affective Relationships among Different Types of Clinical and Experimental Pain." *Pain* 28: 297–307.

Pain, Science; Pain, Affect; Pain, Disassociation

*Barber, T. X. 1959. "Toward a Theory of Pain: Relief of Chronic Pain by Prefrontal Leucotomy, Opiates, Placebos, and Hypnosis." *Psychological Bulletin* 56.

Chapman, C. R., Y. Nakamura, G. W. Donaldson, R. C. Jacobson, D. H. Bradshaw, L. Flores, and C. N. Chapman. 2001. "Sensory and Affective Dimensions of Phasic Pain Are Indistinguishable in the Self-Report and Psychophysiology of Normal Laboratory Subjects." *Journal of Pain* 2(5): 279–294.

*Fernandez, E., and D. C. Turk. 1992. "Sensory and Affective Components of Pain: Separation and Synthesis." *Psychological Bulletin* 112(2): 205–217.

Fernandez, E., and T. W. Milburn. 1994. "Sensory and Affective Predictors of Overall Pain and Emotions Associated with Affective Pain." *Clinical Journal of Pain* 10(1): 3–9.

*Foltz, E. L., and E. W. White. 1962. "The Role of Rostral Cingulumonotomy in 'Pain' Relief." *International Journal of Neurology* 6: 353–373.

Freeman, W., and J. W. Watts. 1946. "Pain of Organic Disease Relieved by Prefrontal Lobotomy." *Proceedings of the Royal Academy of Medicine* 39: 44–447.

Freeman, W., and J. W. Watts. 1950. *Psychosurgery, in the Treatment of Mental Disorders and Intractable Pain*. Springfield, Ill.: Charles C. Thomas.

*Gracely, R. H., R. Dubner, and P. A. McGrath. 1979. "Narcotic Analgesia: Fentanyl Reduces the Intensity but Not the Unpleasantness of Painful Tooth Pulp Sensations." *Science* 203: 1261–1263.

Hardy, J. D., H. J. Wolff, and H. Goodell. 1952. *Pain Sensations and Reactions*. Baltimore: Williams and Wilkins.

Head, H., and G. Holmes. 1911. "Sensory Disturbances from Cerebral Lesions." *Brain* 34: 102–154.

*Ploner, M., H. J. Freund, and A. Schnitzler. 1999. "Pain Affect without Pain Sensation in a Patient with a Postcentral Lesion." *Pain* 81(1/2): 211–214.

*Rainville, P., B. Carrier, R. K. Hofbauer, M. C. Bushnell, and G. H. Duncan. 1999. "Dissociation of Sensory and Affective Dimensions of Pain Using Hypnotic Modulation." *Pain Forum* 82(2): 159–171.

*Rainville, P., G. H. Duncan, D. D. Price, B. Carrier, and M. C. Bushnell. 1997. "Pain Affect Encoded in Human Anterior Cingulate but Not Somatosensory Cortex." *Science* 277(5328): 968–971.

Pain, Science; Pain, Asymbolia; Pain, Disassociation

*Berthier, M. L., S. E. Starkstein, and R. Leiguarda. 1988. "Pain Asymbolia: A Sensory-Limbic Disconnection Syndrome." *Annals of Neurology* 24: 41–49.

Berthier, M. L., S. E. Starkstein, M. A. Nogues, and R. G. Robinson. 1990. "Bilateral Sensory Seizures in a Patient with Pain Asymbolia." *Annals of Neurology* 27(1): 190.

Rubins, J. L., and E. D. Friedman. 1948. "Asymbolia for Pain." *Archives of Neurology and Psychiatry* 60: 554–573.

Weinstein, E. A., R. L. Kahn, and W. H. Slate. 1995. "Withdrawal, Inattention, and Pain Asymbolia." *Archives of Neurology and Psychiatry* 74(3): 235–248.

Pain, Science; Pain, History; Pain, General

Goldscheider, A. 1894. *Ueber Den Schmerz Im Physiologischer Und Klinischer Hinsicht.* Berlin: Hirschwald.

Livingston, W. K. 1943. *Pain Mechanisms.* New York: Macmillan.

*Melzack, R., and P. D. Wall. 1962. "On the Nature of Cutaneous Sensory Mechanisms." *Brain* 85: 331–356.

Noordenbos, W. 1959. *Pain.* Amsterdam, New York: Elsevier.

*Sufka, K. J., and D. D. Price. 2002. "Gate Control Theory Reconsidered." *Brain and Mind* 3: 277–290.

von Frey, M. 1894. "Beitrage Zur Physiologie Der Shmerzsinnes." *Ber. Verhandl. konig. sachs. Ges. Wiss.* 46(185).

Pain, Science; Pain, Imaging

Casey, K. L. 1999. "Forebrain Mechanisms of Nociception and Pain: Analysis through Imaging." *Proceedings of the National Academy of Sciences of the United States of America* 96(14): 7668–7674.

Casey, K. L., and M. C. Bushnell, eds. 2000. *Pain Imaging.* Progress in Pain Research and Management, vol. 18. Seattle: IASP Press.

*Coghill, R. C., C. N. Sang, J. M. Maisog, M. J. Iadarola. 1999. "Pain Intensity Processing within the Human Brain: A Bilateral, Distributed Mechanism." *Journal of Neurophysiology* 82: 1934–1943.

*Coghill, R., J. McHaffie, and Y. Yen. 2003. "Neural Correlates of Interindividual Differences in the Subjective Experience of Pain." *Proceedings of the National Academy of Sciences of the United States of America* 100: 8538–8542.

Jones, A. K. P., W. D. Brown, K. J. Friston, L. Y. Qi, and R. S. J. Frackowiak. 1991. "Cortical and Subcortical Localization of Response to Pain in Man Using Positron Emission Tomography." *Proceedings: Biological Sciences* 244(1309): 39–44.

Ploner, M., F. Schmitz, H.-J. Freund, and A. Schnitzler. 1999. "Parallel Activation of Primary and Secondary Somatosensory Cortices in Human Pain Processing." *Journal of Neurophysiology* 81(6): 3100–3104.

Talbot, J. D., S. Marrett, A. C. Evans, E. Meyer, M. C. Bushnell, and G. H. Duncan. 1991. "Multiple Representations of Pain in Human Cerebral Cortex." *Science* 251(4999): 1355–1358.

Timmermann, L., M. Ploner, K. Haucke, F. Schmitz, R. Baltissen, and A. Schnitzler. 2001. "Differential Coding of Pain Intensity in the Human Primary and Secondary Somatosensory Cortex." *Journal of Neurophysiology* 86(3): 1499–1503.

*Tölle, T. R. T., Kaufmann, T. Siessmeier, S. Lautenbacher, A. Berthele, F. Munz, W. Zieglgänsberger, F. Willoch, M. Schwaiger, B. Conrad, and P. Bartenstein. 1999. "Region-Specific Encoding of Sensory and Affective Components of Pain in the Human Brain: A Positron Emission Tomography Correlation Analysis." *Annals of Neurology* 45: 40–47.

*Treede, R.-D., D. R. Kenshalo, R. H. Gracely, and A. K. P. Jones. 1999. "The Cortical Representation of Pain." *Pain* 79: 105–111.

Pain, Science; Pain, Insensitivity

Baxter, D. W., and J. Olszewski. 1960. "Congenital Universal Insensitivity to Pain." *Brain* 83.

Boyd, D. A., and L. W. Nie. 1949. "Congenital Universal Indifference to Pain." *Archives of Neurology and Psychiatry* 61: 402–404.

Brand, P. W., and P. Yancey. 1993. *Pain: The Gift Nobody Wants.* New York: HarperCollins.

Cohen, L. D., D. Kipnis, E. C. Kunkle, and P. E. Kubzansky. 1955. "Observations of a Person with Congenital Insensitivity to Pain." *Journal of Abnormal and Social Psychology* 51(2): 333–339.

Dearborn, G. V. N. 1932. "A Case of Congenital General Pure Analgesia." *Journal of Nervous and Mental Disease* 75: 612–615.

Dehen, H., J. C. Willer, S. Prier, F. Boureau, and J. Combier. 1978. "Congenital Insensitivity to Pain and the 'Morphine-Like' Analgesic System." *Pain* 5(4): 351–358.

Dubovsky, S. L., and S. E. Groban. 1975. "Congenital Absence of Sensation." *Psychoanalytic Study of the Child* 30: 49–73.

Ford, F. M., and L. Wilkins. 1938. "Congenital Universal Insensitiveness to Pain." *Johns Hopkins Hospital Bulletin* 62: 448–466.

Kane, F. J., A. W. Downie, D. B. Marcotte, and M. Perez-Reyes. 1968. "A Case of Congenital Indifference to Pain: Neurologic and Psychiatric Findings." *Diseases of the Nervous System* 29(6): 409–412.

Larner, A. J., J. Moss, M. L. Rossi, and M. Anderson. 1994. "Congenital Insensitivity to Pain: A 20-Year Follow Up." *Journal of Neurology, Neurosurgery, and Psychiatry* 57(8): 973–974.

Magee, K. R., S. Schneider, and N. Rosenzweig. 1961. "Congenital Indifference to Pain." *Journal of Nervous and Mental Disease* 132: 248–259.

McMurray, G. A. 1955. "Congenital Insensitivity to Pain and Its Implications for Motivational Theory." *Canadian Journal of Psychology* 9(2): 121–131.

McMurray, G. A. 1975. "Theories of Pain and Congenital Universal Insensitivity to Pain." *Canadian Journal of Psychology* 29(4): 302–315.

*Nagasako, E. M., A. N., Oaklander, and R. H. Dworkin. 2003. "Congenital Insensitivity to Pain: An Update." *Pain* 101: 213–219.

Rapoport, J. L. 1969. "A Case of Congenital Sensory Neuropathy Diagnosed in Infancy." *Journal of Child Psychology and Psychiatry and Allied Disciplines* 10(1): 63–68.

Sternbach, R. A. 1963. "Congenital Insensitivity to Pain: A Critique." *Psychological Bulletin* 60(3): 252–264.

Swanson, A. G., G. C. Buchan, and E. C. Alvord. 1965. "Anatomic Changes in Congenital Insensitivity to Pain." *Archives of Neurology* 12(1): 12–18.

Pain, Science; Pain, Modulation

*Fields, H. L., and A. I. Basbaum. 1999. "Central Nervous System Mechanisms of Pain Modulation." Pp. 309–329, in *Textbook of Pain*, ed. R. Melzack and P. D. Wall. Edinburgh: Churchill Livingstone.

Fields, H. L., and J.-M. R. Besson, eds. 1988. *Pain Modulation*. Progress in Brain Research, vol. 77. Amsterdam, New York: Elsevier.

Jones, S. L. 1992. "Descending Control of Nociception." In *The Initial Processing of Pain and Its Descending Control: Spinal and Trigeminal Systems*, ed. A. R. Light. New York: Karger.

Sandkühler, J. 1996. "The Organization and Function of Endogenous Antinociceptive Systems." *Progress in Neurobiology* 50: 49–81.

Pain, Science; Pain, Surgery

Bouckoms, A. J. 1994. "Limbic Surgery for Pain." Pp. 1171–1187, in *Textbook of Pain*, ed. P. D. Wall and R. Melzack. Edinburgh: Churchill Livingstone.

Foltz, E. L., and L. E. White. 1962. "Pain 'Relief' by Frontal Cingulotomy." *Journal of Neurosurgery* 19: 89–100.

Mark, V. H., F. R. Ervin, and T. P. Hackett. 1960. "Clinical Aspects of Stereotactic Thalamotomy in the Human, I: The Treatment of Severe Pain." *Archives of Neurology* 3: 351–367.

White, J. C., and W. H. Sweet. 1969. *Pain and the Neurosurgeon: A Forty-Year Experience*. Springfield, Ill.: Charles C. Thomas.

Pleasure, Science; Pain, Science; Emotion

*Berridge, K. C. 1999. "Pleasure, Pain, Desire, and Dread: Hidden Core Processes of Emotion." Pp. 525–557, in *Well-Being: The Foundations of Hedonic Psychology*, ed. D. Kahneman, E. Diener, and N. Schwarz. New York: Russell Sage Foundation.

Berridge, K. C. 2003. "Pleasures of the Brain." *Brain and Cognition* 52(10): 106–128.

Berridge, K. C. 2004. "Motivation Concepts in Behavioral Neuroscience." *Physiology and Behavior* 81(2): 179–209.

Pancksepp, J. 1998. *Affective Neuroscience: The Foundations of Human and Animal Emotions*. London: Oxford University Press.

Contributors

Colin Allen Department of History and Philosophy of Science and Program in Cognitive Science, Indiana University, Bloomington.

Murat Aydede Department of Philosophy, University of Florida, Gainesville, Florida.

Ned Block Department of Philosophy, New York University, New York City.

Austen Clark University of Connecticut, Philosophy Department, Storrs, Connecticut.

Robert C. Coghill Department of Neurobiology and Anatomy, Wake Forest University School of Medicine, Winston-Salem, North Carolina.

Robert D'Amico Department of Philosophy, University of Florida, Gainesville, Florida.

Fred Dretske Department of Philosophy, Duke University, Durham, North Carolina.

Perry N. Fuchs Department of Psychology, University of Texas at Arlington, Texas.

Shaun Gallagher Philosophy and Cognitive Science, University of Central Florida, Orlando, Florida.

Don Gustafson Department of Philosophy, University of Cincinnati, Ohio.

Christopher S. Hill Department of Philosophy, Brown University, Providence, Rhode Island.

Barry Maund Department of Philosophy, The University of Western Australia, Crawley WA, Australia.

Eddy Nahmias Department of Philosophy, Georgia State University, Atlanta, Georgia.

Paul Noordhof Department of Philosophy, University of Nottingham, United Kingdom.

Morten Overgaard Hammel NeuroCenter, The University of Aarhus, Denmark.

Jaak Panksepp J. P. Scott Center for Neuroscience, Mind and Behavior, Department of Psychology, Bowling Green State University, Bowling Green, Ohio.

Moreland Perkins Department of Philosophy, University of Maryland, College Park, Maryland.

Thomas W. Polger Department of Philosophy, University of Cincinnati, Ohio.

Donald D. Price Oral and Maxillofacial Surgery, University of Florida, Health Science Center, Gainesville, Florida.

Adam Shriver Department of Philosophy, Texas A&M University, College Station, Texas.

Kenneth J. Sufka Department of Psychology and Pharmacology, University of Mississippi, Mississippi.

Michael Tye Department of Philosophy, University of Texas, Austin, Texas.

Hilary D. Wilson Department of Psychology, University of Texas at Arlington, Texas.

Index